The Ecosystem Approach to Fisheries

The Ecosystem Approach to Fisheries

Edited by

Gabriella Bianchi

Fisheries Management Division
Food and Agriculture Organization of the United Nations
Rome, Italy

and

Hein Rune Skjoldal

Institute of Marine Research
Bergen, Norway

Published by
Food and Agriculture Organization of the United Nations

www.cabi.org

CABI is a trading name of CAB International

CABI Head Office
Nosworthy Way
Wallingford
Oxfordshire OX10 8DE
UK

Tel: +44 (0)1491 832111
Fax: +44 (0)1491 833508
E-mail: cabi@cabi.org
Website: www.cabi.org

CABI North American Office
875 Massachusetts Avenue
7th Floor
Cambridge, MA 02139
USA

Tel: +1 617 395 4056
Fax: +1 617 354 6875
E-mail: cabi-nao@cabi.org

A catalogue record for this book is available from the British Library, London, UK.

Published jointly by CAB International and FAO.
Food and Agriculture Organization of the United Nations (FAO)
Viale delle Terme di Caracalla, 00153 Rome, Italy
website: www.fao.org

ISBN: 978 1 84593 414 9 (CABI)
ISBN: 978 9 25105 964 7 (FAO)

Typeset by SPi, Pondicherry, India.
Printed in the UK by the MPG Books Group

Contents

Contributors

C.J. Augustyn, *Marine and Coastal Management, Cape Town, South Africa*

G. Bianchi, *Fisheries Management and Conservation Service, Food and Agriculture Organization (FAO) of the United Nations, Rome, Italy*

A. Charles, *Saint Mary's University, Halifax, Canada*

K.L. Cochrane, *Fisheries Management and Conservation Service, Food and Agriculture Organization (FAO) of the United Nations, Rome, Italy*

C. De Young, *Development and Planning Service, Food and Agriculture Organization (FAO) of the United Nations, Rome, Italy*

W.J. Fletcher, *Department of Fisheries, Perth, Western Australia*

S.M. Garcia, *Fisheries and Aquaculture Management Division, Food and Agriculture Organization (FAO) of the United Nations, Rome, Italy*

A. Graham, *World Wildlife Fund (WWF) International, Surrey, UK*

C. Grieve, *Meridian Prime*

D. Johnson, *Australian Fisheries Management Authority (AFMA), Canberra, Australia*

J. Kolding, *University of Bergen, Bergen, Norway*

R. Hecky, *Department of Biology and Large Lakes Observatory, University of Minnesota, Duluth, USA*

A. Hobday, *CSIRO Division of Marine and Atmospheric Research, Hobart, Australia*

R. McLoughlin, *Australian Fisheries Management Authority (AFMA), Canberra, Australia*

O.A. Misund, *Institute of Marine Research, Bergen, Norway*

O. Mkumbo, *Lake Victoria Fisheries Organization, Jinja, Uganda*

N. Rayns, *Australian Fisheries Management Authority (AFMA), Canberra, Australia*

P. Sandberg, *Directorate of Fisheries, Bergen, Norway*

H.R. Skjoldal, *Institute of Marine Research, Bergen, Norway*

K. Sherman, *National Oceanographic and Atmospheric Administration (NOAA), Washington, DC, USA*

K. Short, *World Wildlife Fund (WWF) International, Surrey, UK*

G. Silsbe, *Department of Biology, University of Waterloo, Ontario, Canada*

M. Sinclair, *Bedford Institute of Oceanography, Dartmouth, Canada*

J. Sígurjónsson, *Marine Research Institute, Reykjavík, Iceland*

T. Smith, *CSIRO Division of Marine and Atmospheric Research, Hobart, Australia*

T. Smith, *Australian Fisheries Management Authority (AFMA), Canberra, Australia*

R. O'Boyle, *Bedford Institute of Oceanography, Dartmouth, Canada*

M.J. O'Toole, *United Nations Development Programme, Windhoek, Namibia*

K. Thórarinsson, *The Federation of Icelandic Fishing Vessel Owners, Reykjavík, Iceland*

G.R. Tromble, *National Oceanographic and Atmospheric Administration (NOAA), Washington, DC, USA*

G. Valdimarsson, *Fish Products and Industry Division, Food and Agriculture Organization (FAO) of the United Nations, Rome, Italy*

M. Vierros, *Institute of Advanced Studies, United Nations University, Tokyo*

I. Winsnes, *Ministry of Environment, Oslo, Norway*

T. Worcester, *Bedford Institute of Oceanography, Dartmouth, Canada*

P. van Zwieten, *Aquaculture and Fisheries Group, University of Wageningen, Wageningen, The Netherlands*

Foreword

Throughout history the seas around Norway have provided an abundant supply of fish as a high-quality and nutritious source of food for the population.

Implementing the ecosystem approach to fisheries management is an important step in the process of establishing a sound basis for sustainable harvest of the marine living resources. Thorough knowledge of the various forces influencing marine ecosystems is fundamental in order to achieve optimal management regimes. In our efforts to understand all the various mechanisms in marine ecosystems we must not forget, however, that the only forces we can control are man's various uses of the sea.

Fisheries management shall in its broadest context include all aspects related to sustainable harvest. When constructing management regimes for fish stocks we have to consider not only the fish stocks themselves and the effects of the particular fisheries on the marine ecosystem, but also the societal effects of the management plan. In my opinion, this is implementation of the ecosystem approach to fisheries management in its broadest understanding.

There has been a lack of a clear and common understanding of what the ecosystem approach to fisheries management means and entails, despite the fact that the term has been widely used in all sorts of publications from political documents to scientific papers. The Bergen Conference and the chapters presented in this book have contributed to explaining and demystifying the concept. The book provides a good overview of different aspects of the concept of ecosystem approach and provides useful examples and experiences from practical implementation. In the concluding remarks from the conference it is stated that although we are converging towards a common understanding, the concept has still to be further clarified and demystified. This can be achieved through learning by the undertaking and sharing of experiences, with road maps and plans being made as we go along.

The Norwegian government has already presented an integrated management plan for the Norwegian part of the Barents Sea and the areas outside

Lofoten. Work is in progress for a similar integrated management plan for the Norwegian Sea to be presented to the Norwegian parliament in spring 2009. Preparatory steps have also been taken to produce an integrated management plan for the Norwegian part of the North Sea, where the principles of the ecosystem approach are fundamental.

We are committed to a clean and rich sea, which can provide healthy food from healthy ecosystems. Used sustainably, the seas and the living resources can provide us with food and other ecosystem goods and services for the future generations to come.

Helga Pedersen
Minister of Fisheries
Norway

Preface

This book presents edited contributions to the Conference on Implementing the Ecosystem Approach to Fisheries, organized by the Nordic Council of Ministers and the Governments of Iceland and Norway, with technical support of the Food and Agriculture Organization (FAO) of the United Nations, held in Bergen, Norway from 26 to 28 September 2006.

Considered as a follow-up to the 2001 Conference on Responsible Fisheries in the Marine Ecosystem, organized jointly by Iceland and the FAO, and with the co-sponsorship of Norway, the Bergen Conference aimed at reviewing concepts and addressing implementation issues related to applying the ecosystem approach to fisheries. Furthermore, experiences made and constraints encountered so far could be exchanged, including strategies and best practices that will facilitate further implementation in practical fisheries management.

The ecosystem approach is central to the implementation of international agreements such as the UN Convention on Biological Diversity. Its principles are also embodied in the FAO Code of Conduct for Responsible Fisheries and in binding law such as the UN Fish Stocks Agreement. A political commitment to implement the ecosystem approach by incorporating ecosystem considerations into fisheries management, resulted from the 2001 Reykjavík Conference. This commitment was reaffirmed and consolidated at the World Summit on Sustainable Development in Johannesburg 2002, where a target year of 2010 was set for its achievement.

A total of about 170 participants from 38 countries and five continents attended the conference, including professionals with different backgrounds and experiences, such as scientists, fisheries management and conservation practitioners, and representatives from the fisheries industry, non-governmental organizations and other interested parties.

The chapters presented in this book have been peer reviewed to help verify factual information and improve the clarity of presentation.

Gabriella Bianchi and Hein Rune Skjoldal
Rome and Bergen
May 2008

Acknowledgements

This book is based on the contributions presented at the Conference on Implementing the Ecosystem Approach to Fisheries held in Bergen (Norway) on 26–28 September 2006, an initiative by the Nordic Council of Ministers and the Governments of Iceland and Norway. Sincere thanks are extended to all those who contributed to the organization and to the success of the conference.

Members of the Organizing Committee:

Dr Per Sandberg, *Directorate of Fisheries (Norway)*
Dr Kristján Thórarinsson, *The Federation of Icelandic Fishing Vessel Owners (Iceland)*
Mr Hein Rune Skjoldal, *Institute of Marine Research (Norway)*
Dr Gabriella Bianchi, *Food and Agriculture Organization (FAO) of the United Nations (Italy)*
Ms Kari Østervold Toft, *Institute of Marine Research (Norway)*
Mr Olav Lekve, *Directorate of Fisheries (Norway)*

The Organizing Committee benefited from the collaboration of several senior staff from FAO including:

Dr S.M. Garcia, *Director, Fisheries and Aquaculture Management Division*
Dr G. Valdimarsson, *Director, Fish Products and Industry Division*
Dr J. Csirke, *Chief, Fisheries Management and Conservation Service*
Dr K. Cochrane, *Senior Fishery Resources Officer, Fisheries Management and Conservation Service*

Special thanks to Ms Kari Østervold Toft for her remarkable contribution in making the conference run smoothly and pleasantly. Ms M.T. Magnan helped with final organization of the documents and figures.

1 The Bergen Conference on Implementing the Ecosystem Approach to Fisheries (Bergen, Norway, 26–28 September 2006): Summary and Main Conclusions

GABRIELLA BIANCHI,[a] PER SANDBERG,[b] HEIN RUNE SKJOLDAL[c] AND KRISTJÁN THÓRARINSSON[d]

[a]*FAO, Rome, Italy;* [b]*Directorate of Fisheries, Bergen, Norway;* [c]*Institute of Marine Research, Bergen, Norway;* [d]*The Federation of Icelandic Fishing Vessel Owners, Reykjavík, Iceland*

Abstract

The Bergen Conference was a follow-up to the Reykjavík Conference in 2001 and was organized by the Nordic Council of Ministers and the Governments of Iceland and Norway, with technical support from the FAO. The aims of the Conference were to review concepts and share experiences from implementation, and to identify strategies and best practices that will facilitate further implementation of the ecosystem approach to fisheries (EAF). The Conference was organized with four sessions on concepts and strategies, knowledge base, approaches and tools, and experiences from case studies, followed by a fifth session on the way forward.

Many terms have been used in relation to the ecosystem approach (EA), but we are converging towards a common understanding of the concept. With respect to fisheries, the EA has two dimensions: a vertical dimension of application of the EAF and a horizontal dimension of integration of fisheries with other sectors into a holistic management framework. The EA is a strategy and not a 'blueprint' action plan, and its application needs to be tailored to the specific ecological, social and cultural conditions in each geographical area. Application of the EA may start with present knowledge, but more focused ecosystem research is needed to make it more effective, and limited knowledge requires added precautions. Ecological risk assessment (ERA) may be an important tool to apply in an EAF, as may the use of marine protected areas (MPAs) in combination with other management measures. An EAF can be kept simple and implemented incrementally from existing measures in fisheries management.

 The Ecosystem Approach to Fisheries (eds G. Bianchi and H.R. Skjoldal)

Introduction

Ecosystem approach (EA) to management is a principle ascribed to, and adopted by, many governments and international organizations and agreements. The World's leaders in Johannesburg in 2002, at the World Summit on Sustainable Development, called for the application of an EA by 2010. The UN Convention on Biological Diversity (CBD) has the EA as a core element of its work programme,[1] and the UN Food and Agriculture Organization (FAO) has developed guidelines for application of the Ecosystem Approach to Fisheries (EAF) (FAO, 2003, 2005).

The Nordic Council of Ministers and the Governments of Iceland and Norway organized, with the technical support of the FAO, a conference on the implementation of the EAF in Bergen, during 26–28 September 2006. This conference was a follow-up from the Conference on Responsible Fisheries in the Marine Ecosystem, held in Reykjavík in October 2001 (FAO, 2002; Sinclair and Valdimarsson, 2003). The aims of the Bergen Conference were to review concepts and share experiences from implementation of the EAF, and to identify strategies and best practices that will facilitate further implementation in practical fisheries management. The Conference was attended by about 170 participants from 38 countries, including scientists, representatives of fisheries administrations, fishermen organizations and environmental non-governmental organizations (NGOs).

The Conference was organized with four consecutive sessions addressing: (i) concepts and strategies; (ii) knowledge base; (iii) approaches and tools; and (iv) experiences from case studies. In each session there were 6–9 invited or submitted oral presentations, followed by a general discussion led by a chairperson. All oral presentations are available at http://www.cieaf.imr.no. There was also a separate poster session with additional presentations related to the topics of the four sessions. Following peer review, the presented papers were published as conference proceedings. A fifth and final session was arranged as a panel discussion, including the session chairs from the previous sessions and supplemented with representatives from the fishing industry, management and research.

The United Nations Open-ended Informal Consultative Process on Oceans and the Law of the Sea (UNICPOLOS) at its seventh meeting held at the UN Headquarters from 12 to 16 June 2006 dealt with the issue 'Ecosystem Approaches and Oceans'. To provide continuity from this meeting, Lori Ridgeway (Canada), one of the two co-chairs, informed the Conference about the outcome of the New York meeting (Document UN GA A761/156). The UNICPOLOS meeting was attended by 101 states, 24 intergovernmental organizations (IGOs) and 16 NGOs with the aim of building a common understanding on EA and closing implementation gaps. While the approach has a broad international buy-in, many participants claimed that there was not enough knowledge to get started. The review on the implementation of the UN fish stock agreement showed that many countries would not take action because of lack of information. An

[1] http://www.biodiv.org/programmes/areas/marine/ecosystem.asp

important objective of the New York meeting was to demystify the concept of EA and to share experiences from its implementation from both developed and developing countries.

Ridgeway stressed that the lack of a clear agreed definition of EA should not be an issue delaying its implementation. Furthermore, EA is about managing human activities and should be implemented also where knowledge is incomplete. However, there is an inverse relationship between knowledge and precaution, and the more limited the knowledge, the more conservative (precautionary) the management measures should be. It is important to get started and improve understanding over time. Integrated management of human activities should still be based on sound sectoral management. Major challenges will be faced at the regional level, as regards for instance fitting the work of Regional Fisheries Management Organizations (RFMOs) into the cross-sectoral approach to management. Basic issues such as overcapacity need to be resolved regardless of whether EA is implemented or not. The main steps that should be taken to implement EA should cover including EA in national policy, increasing research funding, improving coordination among ministries and management bodies, and identifying stakeholders.

The authors of the present conference summary constituted the Steering Group for the Bergen conference. Here we provide a short summary of the presentations in the four sessions, and a summary of the main items arising from the discussions. There was no conference declaration or statement prepared from the meeting. However, based on the conference outcome, the Nordic Council of Ministers presented a statement to the FAO Committee on Fisheries (COFI) that considered the issue of EAF in connection with its 27th session in March 2007.

Concepts and Strategies

The first session dealt with concepts and strategies for the EA. The session was chaired by Michael Sinclair (DFO, Canada). He introduced the session by referring to other relevant meetings, such as the ICES/SCOR symposium on 'Ecosystem Effects of Fishing' (which took place in Montpellier, 1999); the Iceland/FAO/Norway Conference on Responsible Fisheries in the Marine Ecosystem (Reykjavík, 2001) and the 7th meeting of the UNICPOLOS (New York, June 2006).

Based on the documentation and the conclusions from these meetings, it was noted that the EAF encompasses diverse concepts creating confusion among stakeholders (the fishing industry, managers, policy makers and scientists). Three main conceptual pillars were therefore proposed as main components:

1. The effects of fisheries on ecosystems (e.g. trawling impacts and incidental mortality of vulnerable species).
2. The effects of ecosystems on fisheries (such as climate change impacts on abundance and distribution of commercially important stocks).
3. Attempts to 'manage ecosystems' through manipulation (e.g. to generate enhanced cod and shrimp biomass levels by limiting fishing of capelin, and fishing sea urchins to enhance kelp production).

Different stakeholders give different emphasis to each of the components, with the fishing industry most interested in the second and the third pillars, while conservationists and some NGOs are most interested in the first.

The main results from the Reykjavík Conference included a set of necessary conditions and strategies for the implementation of EA to fisheries. Among the main necessary conditions, reduction of fishing capacity and rights-based fishing were mentioned. Relevant strategies included:

- Integrated management of multiple fisheries and other ocean uses within a geographic context.
- Definition of a broader set of conservation objectives to sustain target species and ecosystem structure and functioning.
- Definition of management areas based on ecological boundaries adjusted to areas of administrative convenience as appropriate, recognizing that a nested approach will be required.
- Initiate an evolutionary rather than a revolutionary process.

An important result of the Bergen Conference should be that of providing a balanced and converged perspective of the diverse concepts of EA to fisheries, including the three pillars mentioned earlier.

Six presentations were included in this session. Gabriella Bianchi (FAO) noted that the principles that underlie the EAF, as presented in the FAO guidelines, are not new. They can all be found in the FAO Code of Conduct for Responsible Fisheries (CCRF)[2] which, in turn, was drawn up so as to be consistent with policy developments at the international level within the United Nations Convention on Environment and Development (UNCED) and the United Nations Convention on the Law of the Sea (UNCLOS). The EAF concept has drawn attention to these principles and on the need to put them into practice. Reference was made to the many denominations related to holistic approaches to management (e.g. EAF, Sustainable Livelihood Approach (SLA) and Integrated Coastal Zone Management (ICZM)). It was noted that these are largely consistent with each other in terms of broad sustainability objectives, but differ in the emphasis they give to the various dimensions of a management system, i.e. the human, ecological and institutional dimensions, and that their relevance depends on the context. A major distinction in approaches is perhaps between those that are cross-sectoral, integrating multiple ecosystem uses, and those dealing with a specific sector, like the EAF. Both cross-sectoral and sectoral approaches are relevant, they are complementary and could be implemented in parallel. Despite the progress made in embracing the EAF principles, it was concluded that reconciling short-term economic and social gains with long-term sustainability would still prove to be a major challenge.

The CBD adopted the EA in 1995. This is described as a strategy for integrated management of land, water and living resources, and is underpinned by 12 principles and 5 points of operational guidance (CBD COP Decision V/6[3]).

[2] http://www.fao.org/DOCREP/005/v9878e/v9878e00.htm

[3] http://www.biodiv.org/decisions/default.aspx?m=COP-05&id=7148&lg=0

Marjo Vierros (CBD) explained that these principles are interlinked and that their application should be balanced according to the local context. As a strategy, EA promotes conservation, sustainable use and equity. Although very similar in the basic principles, the CBD EA differs from the EAF in that it entails integrated (cross-sectoral) management. It was stressed that for ocean areas, the main challenge lies in integrating the various management approaches into a comprehensive and cohesive plan. She noted that there is no single way to implement the EA and that the scale has to fit the problem. The 7th meeting of the Conference of the Parties (2004) has developed implementation guidelines to facilitate further implementation of the EA.[4] Important challenges include integration between sectors and participation of stakeholders.

Kristján Thórarinsson (Iceland) noted that despite the overall agreement on its basic principles, perceptions still seem to be quite different as regards what EAF really entails. In Nordic countries such as Greenland, Iceland, the Faroe Islands and coastal Norway, fisheries are extremely important, both for economic and social reasons. Fisheries management is therefore very important and advanced management procedures have been developed. The need for decentralizing decision making was underscored, including the need to use existing institutions and mechanisms to incrementally add ecosystem considerations. The character of EAF concepts was further defined as *didactic* as opposed to *normative*, meaning that EAF should be a reference framework rather than a detailed plan of action advocating specific solutions. Furthermore, priorities for action should be set locally, to optimize the use of limited resources and capacity and reflect the needs at the local level. Possibly, a reduction in fishing effort, also advocated under the conventional fisheries management framework, would result in achieving broader ecosystem objectives. Finally, EAF would require a learning process of the various stakeholders. Managers should understand the implications of the new concepts, government authorities should reflect on the commitments made and the fishing industry needs to know what is expected.

Integrated assessment and management of marine resources and ecosystems in the respective regions are the main objectives of the comprehensive Large Marine Ecosystem (LME) programme network consisting of 121 countries involved in 17 LMEs, supported by funding from the Global Environmental Facility (GEF). Ken Sherman explained how the LME programmes are implemented following a five-module methodology (productivity, fish and fisheries, pollution, socio-economics and governance) that helps countries towards adopting practical joint governance. Indicators have been identified for each of the five modules. The LME programmes are based on an extensive collaboration with many international governmental organizations and NGOs including UNEP and FAO. It was noted that the application of an EA will entail a need for more funding. For example, the USA was expected to have to double their marine research budget to meet the knowledge requirements for EA.

Kathrine Short (WorldWide Fund for Nature (WWF)) presented WWF's work related to Ecosystem-based Management (EBM). Drawing from the FAO CCRF and other relevant international processes, WWF's strategy includes integrated

[4] http://www.biodiv.org/decisions/default.aspx?m=COP-07&id=7748&lg=0

management of high sea areas, and aims at sectoral engagement, mitigating fishing impacts, protecting areas and species, and providing alternative livelihoods. According to WWF's view, EBM principles relate to maintenance of ecosystem structure and functioning, and consider that human needs and values are based on a shared vision of all stakeholders and on scientific knowledge. Maintaining the structure and function of ecosystems should be the main purpose of management. Ecosystem manipulation in the sense of altering ecosystem structure by, for example, removing predators, should be avoided. WWF has developed guidelines for the practical implementation of EBM that lead through 12 operational steps.The desirability of developing an international toolkit to facilitate the application of ecosystem-based approaches was highlighted. This toolkit could include, for example, case studies, be related to policy and legislation (including incentives and enforcement), a minimum suite of indicators, social, ecological and economic aspects and examples of industry voluntary codes of conduct. A proposal in this direction was raised in connection with the forthcoming session of the FAO COFI in early 2007.

An expert consultation on the social, economic and institutional implications of implementing an EAF was convened by FAO in June 2006 and included 15 experts with natural and social scientific backgrounds, representing a wide range of interests. Cassandra de Young (FAO) explained how the meeting had been run and the main themes that were dealt with, such as human values, ecosystem services, benefits and costs of applying EAF, creating incentives, financing its application and the necessary policy and institutional frameworks. One of the main recommendations stemming from the June meeting was the development of supplemental FAO Technical Guidelines for Responsible Fisheries on the economic, social and institutional considerations of applying the EAF, providing a concise document highlighting how economic, social and institutional considerations can be integrated into the application of EAF.

The Knowledge Base

This session was chaired by Poul Degnbol (EC/DG Fisheries, Denmark). He opened the session with the key message that knowledge, in order to be useful for fisheries management, must first of all relate to objectives. Second, management decisions need to be simple and the knowledge supporting them must be communicated clearly so that it is understood, despite describing the complexities of ecosystems. Therefore, knowledge of EA to fisheries should aid understanding of the complexities of marine ecosystems and human interactions with them, while delivering information that can be easily understood and fed into the decision making process. Progress made so far is mainly related to the acceptance of the need to move from a predictive to a more adaptive approach to the knowledge–policy interaction. The knowledge scope for EA was presented based on CBD Decision V/6. Knowledge should be related to the objectives of conservation, sustainable use, and fair and equitable sharing of benefits, and to how to address these three objectives simultaneously. Another important research area, related to implementation, is the development of adaptive management.

Six presentations were included in this session, but it was noted that research needs for EA to fisheries were also dealt with under other sessions.

Robert O'Boyle (Canada) noticed how much attention so far had been devoted to the overall EAF framework, including overall and operational objectives and stakeholder participation, and how there was an urgent need to evaluate the scientific research requirements for effective implementation of the EA. These should include not only specific fishing impacts but also cumulative impacts across fleets and sectors. Based on experiences made from the Scotian Shelf, he illustrated the approach used to identify research priorities in Canada. Starting from each management objective or question posed, associated research is identified and its 'tractability' (probability that the issue can be resolved within 3–5 years) evaluated. The best venue for carrying out the research is also identified. The types of issues considered included the impacts of fisheries on marine ecosystems and the impact of the ecosystem on fisheries. In this context he underscored the challenge of interpreting causality and cumulative effects. Some examples were given of research questions and how these had been dealt with, including biodiversity, productivity and habitat issues.

Gunnar Stefansson (Iceland) presented the results of a study that compared the efficiency of control measures in relation to management objectives in an ecosystem context. The comparison included quota systems, effort controls and marine protected areas (MPAs). Some key results seemed to depart from general perceptions. For example, the results showed that, contrary to what is usually thought, MPAs may have the same effect upon a fish stock and its productivity as do conventional output regulations. On the other hand, MPAs do not guarantee enhanced fishing outside the protected area and they help only if they are large enough to cover most of the resources. MPAs can provide significant benefits if they are combined with other management measures. They can also represent a buffer against uncertainty under catch and effort control systems. Stefansson drew attention to the need for carefully designed MPAs in order to maximize their efficiency.

Greenland is a large country characterized by complex and dynamic marine coastal systems. Helle Siegstad (Greenland) described how in her country these complexities were dealt with in terms of establishing the scientific basis for EBM. She presented the ECOGREEN Programme, intended to improve the understanding of the physical and biogeochemical interactions of the marine ecosystems around Greenland, the ecosystem structure and functioning, human behaviour (drivers) and human activities. Two main research lines are related to natural sciences and social sciences, respectively, feeding into a third area, i.e. the interactions between social and natural systems. These, in turn, feed into management recommendations. ECOGREEN includes a monitoring programme and provides a framework for prioritizing research.

A programme that has come about in response to EAF is ECOFISH, presented by Kjellrun Hiis Hauge (Norway). The approach is multidisciplinary, within the natural science domain, and aims to develop an integrated system of models describing ecosystem functioning, focusing on processes of importance to harvestable stocks. The programme will revise the ecosystem observation system and develop a set of indicators covering a wide range of ecosystem properties. A feedback loop from models will be used to improve sampling schemes.

Another presentation within the natural science domain, by Erik Olsen (Norway), described an ecosystem monitoring programme for the Barents Sea based on extensive ecosystem surveys. These surveys included many components such as oceanography, pollution, pelagic, demersal and 0-group fish, invertebrates, benthos, plankton, marine mammals and seabirds. The main advantages of this programme were related to being able to obtain a synoptic view of the ecosystem, while an obvious challenge was reconciling different survey objectives and strategies.

Ecosystem-based approaches may be different in various ways, but they all share the fact that they deal with risk management. Thus, the approach taken by Australia, as presented by Rick Fletcher (Australia), provides a framework for implementing an EA, including a risk assessment process to systematically identify issues of priority, develop management objectives and identify indicators and management measures needed. The outcomes may be very different in different situations/fisheries. Based on experiences gained in Australia and Pacific Island States, it has become clear that the less industrial the fishery, the more community focus will be necessary. Furthermore, it seems that the main issues are related to governance, while 'ecosystem issues' have not been considered as a main problem. Final recommendations were related to not letting scientists run the process while encouraging a strong participation by those involved in the management of the fishery.

Svein Sundby (Norway) provided an overview of the main climatic processes that affect marine ecosystems. Climate affects various ecosystem components in various ways, both at the individual and population levels, and at different time and space scales. Examples of good correlations between annual temperature fluctuations and abundance of 0-group or juvenile fish are many. These can be related to interannual, decadal and multidecadal processes. The effects of large-scale, decadal climatic variability such as the North Atlantic Oscillation (NAO) index have been documented for many living organisms, ranging from trees to birds and marine mammals. Examples of multidecadal fluctuations (e.g. Atlantic Multidecadal Ocillation, AMO) can also be found, and the Norwegian spring-spawning herring is one of the fish stocks showing a response to such fluctuations. It was noted that, because of the strong relationship between zooplankton and pelagic fish biomass, measuring zooplankton biomass should be a priority.

Approaches and Tools for Managing Fisheries as Part of the Ecosystem Approach

This session was chaired by Serge Garcia (FAO). He introduced the session by illustrating the conditions needed or desirable for a successful application of an EA. An enabling environment, with political commitment, appropriate legal framework and rules, ministerial coordination, etc., is one of the prerequisites.

Effective implementation depends on additional factors such as adequate administration, common understanding of the framework, participation, availability of relevant information and successful integration among various interrelated programmes. Implementation can be facilitated thorough programmes of awareness raising and by various incentives/disincentives and capacity building, and can be assessed periodically. Conservation measures (such as establishment of MPAs) can also be implemented as part of the EA.

Grimur Valdimarsson (FAO) presented the industry perspective on EAF. There seems to be some scepticism on new demands and worries about extremisms. The industry perceives management objectives as often being complicated and sometimes contradictory. Overall, they look for clearer rights, shared responsibilities and simplicity in the new framework. Successful EAF would require a shift in approach, with the industry *playing a significant role* in its implementation.The importance of fishing rights for achieving sustainability objectives was underscored.

One of the tools often proposed for an effective implementation of EAF is the use of MPAs. Peter Gullestad (Norway) illustrated the Norwegian experience on the use of MPAs or area closures and regulations used in fisheries management, and examples were provided of their use in the management of redfish, lobster and seaweeds. He concluded that MPAs have been a management tool for many decades already, but that their use will become more extensive under an EA framework.

Modelling and simulations have been important tools to identify and assess/compare management options. Under an EAF, the work needs to be expanded to include multi-species interactions and environmental impacts on these. A joint project between Norway and Russia, aiming at developing such a modelling tool for the Barents Sea, was presented by Sigurd Tjelmeland (Norway). Preliminary results have shown the great importance of recruitment variability. The choice of model for the recruitment function is of critical importance for the simulation results, and the mechanisms behind the recruitment variability and function need to be more closely investigated.

Henning Winker (Germany) presented a single-species approach based on the concept of L_{opt}, suggested as an alternative to the approach adopted by the European Union that uses maximum sustainable yield (MSY) as a target reference. L_{opt} is the length of fish corresponding to the maximum cohort (year-class) biomass. This approach is more conservative than the MSY target in that it implies letting more fish grow to a large size. Thus, it favours the reduction of catches and discards of juvenile fish and pre-spawners.

The implementation of the EAF entails costs and benefits. Anthony Charles (Canada) emphasized that these should always be taken into account when considering alternative management strategies. Costs and benefits can be grouped according to categories such as ecological, economic, social and management. Costs and benefits should be assessed at different timescales. He underscored the importance of assessing the issue of distribution of costs and benefits among fishers and between them and society, which are central issues behind perceptions and social responses.

Communication was the main focus of the following two presentations. Wojciech Wawrzynski (Poland) highlighted the importance of marine science communication to the public, using media such as video programmes (an example was given at the conference). Scientific results are translated into a more comprehensible language and, in this way, become more easily available to the public. The importance of another aspect of communication, that between scientists, policy makers and stakeholders, emerged in the results of a study by Dorothy Jane Dankel (Norway), who had analysed the reasons for success and failure in a number of fisheries. The importance of communication would become even greater under an EA, mainly because this is characterized by a stronger emphasis on bottom-up approaches.

Hein Rune Skjoldal (Norway) reported on the important developments in the North Sea towards an EA to management. At an intermediate ministerial meeting in 1997, ministers and EU commissioners responsible for North Sea fisheries and the environment agreed to develop and apply an EA in order to integrate fisheries and environmental protection, conservation and management measures. This culminated in the Bergen Declaration from the 5th North Sea Conference in 2002, where a political commitment was made to implement an EA. The ministers agreed to a conceptual framework for the EA including an integrated set of Ecological Quality Objectives. The European Union has developed a proposed European Marine Strategy Directive that focuses on the implementation of the EA at the scale of geographically defined marine ecosystems (e.g. the Baltic Sea and the North Sea). In Norway, because of these developments, the Institute of Marine Research, which provides most of the scientific advice for fisheries management, has recently changed its structure to strengthen an ecosystem focus in its research and advisory work.

Serge Garcia (FAO) focused on the interface between fisheries assessments and decision making and the challenges under an EA. Decision making will have to balance tensions and reconcile different interests and management objectives, and the interface between science and policy has a key role to play in this respect. A key issue, however, is related to the complexity of the systems to be assessed and the validity of different approaches to address this complexity. The conventional scientific approach is related to positivism (aiming at unravelling the true laws of nature). Its adequacy to provide, in the short term, knowledge that can be usefully applied to policy making and management is questioned. An alternative approach, related to constructivism, questions the existence of such laws and aims at social construction of knowledge. Garcia concluded that the change towards a more constructivist (post-normal) approach was already taking place, and this was seen as being justified and necessary. An integrated advisory process (IAP) that combines the analytical process with a participatory process could provide a platform for this change to take place towards a system that utilizes both approaches.

Experiences from Case Studies

The chair, Lori Ridgeway (Canada), introduced the session by referring to the UNICPOLOS meeting in New York in 2006 that she had co-chaired (see Introduction). At that meeting, emphasis was given to demystifying the concept of EA, and the progress of implementing an EA was presented for selected countries and regions. The examples included both developed and developing countries. At the present conference, some of the same examples were presented as case studies. Some dealt with management at the cross-sectoral level, while others were related to the application of an EA within the fisheries sector.

The Barents Sea is rich in natural resources, both living (e.g. fish) and non-living (e.g. oil), and is the basis for considerable economic activity. As an example of an integrated, cross-sectoral approach, Inger Winsnes (Norway) presented a management plan for this region. The Government of Norway has adopted the EA to ocean management and the management plan for the Barents Sea is a step towards its practical implementation. The plan was developed to reconcile different uses by providing a framework that allows the exploitation of the various resources while maintaining the ecosystem structure and function. Goals and targets were set and agreed for this region. Governance is based on the establishment of a steering committee under the Ministry of Environment that includes representatives of relevant government agencies. The committee agrees on management measures. Advice is provided by a 'Management Forum', which receives input from research, monitoring and from the users. The management plan is a dynamic document and will be updated regularly, the first update foreseen for 2010. It is recognized that cooperation with Russia is important in order to include the whole Barents Sea as an ecosystem.

Jóhann Sigurjónsson (Iceland) recalled that the Reykjavík conference had concluded that many of the measures implemented under single-species management schemes were also useful under an EAF scheme. What was needed was their successful implementation. The adoption of an incremental and pragmatic approach to EAF was therefore seen as the way to go. Examples of how Iceland was incorporating ecosystem considerations in this pragmatic way were presented. It was concluded that this approach would eventually also contribute to a more holistic management approach.

Three countries from southern Africa (Angola, Namibia and South Africa), making up the coastal states of the Benguela Current Large Marine Ecosystem (BCLME), are committed to the implementation of the EA. Michael O'Toole explained how these countries use the opportunity provided by the GEF BCLME programme to strengthen progress towards this end through a project that consists of a cooperative effort by the management agencies of the three countries, the BCLME programme and FAO. Focusing on several of the main fisheries in the respective countries, the project has pursued a structured and participatory approach to identify gaps in existing management approaches and to prioritize measures to address these gaps. Costs and benefits are being measured in terms of the broad objectives applicable in each fishery. The results of this project provide a valuable framework for future refinement and implementation of the EAF as part of a wider cross-sectoral framework. The three countries have recently signed an agreement to establish a management commission for the Benguela Current LME.

Since the Reykjavík Conference in 2001, Australia has made good progress in implementing many of the elements of an EAF management. Richard McLoughlin

(Australia) explained that by 2007 all the elements relevant to the EA were to be fully integrated and implemented.The decision to move in this direction was taken by the Australian Government, with inputs from science, management and the industry.The main elements of the approach include implementing formal harvest strategies for target and by-product stocks in every fishery; undertaking ecological risk assessment (ERA) and developing a risk management response; implementing large-scale spatial management; enhancement of fishery data collection; and enhancing liaison and communication capacity.

Protecting, restoring and managing the use of coastal and ocean resources is one of the strategic goals of the National Oceanographic and Atmospheric Administration (NOAA), and NOAA Fisheries work to achieve this goal. Galen Tromble (USA) presented data from the USA which showed, despite the perceived challenges in implementing the EA, progress had been made in several regions where Fisheries Ecosystem Plans (FEPs) had been developed. FEPs describe the known components of ecosystems and main interactions for a given region. They increase the managers' awareness of how their decisions affect the ecosystem. Progress is also being made in the science needed to support the implementation of EAF. However, important challenges still remain and are related to providing management with decision support tools to help deal with increasing complexity of objectives and information, to the need for better communication and outreach to the public and to policy makers, and to the need to strengthen the statutory basis for the EA. Finally, governance issues are seen as very challenging.

Qisheng Tang (China) considered the Yellow Sea and illustrated some of the main management issues that need to be dealt with in this ecosystem. Major changes in species composition have been witnessed since the 1950s with an increase in small pelagics compared to bottom-dwelling and long-lived species. This is also reflected in a drop in trophic level during that period.The Yellow Sea LME is characterized by multiple uses and a major challenge is to look at the combined effects of these on the ecosystem.Another major challenge is how to deal with increasing demand for seafood while the ecosystem has a limited carrying capacity.Tang described how the science to support EA was developed in several GLOBEC (global ocean ecosystem dynamics) research programmes in China.

Jorge López (Nicaragua) provided information on the political commitment expressed by the Central American countries belonging to the organization OSPESCA (Organization del Sector Pesquero y Aquicola del Istmo Centroamericano) about the principles of sustainability and precaution that are consistent with an EA.

Implementing the Ecosystem Approach to Fisheries: The Way Forward

Panel discussion

The final session of the Conference was a plenary discussion assisted by a panel and moderated by Mike Sinclair (Canada). A bullet point summary of the preceding four sessions was prepared by a group consisting of the session chairs and the Conference Steering Group assisted by Robert O'Boyle (Canada).This summary was distributed to the meeting and presented to the plenary by O'Boyle.

The panel consisted of the session chairs (Poul Degnbol, Serge Garcia and Lori Ridgeway) supplemented with representatives from the fishing industry (Inge Halstensen), management (Peter Gullestad) and research (Jóhann Sigurjónsson). As backdrop to the discussion, each panelist briefly commented on what she/he perceived to be the main issues.

Inge Halstensen (Norway) gave a brief sketch of the history of the relationship between the fishing industry and fisheries science in Norway during the last 50 years.This history had gone through a time of confrontation and distrust, to the present situation with much improved communication and a feeling from the fishing industry that they are included in the process. Halstensen emphasized that modern fishing vessels are well equipped and can provide valuable information on the fish stocks to the fisheries scientists.

Peter Gullestad (Norway) is the Director of Fisheries in Norway. He emphasized that the EA is incremental as a process, but represents a revolutionary change in the way we need to think about fish and fisheries in the marine ecosystem. Multi-species interactions, climate forcing, recruitment variability, bottom habitats and genetic effects are keywords for a broadened ecological context for fisheries. He challenged the scientists to coordinate their work better to meet the need for a broader cross-disciplinary approach, and pointed to the need for more integration in the scientific advisory process supporting management.

Jóhann Sigurjónsson (Iceland) is Director of the Marine Research Institute in Iceland. He said that the EA should not be seen as a threat, but as an opportunity to do better and to avoid mistakes of the past. Management actions still need to be taken within the fisheries sector, and effort reduction is one measure that will lead to less environmental impacts by fisheries.The increased information needed on different aspects of the marine ecosystems may mean that we have to be satisfied with qualitative assessments where quantitative assessments are difficult to perform. The cost of science will inevitably increase, and one issue is to secure the motivation of the fisheries scientists who may see the EA as just another burden put on their shoulders.

Qisheng Tang (China) is Director of the Yellow Sea Fisheries Institute. Since the late 1980s there has been a move to ecosystem focus in the management of this sea. At the same time, major changes in the ecosystem have been taking place that are not well understood.There is a need for better ecosystem knowledge to advise the government about management of the Yellow Sea to secure long-term food production from capture-fisheries and aquaculture. Elements in the implementation of the EA for the Yellow Sea include basic research in China-GLOBEC, monitoring in the GOOS (Global Ocean Observing System) framework, and management as an LME.

Poul Degnbol stressed that the bottom line is to regulate fishing activities. The EU has issued a Green paper on its maritime policy,[5] including the proposed

[5] http://ec.europa.eu/maritimeaffairs/pdf/com_2006_0275_en_part2.pdf

Marine Strategy[6] as the environmental sustainability pillar. In 2003, the status of European fisheries was described as being generally poor and there has been little improvement in this bleak situation. The Precautionary Approach was seen as part of the problem since this builds upon limits to be avoided. With a lack of clear targets, staying out of real trouble has come to be seen as an acceptable and a *de facto* target. There is the need for political will to move from bad to good. This should include a move to an adaptive approach, where the fishing industry is brought into the management process. The establishment of Regional Advisory Councils (RACs) was seen as one step in moving away from a top-down to a more inclusive approach.

Serge Garcia considered the reliance on limit-based precautionary reference points and the lack of targets as a serious drawback for fisheries management. He said it was like driving at high speed along roads without knowing where one was going, and pointed to the tensions between holism and reductionism in science and management, and to the need to bring socio-economic aspects more strongly into play in the management process.

Lori Ridgeway spoke as a policy maker or 'integrator', with a focus on the EA as a framework for planning and decision making. She pointed to specific challenges that had to be addressed for successful implementation and application of the EA. These included identification of core fisheries management issues that will have to be tackled, irrespective of an EA. Another issue was the question of how to build the appropriate buy-in from industry, government and other stakeholders for the difficult decisions that an EA can entail. The multiplicity of risks and benefits that must be taken into account, as well as the need for increased precaution, may mean lower activities in all sectors including fisheries. How then to devise win–win outcomes that may secure buy-in and provide incentives for cooperation and compliance by industry and other stakeholders? The EA is an inclusive approach to planning and decision making, and governance therefore matters. Ridgeway raised the question of how to create inclusive stakeholder processes without bogging down the whole decision making process itself. Which institutional arrangements or 'tables' do we need, which decisions are to be made at these 'tables', and by whom? She finally raised the question of whether there is need for changes to policy and legal frameworks. Are there policy gaps that hinder the implementation of the EA?

The moderator invited brief interventions by representatives from the fishing industry and environmental NGOs. Other conference participants also gave brief interventions during the panel discussion.

Main items from conference discussions and presentations

A summary of the discussions during each of the first four sessions and the final panel discussion is provided here. We have combined what we consider the main points made during discussions or presentations under four subheadings

[6] http://ec.europa.eu/environment/water/marine.htm

corresponding to the four themes addressed by the Conference. This is followed by some concluding remarks.

Concepts and strategies

MANY TERMS BUT CONCEPTUAL CONVERGENCE A wide range of terms are being used related to EA, such as *ecosystem management*, EBM, *ecosystem-based fisheries management* (EBFM) and EAF. While some of these terms may lack clear definitions and be used in different ways, there are core elements in common across the different terminologies. We are therefore converging towards a common understanding of the concept of EA and can move forward despite the differences in terminology. However, the different terms and their different uses in different contexts still contribute to confusion and a lack of clarity.

APPLICATION WITHIN AND ACROSS SECTORS Perhaps the most important distinction in concepts is between application of the EA within the fisheries sector (as well as in other sectors) and across multiple sectors including fisheries. The latter is the truly holistic approach as used, for instance, in CBD, while the EAF of FAO is an example of the former. These two dimensions (vertically within a sector and horizontally across sectors) should not be seen as opposing, but rather as complementary to each other. Sound sectoral management is likely to be a prerequisite to achieve successful integrated management of human activities across sectors. The need for cross-sectoral integration will vary depending on the specific circumstances and is likely to be the greatest in the coastal zone where pressures from different human activities are most expressed.

STRATEGY, NOT 'BLUEPRINT' ACTION PLAN The EA is a strategy for integrated management that builds on a number of general principles. The application of the EA needs to be tailored to the specific ecological, social and cultural conditions in each specific geographical area. There are therefore many different ways to implement the EA that are consistent with the strategy and its general principles. The EA is not a detailed and prescriptive action plan to be applied everywhere without adaptation to local conditions.

MULTIPLE OBJECTIVES: MANY STAKEHOLDERS The broadening to more ecosystem considerations in fisheries management, as well as the need to coordinate with other sectors, means that multiple objectives are a key feature of the EA. This implies extensive communications between different stakeholder interests, researchers and managers. New mechanisms of interaction need to be developed, which are interactive and exploratory of options and not based on a one-way process from predictions through management proposals to consultations. The objectives are policy objectives, and their translation into operational management objectives. An important part of the consultations is to reach a common understanding and agreement on the objectives.

ECOSYSTEM MANIPULATIONS There is a two-way interaction between fisheries (and any other relevant human activities in other sectors) and marine ecosystems in that fisheries impact the ecosystem and the ecosystem conditions affect fisheries.

Fisheries reflect *de facto* a human modification of the ecosystem. Methods for stock rebuilding or habitat rehabilitation are indeed manipulations, aiming to reverse excessive stress. However, the complexity of the two-way interaction between fisheries and ecosystems limits our ability to assess impacts and predict consequences of remedial measures. The principle of sustainability is to use nature within its own limits so that its productive and regenerative capacity and biodiversity are not reduced or threatened in the long term.

Knowledge base

START WITH PRESENT KNOWLEDGE Lack of knowledge should not be used as an excuse to delay implementation of the EA. There is always some knowledge of any area and we can start from that basis. While good knowledge about the ecosystem is an advantage for effective application of the EA, ecosystem considerations in fisheries and integration across sectors can start with present knowledge and be improved as we go along. There is however an urgent need to improve knowledge and understanding of social aspects and institutional frameworks required for adaptive change.

LIMITED KNOWLEDGE MEANS MORE PRECAUTION There is an inverse relationship between the degree of scientific certainty and the degree of caution we need to exert in order not to adversely affect nature. The better our knowledge, the more precisely we can predict impacts and advise on management measures. In contrast, poor knowledge entails limited ability to predict and consequently the need to exercise considerable precaution in our measures. This is one of the main challenges of the EA: how to balance knowledge and precaution and how to communicate this balance to achieve broad consensus among stakeholders.

MORE FOCUSED ECOSYSTEM RESEARCH IS NEEDED 'More research is needed' is a common statement from scientists, who have limited credibility in this context since they are stakeholders in the activity of science. However, there is general acceptance that the broader ecosystem considerations that are needed in fisheries, and cross-sectoral in other sectors as well, will require more knowledge and information, which must be supplied through increased research and monitoring. Understanding the biodiversity–productivity linkage, trophic processes, habitat resilience to human disturbance and impacts of climate variation and change are key natural science themes that need to be further addressed and explored. A part of the increase can no doubt be achieved through better coordination and use of current knowledge and resources spent in different sectors, as well as by stricter prioritization of relevant research that provides us with better insight into the workings of the marine ecosystems.

Approaches and tools

ERA CAN BE AN IMPORTANT TOOL Risk assessment is a common tool in business and industry at large. A similar approach can be usefully applied within an EA, where ERA related to human well-being, ecosystem conservation and sustainable use can be a core tool. ERAs need to be carried out for all fisheries where relevant, and can be applied both in data-rich and data-poor situations. Risk assessments should be linked with other broader assessments of environmental or ecosystem status and of impacts from other human activities on the marine ecosystem. ERA is a tool that can help to identify critical issues for implementing EAF, as well as to sort out which issues can and cannot be influenced by management actions.

MPAS AND AREA CLOSURES IN FISHERIES Area closures and fishing restrictions in MPAs have been widely used as measures in fisheries management for many decades. They have been used to protect juvenile or spawning fish and important fish habitats or to regulate different fisheries. Area restrictions established for fisheries management purposes can also serve broader conservation objectives. MPAs can be an important tool but they are not a panacea and are most useful if used in combination with other management tools. In this respect, fleet behaviour needs to be carefully considered when using this management tool.

MORE EMPHASIS ON THE HUMAN DIMENSION The EA is primarily about managing human beings. It is therefore important to include socio-economic and institutional considerations in EA planning needed for adaptive change to achieve the dual objectives of socio-economic benefits and environmental sustainability. Fair and equitable sharing of benefits is also an element that needs attention. Knowledge and tools are needed to facilitate inclusion of equity and social aspects, and to strengthen the human dimension of the EA. People tend to respond more to incentives than to commands. Therefore, objectives and incentives need to be aligned in order to facilitate successful implementation of EA to fisheries. Cost-benefits analysis should always be undertaken when considering alternative management strategies. The issue of distribution of costs and benefits among fishers and between them and society and between generations is a central issue behind perceptions and social responses.

INCREMENTAL IMPLEMENTATION: REVOLUTION IN THINKING Many of the elements of today's fisheries management, like effort or fleet control, harvest control rules, modelling and simulations, will continue to play important roles under an EA to fisheries. If successfully implemented, they can contribute substantially to sustainable use and ecosystem conservation. Thus, the EA can build on existing elements and be further implemented and improved in an incremental or step-wise manner. However, what may be required is a radical change or revolution in our thinking and attitudes towards ecosystems, ecological relationships, stakeholder involvement and collaborative frameworks.

WE NEED TARGETS SO THAT WE KNOW WHERE TO GO Staying out of real trouble and avoiding falling off the cliff are not really good targets. Sadly, this is the current situation where avoiding limits, often with limited success, is the common practice in fisheries management in many places. We need to develop and apply ecologically based targets that help us know where to go so that we can achieve the dual objectives of sustainable use and ecosystem conservation. This will help us towards achieving the commitment from the WSSD (2002) to rebuild fish stocks to MSY levels by 2015[7].

[7] http://www.un.org/esa/documents/WSSD_POI_PD/English/POIChapter4.htm

Experiences from case studies

IMPLEMENTATION IS UNDERWAY The EA to management in general, and to fisheries in particular, is underway in many nations and in international contexts. Many governments have adopted the principle and EA is being implemented nationally. There is a wide range of cases where two or more countries collaborate across exclusive economic zones (EEZ) borders to implement EA to the management of LMEs. Learning by doing is important, and the range of national and international projects is providing important lessons that should be broadly shared as a basis for improvement as we go along.

KEEP IT SIMPLE: EA IS NOT MYSTICAL While ecosystems may seem complex with their diversity of species, populations and habitats, the EA is fairly straightforward. There is nothing mystical about either ecosystems or EA. Experiences that are emerging from case studies suggest that the EA can be kept simple, starting with existing institutional structures, and modified and improved as we go along. The most important thing is perhaps a change in mindset to be more open to collaboration and to stakeholder involvement.

GET INVOLVED, PLEASE Stakeholder involvement is important, as is the need for improved communication between science, policy making and society. Stakeholder involvement and the need for broader considerations both within the fisheries sector and across the different sectors require new approaches such as an integrated advisory process (IAP). Such processes already operate in a few countries and should be strengthened and generalized, although their application may require additional costs when compared to conventional management.

Concluding Remarks

Implementing the EAF is often perceived as a very challenging goal and the concept has intimidated government institutions worldwide. The Bergen Conference clearly resulted in the recognition that EA core issues are not new. While a number of holistic approaches are being proposed that may differ in emphasis, it was recognized that they largely converge conceptually by aiming to implement principles of sustainable development by harmonizing ecosystem sustainability with human well-being. EAF is only the consolidation and actual implementation of principles and policies that are already agreed, such as the UNCLOS, the CBD and the CCRF. EA is a strategy that should promote conservation, sustainable use and equitable sharing of ecosystem services. There are many ways in which it could be implemented, depending on context, means, culture, etc.

The Reykjavík Conference in 2001 can be considered as a milestone in terms of putting into focus the issue of ecosystem considerations in fisheries management. The World Summit on Sustainable Development (WSSD) (Johannesburg, 2002) specifically refers to Reykjavík and sets 2010 as the time frame for the application of the EA. Although ambitious, this target date has urged countries to take initiatives towards the realization of an EA; the implementation phase towards this end, as shown by this conference, seems to be well on its way.

The UNICPOLOS meeting in June 2006 focused on demystifying the concept of EA by sharing experiences from its implementation around the world. The Bergen Conference has followed this path and contributed to the convergence of various perceptions towards a common understanding of the EA concept. We are not quite there yet as some differences in perception still persist. Provision of clear definitions and explanations of terminology is one way to improve clarity and avoid misunderstandings on semantic grounds. Through learning by doing and sharing experiences as we go along in the further implementation of the EAF, we will no doubt contribute to a common understanding.

Concluding remarks at the Conference were provided by Hein Rune Skjoldal (Norway). He underlined three points in his summary:

1. The EA is an approach to management and not to science. It has implications and requirements to science as one of the supporting elements of an EA framework.
2. The EA requires ecosystems. These should be defined geographical entities as increasingly recognized, for instance, by UNICPOLOS 2006. The LMEs identified worldwide are good examples. Once ecosystem boundaries are defined, it becomes obvious who are the competent authorities and relevant stakeholders for its management.
3. The EA has two main dimensions - vertically within a sector (e.g. fisheries) and horizontally across sectors. Both dimensions should be seen as relevant and complementary. This distinction may help clarify the EA concept.

References

FAO (2002) Report of the Reykjavík conference on responsible fisheries in the marine ecosystem. Reykjavík, Iceland, 1–4 October 2001. FAO Fisheries Report No. 658. FAO, Rome, Italy, 128 pp.

FAO (2003) The ecosystem approach to fisheries. FAO Technical Guidelines for Responsible Fisheries. No. 4, Suppl. 2. FAO, Rome, Italy, 112 pp.

FAO (2005) Putting into practice the ecosystem approach to fisheries. FAO, Rome, Italy, 76 pp.

Sinclair, M. and Valdimarsson, G. (eds) (2003) *Responsible Fisheries in the Marine Ecosystem*. CAB International, Wallingford, UK and FAO, Rome, Italy, 426 pp.

2 The Concept of the Ecosystem Approach to Fisheries in FAO

GABRIELLA BIANCHI

FAO, Rome, Italy

Abstract

The Reykjavík Conference on 'Sustainable Fisheries in the Marine Ecosystem' in 2001, and the commitment made by FAO members in the Reykjavík Declaration to responsible and sustainable fisheries in the marine ecosystem, strengthened and legitimized the ecosystem approach to fisheries (EAF) as the reference framework for managing the fishery sector. This direction was further reinforced by the World Summit on Sustainable Development (WSSD) (Johannesburg, 2002) that recommended the implementation of an ecosystem approach to aquatic resources management by 2010.

The requirements implicit in the EAF, such as addressing more complex and poorly understood systems and the associated uncertainty, increasing data requirements, consideration of several timescales, and the recognition of the importance for a broader stakeholder participation at various stages of the fisheries management process, have initially intimidated many and fostered a perception of the EAF as a difficult and perhaps impossible task. Furthermore, the understanding of the basic principles of what an ecosystem approach actually implies are still not always understood or agreed upon.

Attitudes are however changing, both at the international and at the national levels, and a pragmatic approach has been adopted in many places to see how conventional fisheries management can be improved to incorporate ecosystem considerations and more properly deal with the social dimension.

While it could be argued that a large proportion of FAO's work is either directly or indirectly promoting the application of an ecosystem approach, FAO has also specifically addressed EAF by developing guidelines for its implementation, following the mandate issued in connection with the Reykjavík Conference. Promotion has been conducted in a number of conferences, regional and national initiatives have been monitored informally and specific case studies have been implemented through field projects.

This contribution will summarize the developments in the conceptual framework that have taken place in FAO since the Reykjavík Conference, and try to put emphasis on the basic principles that should underpin the application of the EAF. Despite the progress made, important challenges still need to be faced. These are not only related to the direct drivers of marine ecosystem change, such as fisheries and other sectors utilizing goods and services from the marine ecosystem, but also related to the indirect drivers such as

 The Ecosystem Approach to Fisheries (eds G. Bianchi and H.R. Skjoldal)

changes in human population coupled with a widespread aspiration for an improved standard of living, and global economic policies.

Introduction

The political commitment to an ecosystem approach to fisheries (EAF) formally materialized in connection with the 'Reykjavík Conference on Sustainable Fisheries in the Marine Ecosystem' in 2001, when 45 participating countries signed a declaration and a pledge to incorporate ecosystem considerations in fisheries management (http://www.fao.org/docrep/meeting/004/Y2211e.htm). This commitment was reinstated in connection with the World Summit on Sustainable Development (WSSD) in Johannesburg in 2002, where 2010 was agreed as target for its application (WSSD, Plan of Implementation, Paragraph 29d). Today, while there is broad acceptance worldwide that an ecosystem approach is the most appropriate framework for managing exploitation of renewable resources, including fisheries, many are still grappling with the interpretation of the concept or with defining strategies for its implementation.

This contribution will briefly outline the developments in the conceptual framework that have taken place in the international arena, relevant to sustainable fisheries, that have lead to the formulation of the EAF and the basic principles that underpin its application, based on FAO's perspective. Guidance to practical implementation is provided by FAO and the highlights will be presented here, together with examples of practical implementation.

Development of International Instruments Relevant to the EAF

Awareness of the negative effects regarding society's use of natural resources has been reflected in international instruments since the late 1950s. The UN Conference on the Human Environment, held in Stockholm in 1972, recognized that:

> [i]n the long and tortuous evolution of the human race on this planet a stage has been reached when, through the rapid acceleration of science and technology, man has acquired the power to transform his environment in countless ways and on an unprecedented scale.

These concerns were very relevant also for fisheries, and a number of fish stocks had been already heavily or overexploited in immediate post-war times, particularly in the northern hemisphere. The Stockholm Conference also stressed that '[e]conomic and social development is essential for ensuring a favourable living and working environment for man and for creating conditions on earth that are necessary for the improvement of the quality of life'.[1] In coupling the need for

[1] http://www.unep.org/Documents.Multilingual/Default.asp?DocumentID=97&ArticleID=1492&l=en

sustainable use with that of human development, the Stockholm conference laid the basis for the concept of ecologically sustainable development, the foundation of EAF. Principle 25 of the Stockholm declaration calls for states to 'ensure that international organizations play a coordinated, efficient and dynamic role for the protection and improvement of the environment'. Consistent with the above call, sustainable development principles in fisheries have been reflected in international instruments and in the work of international intergovernmental organizations at least along three main strings of the international policy arena (Turrell, 2004). These are related to legal, environmental and fisheries management aspects, respectively (Fig. 2.1).

The legal string has as milestone the UN Convention on the Law of the Sea (UNCLOS) of 1982. The resulting United Nations Law of the Sea (UNLOS) has provisions for sustainable use of target stocks, also taking into account non-target species and species interactions. The UN Fish Stock Agreement (1995) notes the importance of preserving biodiversity, maintaining integrity of marine ecosystems and minimizing risks. As part of this string, the United Nations open-ended

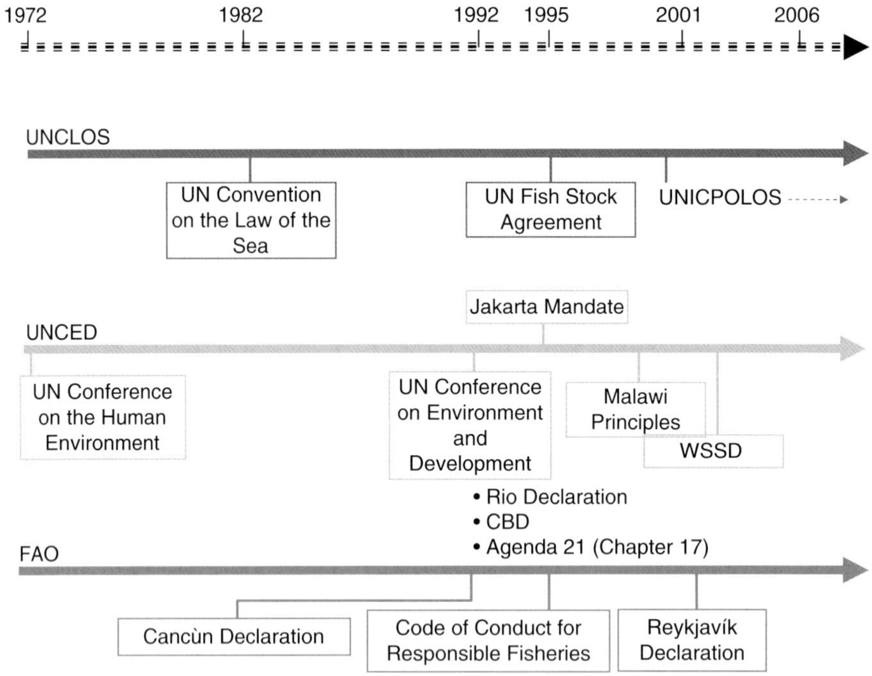

Fig. 2.1. Schematic representation of developments in international instruments relevant to the ecosystem approach to fisheries. UNCLOS = United Nations Convention on the Law of the Sea; UNCED = United Nations Commission on Environment and Development; WSSD = World Summit on Sustainable Development; CBD = Convention on Biological Diversity; UNICPOLOS = United Nations Informal Consultative Process on the Law of the Sea.

Informal Consultative Process on Oceans and the Law of the Sea (UNICPOLOS) was appointed by the United Nations General Assembly (UNGA) in 1999 to 'deal specifically with developments in ocean affairs and the law of the sea'. In 2006, in connection with its 7th meeting, UNICPOLOS dealt specifically with 'ecosystem approaches and oceans', with the aim of building a common understanding on EA and to close implementation gaps (UNGA, 2006).

The environmental string has its origins in the Stockholm Conference. In 1983, the UN set up the World Commission on Environment and Development, led by Gro Harlem Brundtland of Norway, that put forward more clearly the concept of sustainable development as an alternative approach to one simply based on economic growth - one 'which meets the needs of the present without compromising the ability of future generations to meet their own needs'. In 1987, the Commission delivered its report and, based on its findings and recommendations, the UNGA called for the UN Conference on Environment and Development (UNCED, also known as the Earth Summit) to come to an understanding of 'development' that would support socio-economic development and prevent the continued deterioration of the environment. In 1992, the Earth Summit resulted in agreements, such as Agenda 21, and legally binding Conventions, including the Convention on Biological Diversity (CBD) having an overarching significance for all human activities. The CBD marked the beginning of a new era, and its success would be measured by the implementation - locally, nationally and internationally - of its agreements. Following on from the CBD, and of direct interest for aquatic resources use, were the Jakarta Mandate on coastal and marine biodiversity (Jakarta, 1995) that specifically linked issues of biodiversity and conservation to fishing activities, and the Malawi Principles (1998) establishing the basic requirements for an ecosystem approach. Ten years after the Earth Summit, at the WSSD (2002), a commitment was made to implement an EA to fisheries by 2010 (WSSD, Plan of Implementation, paragraph 29d).

FAO is the mandated UN agency for Fisheries Management. FAO's normative work is largely concerned with developing intergovernmental understanding on key issues for policy coherence, development of global norms to be agreed on and implemented by member countries, monitoring compliance, collecting global and regional information and statistics to be made available in the public domain and for global and regional analyses and trends, and promoting best practice. While the broad lines of the FAO Fisheries Department areas of work were laid down already at its establishment in 1945, the work focus has been changing over time and it has largely been consistent with and reflecting international developments in the area of natural resources conservation and management (Garcia, 1992). The FAO Code of Conduct for Responsible Fisheries (CCRF, 1995) can be considered as the first milestone within fisheries in terms of capturing principles of sustainable use found in the overarching legal and environmental strings, and establishing principles and standards for the conservation, management and development of all fisheries. These principles are captured and their implementation given more impetus through the EAF (FAO, 2003).

Yet the above developments at the international policy level did not immediately translate, in most cases, into the necessary fundamental changes at national policy and management levels. The Reykjavík Conference and related declaration (2001) can be considered as an attempt to build a bridge between the commitments on sustainable use that countries had agreed to over the years and their actual implementation within the fisheries sector. In this sense, Reykjavík 2001 can be considered as a major step towards operationalization of the principles of sustainable development in fisheries.

Based on the above, and in relation to the incorporation of environmental concerns in fisheries management, three main phases can be detected at the global level:

- The phase of raising awareness, with its roots in the Stockholm Conference (1972) and culminating with the Earth Summit (1992), with principles of sustainable use of living resources being reflected in the UNLOS of 1982.
- Harmonization of fisheries management with environmental objectives, with the development of international instruments at sectoral level, such as the CCRF (1995).
- A third phase, that of commitment to implementation, as stated in the Reykjavík Declaration, and made even more urgent through sensitization of the public, largely by NGOs.

The Ecosystem Approach: Definition and Basic Principles

There are various definitions and denominations in use to indicate holistic approaches to management that ultimately aim at the implementation of sustainable development concepts in fisheries, to be achieved through democratic and transparent practices that take account of diverse societal interests and using mechanisms that allow participation of stakeholders in the planning and decision making processes. These, despite distinct wording or emphasis, refer to approaches that share more commonalities than differences. There has been great uncertainty so far, however, on what the ecosystem approach actually entails. Perceptions have ranged from that of a purely natural science-driven endeavour, based on detailed understanding of ecosystems' structure and functioning, to the perception that the use of marine protected areas (MPAs) is synonymous with EAF. In the FAO definition, the word 'ecosystem' is used to emphasize the holistic nature of the approach, addressing the fishery system as an integrated social-ecological system. Human beings are an integral part of the ecosystem. FAO's definition reflects this notion:

> An Ecosystem Approach to Fisheries strives to balance diverse societal objectives, by taking account of the knowledge and uncertainties about biotic, abiotic and human components of ecosystems and their interactions and applying an integrated approach to fisheries within ecologically meaningful boundaries.
>
> (FAO, 2003)

The above clearly addresses both human and ecological well-being, thus combining two concepts: that of conserving biodiversity, ecosystem structure and functioning, and that of fisheries management dealing with providing food, income and livelihoods for humans. The definition therefore provides the basis for mainstreaming sustainable development into fisheries policy frameworks and decision making at national, regional and global levels.

The main principles that characterize the EAF are reflected in the above definition. Most appear in international instruments such as the UNLOS, the Fish Stock Agreement and the CCRF. They have been described and commented on in various documents (FAO, 2003; Garcia *et al*., 2003; Garcia and Cochrane, 2005). Here some key principles are presented and reorganized according to three main frameworks, the normative, the operational and the cognitive, that underpin societal organizational processes.

The *normative framework* consists of the agreed high-level conceptual objectives as reflected in international and national legislation. At this level, political commitments and high goals are expressed such as ensuring food security and safety, promoting development of rural areas or ensuring maintenance of biodiversity. Application of EAF will require that commitments are explicitly envisaged regarding:

- Maintaining *ecosystem integrity*: there is no agreed definition, but it is usually taken as implying maintenance of biodiversity in all its aspects (habitat, species and genetic), and maintenance of ecological processes that support biodiversity and productivity. The objective of EAF is sustainable use of aquatic resources for efficient delivery of food and services, with the view of improving human well-being.
- The principle of *equity* (balancing diverse societal objectives) implies both the intra-generational equity, i.e. fair distribution of rights between various sections of society at present, and inter-generational equity, and thus the need to make sure that future generations will be able to draw the same benefits from aquatic ecosystems as we do.

Implementing these principles will, in the first instance, imply that sustainability and equity goals be reflected in relevant policy documents at the international, regional (e.g. mandate of regional fisheries management organizations, RFMOs) and national (e.g. national fisheries law) levels, as a reference for operationalization.

The *operational framework* relates to the institutions, processes and resources necessary for achieving the high goal objectives. Guiding principles in operationalizing the EAF include:

- Application of the *precautionary approach*, implying that where there are threats of serious irreversible damage, the lack of full scientific knowledge shall not be used as a reason for postponing or failing to take measures to prevent environmental degradation (CCRF, Article 6.5, FAO, 1995a). Interpretations of this approach indicate that it implies the reversal of the burden of proof for legislation, shifting the onus of proving non-reversible negative impacts onto the responsible entity before a potential disturbance is allowed, i.e. assuming that human actions are harmful unless proven otherwise. FAO has produced guidance in the application of the precautionary approach (FAO, 1995b, 1996) that recognizes the following:
    - all fishing activities have environmental impacts, and it is not appropriate to assume that these are negligible until proved otherwise;
    - although the precautionary approach to fisheries may require cessation of fishing activities that have potentially serious adverse impacts, it

does not imply that no fishing can take place until all potential impacts have been assessed and found to be negligible;
    - the precautionary approach to fisheries requires that all fishing activities be subject to prior review and authorization; that a management plan be in place that clearly specifies management objectives and how impacts of fishing are to be assessed, monitored and addressed; and that specified interim management measures should apply to all fishing activities until such time as a management plan is in place; and
    - the standard of proof to be used in decisions regarding authorization of fishing activities should be commensurate with the potential risk to the resource, while also taking into account the expected benefits of the activities.
- The second guiding operational principle is related to the need of moving towards *adaptive management systems*, given the complexity and dynamics of ecosystems and society and the difficulty in predicting outcomes of different management measures. The emphasis, as compared to conventional practice, is on the need for mechanisms that from observation and experience feed back into policy and management decisions. Obviously, learning is an important aspect of adapting and in addition to individual fishermen learning, processes are needed to ensure institutional and societal learning, for which documentation of decisions and experiences made is essential (e.g. http://nsgl.gso.uri.edu/conn/conng03002.pdf).
- The principle of *compatibility* stresses the importance of coherence of management measures across the resource/ecosystem range. Related to this is the need to collaborate at the regional level, when resources and ecosystems are transboundary (CCRF, Article 6.12, FAO, 1995a). Related to this is the desirability of clearly defining the appropriate scale for resource management, from local (as in the case of locally distributed resources and processes affecting them) to regional and global.
- The principle of *participation* is reflected in most recent international instruments, requiring that stakeholders be more closely associated with the management process, data collection, knowledge building, option analysis, decision making and implementation. This results from the recognition that decisions will be considered as having greater legitimacy by stakeholders, and also that greater participation in decision making will bring important additional information and insights into the fishery system that will enhance the probability of achieving agreed objectives. The CCRF does not address participatory approaches directly, but encourage states to 'ensure transparency' (CCRF, Article 7.1.9, FAO, 1995a) and to 'explain the bases and purposes of management measures to stakeholders' (CCRF, Article 7.1.10, FAO, 1995a). A broader participation of stakeholders implies specific institutional arrangements, mechanisms and resources (see also Garcia, this volume).
- Using *incentives*, as compared to being prescriptive, is another guiding principle in the application of EAF. Conventional fisheries management is largely built on developing norms and punishing those who do not comply (negative incentives). These deterrents are obviously necessary but can be disregarded in

situations where monitoring and control are not optimal, which is a prevalent situation. These can be complemented by measures that support positive behavioural change (positive incentives) and could be social, economic, legal or institutional in nature (De Young and Charles, this volume).

- Coordination and harmonization across sectors (*sectoral integration*) are needed for a successful application of EAF. Mechanisms or institutional changes are therefore required that allow interaction across sectors, recognizing that resources and ecosystems may be subject to human activities other than fisheries.

The principles related to the operational framework are those that will probably be the most challenging ones as they require a fundamental change in the way management systems work. For example, important institutional changes will be needed to address increased stakeholder participation, for implementing adaptive management or for sectoral integration (Charles and De Young, this volume). Furthermore, successful application requires a shared vision as regards global sustainability objectives by society and the users, which can be achieved only if a profound political commitment exists.

The *cognitive framework* relates to the acquisition of information, analysis and translation into knowledge usable by society. Knowledge requirements have been perceived by many as the greatest barrier to the application of EAF, based on the assumption that the knowledge needed will be broader in scope and, at the same time, a more detailed understanding of the functioning of complex ecological and social systems would be required. This perception has led many to believe that the application of EAF would be virtually impossible to realize. This, however, is not the case as will become clear through the following points:

- Improving *scientific understanding* of ecosystems in all their components is required under the EAF. The scope of fisheries research will be broadened and layers of complexity added as compared to the conventional fisheries management context. Furthermore, in recognition of the fact that problems experienced in fisheries management so far are due largely to poor governance and to social and economic factors, fishery research should be expanded to deal with these aspects. Increased funding will therefore be needed for integrated (both biotic and abiotic) ecosystem research, as well as for a wide range of research to address the human part of the system, and for integrating social and ecological knowledge. However, recalling the Precautionary Principle, fisheries management is required explicitly to take decisions when there is a lack of complete scientific knowledge. Strictly related to this is the principle that decisions are to be based on 'the best scientific evidence available' (CCRF, Article 7.4.1, FAO, 1995a), including traditional knowledge.
- Encourage research on *selective and environmentally safe fishing gear* and practices. Key ecosystem impacts of fishing depend on the use of fishing gear that results in collateral damage on non-target species and habitats. In most cases this damage can be substantially reduced by appropriate gear modifications. These in turn, often require extensive and costly experimentation before they can be successfully implemented.

- *Move from a predictive to an adaptive science framework.* One difference between the role of research as perceived under conventional fisheries management and under an EAF framework is that in the latter there is acceptance that management measures may be taken in a situation of high uncertainty and lack of predictive models. In these situations, for example, development and use of a set of meta-indicators and regulation of overall fishing pressure, through spatial management, may be considered valid management interventions. While methods to support decision making in conditions of high uncertainty exist, their application to fisheries is not yet mainstream and their utilization is still largely confined to the academic world. These methods facilitate utilization of computer tools that help assess risks and account for decisions, and include the use of fuzzy logic (Paterson *et al.*, 2007), and Bayesian Belief Networks (Pakarinen *et al.*, 2001; Kjærulff and Madsen, 2006). To the author's knowledge, while application of these methods in decision making has taken place in other fields such as medicine and agriculture, no example of their actual application in decision making in fisheries exists. It may be worthwhile to further explore the usefulness of these methods for decision making under high levels of uncertainty.

Finally, it should be noted that none of the principles that underlie the EAF is new. They can all be traced in earlier instruments, agreements and declarations, while it is their implementation that lags behind. The reasons for the delay in implementation are many and complex (Aqorau, 2003), including the fact that many States are not party to the instruments and agreements, thereby limiting their efficacy. Another aspect is related to the distance between the discourse at the international level and the actual situation on the ground in terms of societal values and socio-economic conditions. The application of the principles described above is based on the assumption of transparent, democratic and well-informed societies, with a large portion of citizens supporting sustainability and equity values. These assumptions may be too optimistic as, for example, only about 17% of the countries worldwide enjoy a functional democracy (Kekik, 2007).

Implementation: a Pragmatic Approach

The EAF reorganizes the principles of sustainability and equity described above, making their implementation more compelling and provides a framework for a comprehensive implementation of these principles (as compared to a piecemeal, non-systematic implementation). This entails going through a systematic process of assessing fishing activities (both ongoing and new) along the three main dimensions of a fishery system, i.e. the environmental, the socio-economic and the governance dimensions. Suitable management strategies are identified and agreed upon and consolidated in fisheries management plans.

Development of fisheries management plans is a key step in the implementation of EAF. It should be noted that the CCRF (FAO, 1995a) also explicitly requires that 'long-term management objectives should be translated into management actions, formulated as a fisheries management plan or other management framework'. Implementation of an ecosystem approach requires, perhaps more explicitly than under conventional fisheries management and the CCRF, that management plans be developed consistent with the EAF principles presented above.

The above process is described in detail in the FAO guidelines (FAO, 2003, 2005a). Inspired by Australia's experiences in developing fisheries management practices consistent with the principles of ecological sustainable development (Fletcher *et al.*, 2002), the FAO guidelines provide a framework for planning and managing fisheries in a way that is consistent with EAF, including being participatory and transparent.

The planning process consists largely of examining existing or developing fisheries to identify key priority issues to be dealt with by management in order to be consistent with an ecosystem approach. The main result of this planning process is the backbone of EAF fisheries management plans. Figure 2.2 shows the management cycle that includes initial planning, implementation and feedback loops that are essential under an adaptive framework. Implementation of EAF will require an initial planning exercise (including 'Scoping', 'Setting objectives' and 'Formulating actions and rules'), to revise existing or developing new management plans for a given fishery, a sub-sector (e.g. small-scale fisheries) or a given region. The steps of the planning and management cycle of Fig. 2.2 are very similar to those undertaken under conventional fisheries management. There are, however, a number of additional mandatory elements under an EAF. These include stakeholder participation at all steps of the planning, management and

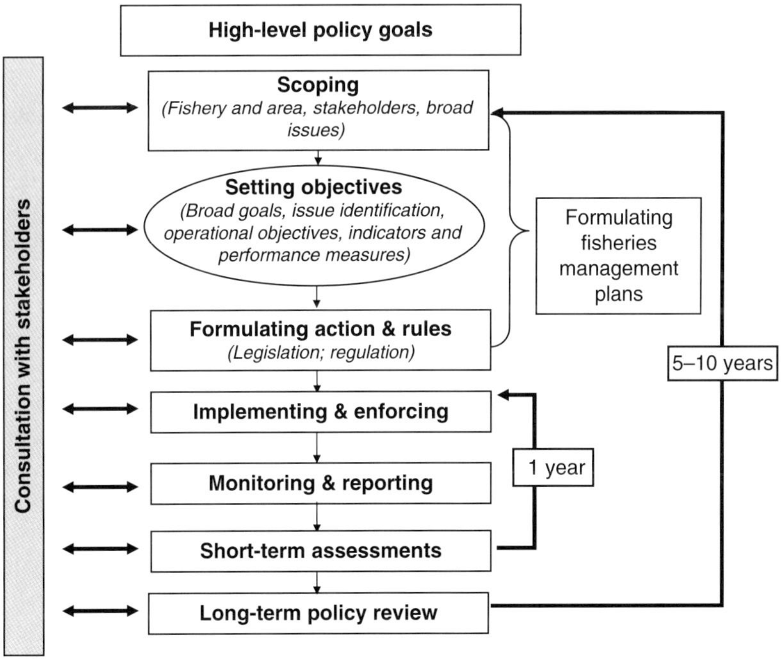

Fig. 2.2. Management cycle with feedback loops characteristic of adaptive strategies. Scoping, setting objectives and formulation of action and rules represent the key steps for developing fisheries management plans. The following steps regard implementation. (Modified from FAO, 2003.)

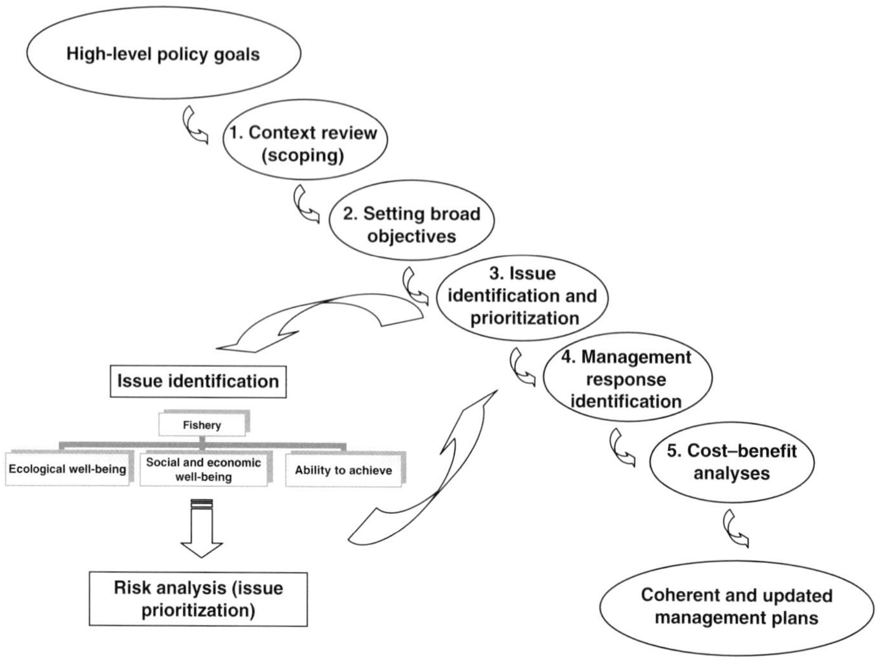

Fig. 2.3. Developing fisheries management plans under an EAF. (From Bianchi *et al.*, in press.)

decision-making process, use of best available knowledge, which also implies that the planning and decision making should take place without being postponed until improved knowledge is available. Another innovative element of the EAF framework is to consider the priority of actions along the three main dimensions of fisheries systems, i.e. the ecological, human and institutional dimensions.

Figure 2.3 shows the planning process in greater detail, a process that is key to translating high-level policy goals into operational objectives, and that results in the formulation of agreed coherent management plans (FAO, 2003; Cochrane *et al.*, 2007; Fletcher, this volume).

Identification of the key stakeholders is fundamental to the successful development and implementation of the management plans. Although stakeholder identification can take place informally, more formal ways can be used (Renard, 2004; Vierros *et al.*, 2006). In addition to ensuring stronger legitimacy and transparency, a good process for stakeholder identification and analysis also provides the basic understanding of the social and institutional context relevant to the planning process.

Comparison with Other Approaches

As described in the preceding section, there are a number of overarching principles that have been emerging as important in modern natural resources management. At the same time, a number of approaches have developed that incorporate these principles in a holistic manner, however, each with different emphasis. Examples are ecosystem-based management (EBM),[2] ecosystem-based fisheries management (EBFM, Pikitch *et al.*, 2004), EAF (FAO, 2003), integrated coastal zone management (ICZM) or integrated coastal area management (ICAM),[3] territorial user rights in fisheries (TURFS, Christy, 1992), sustainable livelihoods approach (SLA),[4] etc. The feasibility of applying ecosystem-based management in a tropical context is analyzed and assessed by Christie *et al.* (2007).

Here the intention is not to detail the characteristics of each approach and the differences among these but only to look at broad categories. One major distinction is between approaches specific to a sector as compared to those that are fully holistic (cross-sectoral). In the first case, a given sector or sub-sector is placed into an ecosystem context (e.g. the EAF, as defined by FAO, 2003), where fisheries (as other sectors operating in a given area) are managed consistently with the overall principles and objectives set for a given ecosystem. Interactions with other sectors are considered and, where relevant, are established for harmonization purposes. The other, the cross-sectoral approach considers, in an integrated and holistic manner, all human activities impacting on a given ecosystem simultaneously. This approach has been adopted, for example, by Australia (Integrated Ocean Policy, 1998), by Canada (Canada's Ocean Strategy, 2002) and is also proposed by the European Union in the recently developed European Marine Strategy (EU, 2004). Both the sectoral and cross-sectoral approaches are relevant; they are complementary and should be implemented in parallel. The implications for science and management are different. The science for an integrated approach across various sectors and impacts has to address a new layer of complexity as compared to the sectoral approach, including assessments of the combined effects of various human activities on the ecosystem or developing common reference points. Also institutional implications are different, as at the cross-sectoral level mechanisms or institutions are required to allocate rights to different users, determine common goals or reconcile conflicts.

In some cases, differences in emphasis may be due to a discipline-centred perspective. For example, EBFM stems from a biodiversity/ecological perspective, i.e. having as a primary objective the maintenance of healthy marine ecosystems, while TURFS focuses on the governance aspects, from the perspective that allocation of territorial rights is a way of achieving sustainable development. The livelihood approach, particularly relevant for communities where poverty is a major issue, stems from the human sustainability perspective. The sustainable livelihood approach focuses on the human dimension, placing people's social and economic activities at the centre, while less emphasis is given to ecological sustainability, which, according to this approach, is closely linked to and depends on sustainability of livelihoods. The approach is believed to be particularly useful for poverty reduction in fishing communities. Figure 2.4 shows how various

[2] http://www.ices.dk/icesworK/asc/2005/KeithSainsbury_talk.pdf

[3] http://www.oceansatlas.org/servlet/CDSServlet?status = ND0xMjc2MiZjdG5faW5mb192aWV3X3NpemU9Y3RuX2luZm9fdmlld19mdWxsJjY9ZW4mMzM9KiYzNz1rb3M~

[4] http://www.fao.org/fishery/topic/14837

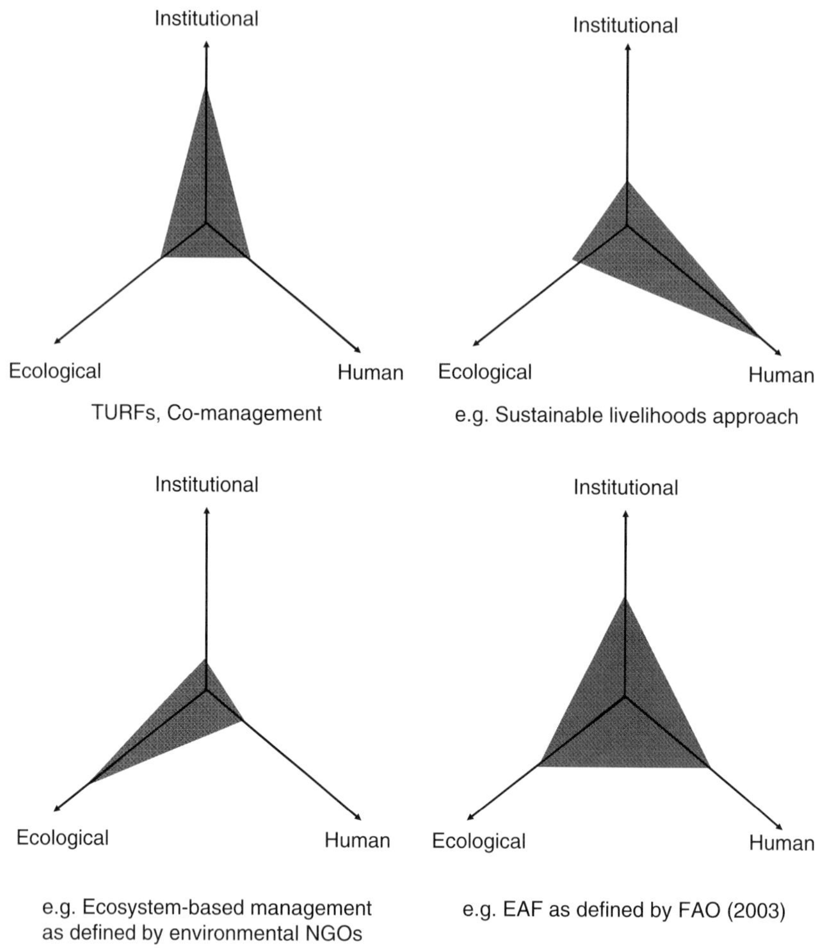

Fig. 2.4. Simplified representation of different emphasis given by different approaches along the three main dimensions of a fishery system.

approaches can be represented along the three main dimensions of a fishery system, i.e. the ecological, the socio-economic and the governance dimensions. The EAF, as defined in the FAO guidelines, gives equal prominence to the three dimensions through a planning process that explicitly takes these into account.

The main point here is that different approaches have different emphasis, their applicability and relevance depending on the context.

From International Policy Developments to Implementation: FAO's Work

The requirements implicit in the EAF, such as addressing more complex and poorly understood systems and the associated uncertainty, increasing data requirements, consideration of several timescales and the recognition of the importance of broader stakeholder participation at various stages of the fisheries management process, have initially intimidated many and fostered a perception of the EAF as a difficult and perhaps impossible task.

Attitudes are however changing, both at international and at national levels, and a pragmatic approach has been adopted in many places to improve conventional fisheries management by incorporating ecosystem considerations and more properly dealing with the social dimension. In a few cases, this has been done in a comprehensive way (e.g. in Australia, see McLoughlin *et al.*, this volume), while in most cases ecosystem considerations are being incorporated piecemeal in response to specific concerns (e.g. Norway's closure of known cold-water coral reef areas to bottom trawling because of impacts on cold-water corals).

In fact any initiative taken to implement FAO's CCRF is fully consistent with and contributes to a comprehensive application of the EAF.

FAO is committed to continue the work with further development of the framework and of the tools for the application of EAF and to provide technical assistance to member countries in its realization. Garcia (2006) summarized the progress made towards implementation of EAF and the work done by FAO in this direction, showing how the work programme of the FAO Fisheries Department is dedicated to the active promotion and monitoring of responsible fisheries development and management. As a consequence, a very substantial part of FAO's budget is used for activities that contribute to the establishment of a better balance between resource use and conservation (both in terms of biodiversity and ecosystems), through improvement of fisheries governance. This is reflected in the international collaboration with UN agencies and a number of environmental NGOs. It is also reflected in the type of guiding documents and policy instruments developed to facilitate implementation of EAF - such as the guidelines produced in 2003 - completing the guidance already available in technical guidelines in support of the CCRF, such as those on the precautionary approach (1995b) and the integration of fisheries in coastal areas management (1996), just to mention a few. A number of International Plans of Action (IPOAs) have also been adopted on the control and reduction of fishing capacity (1999), to reduce the incidental catch of seabirds in long-line fisheries (1999), for the conservation and management of sharks (1999), and to deter illegal fishing (2002). Of direct relevance to the EAF is the work initiated in 2006 on the social, economic and institutional implications of applying the EAF, with an Expert Consultation focusing on the human side of the EAF, held at the FAO headquarters in Rome from 6 to 9 June 2006 (De Young, this volume). This process will result in the production of technical guidelines covering these aspects. FAO has also initiated the process of developing a toolbox that provides detailed guidance on available methods and tools to facilitate application of EAF at all levels, from policy formulation and planning to day-to-day application (FAO, in preparation).

A recent restructuring of FAO has resulted in the change of name from 'Fisheries Resources Service' (a service that traditionally has dealt with fishery resources assessments and management) to 'Fisheries Management and Conservation Service', thus reflecting a broadening in emphasis consistent with EAF as regards the scope of the Organization's activities in this area.

Promotion of EAF has been conducted in a number of conferences, regional and national initiatives have been monitored informally, and specific case studies have been implemented through field projects, all in order to promote its implementation.

Several FAO projects and activities address EAF in a comprehensive way and aim at simultaneously achieving progress in several relevant aspects in selected locations or ecosystems. One project aims to examine the feasibility of implementing EAF in the Benguela region in cooperation with the Benguela Current Large Marine Ecosystem programme (BCLME) and the fisheries management agencies of Angola, Namibia and South Africa (Cochrane *et al.*, 2007; Cochrane *et al.*, this volume).This project pursues a structured and participatory approach based on the FAO Guidelines, to identify and prioritize the gaps in the existing, largely conventional, approaches to fisheries management in those countries and to consider potential management actions to address them.

Through another project, with the support of the Government of Japan, technical assistance is provided to fisheries institutions of selected countries and regions to develop the information tools (including ecosystem modelling, the use of GIS and collection of standard fisheries data) to improve management of their pelagic resources and fisheries in accordance with EAF. Another wide encompassing project, funded by the Government of Norway, is being implemented in partnership with various GEF (global environment facility)-funded large marine ecosystem (LME) projects to strengthen the knowledge base for implementing the EAF in developing countries.The initial focus is in the African region, providing capacity building, standardized data collection and monitoring of marine fisheries and related ecosystems, and supporting policy development and management practices consistent with EAF principles.

The above is not meant to be a comprehensive overview of FAO's work of direct relevance to EAF, but to give a flavour of the type of activities and of the commitment of FAO to this approach.

Challenges

Important challenges to the realization of sustainable development, and, in particular, to the application of EAF exist, beyond the technical aspects of practical day-to-day implementation. The challenges are not only related to controlling the direct drivers of marine ecosystem change, such as fisheries and other sectors utilizing goods and services from the marine ecosystem, but are also related to the indirect drivers such as changes in human population coupled with a widespread aspiration for an improved standard of living, and global economic policies, governance and social and economic conditions.

Globalization and international trade can have both positive and negative effects on fisheries sustainability. Globalization has been promoted with the vision that an increased flow of goods, money and information across national boundaries would have mainly positive effects on the environment, including the regulation of international trade based on sustainability standards, and increased well-being favouring environmental and social programmes (OECD, 1997). On the other hand, environmental negatives such as increased consumption and economic return leading to depletion of natural resources, particularly in a situation of poorly defined user rights, have also been recognized. Important efforts have taken place to develop sustainability requirements for products being traded in the international market. For example, the Marine Stewardship Council (MSC), an independent, global, non-profit organization, contributes to sustainable fish trade through a certification programme for well-managed fisheries.[5] FAO has also developed eco-labelling guidelines (FAO, 2005b), and there are ongoing efforts to better define sustainability requirements to take account of broader ecosystem concerns.

Perhaps the key issue is that the *de facto* value system that has dominated the last few decades has been that of economic growth. Prospects of increased well-being and of poverty reduction have been powerful forces in motivating decision making, often downplaying concerns of environmental sustainability, and reflecting more the interests of large capital investment as compared to ensuring local food security. During the past two decades, policy reforms designed to promote growth and liberalization have been encouraged with little regard to the environmental consequences (Arrow *et al.*, 1995). In other words, there is usually very limited harmonization between economic and social policies and environmental sustainability commitments. Harmonization between these domains would be the proof that the expression 'sustainable development' is not an oxymoron.

Another major concern is the suitability of present forms of governance to the application of the EAF, considering that this approach requires democracy, transparency and a vision of fairness, equity and sustainability shared among the various stakeholders and within the society. Furthermore, overall economic well-being is needed to prioritize long-term sustainability concerns as compared to short-term economic or survival pressures. A vast majority of nations are in a poor governance situation (Kekik, 2007), aggravated in most cases by suboptimal economic conditions or poverty. These conditions may hinder the application of EAF.

The ecosystem approach requires appropriate institutional arrangements ensuring balanced, transparent and legitimate decision making in relation to different possible trade-offs and stakeholders. Sustainable development and derived principles are largely defined at the interface between productive sectors and conservation. This makes the decision making process aimed at optimizing both production and conservation challenging at the institutional level. For example, conservation is often dealt with by a Ministry of Environment, while fisheries management is the responsibility of a specific Ministry of Fisheries (or of a directorate that is part of a larger Ministry, e.g. that of Agriculture), with limited collaboration between them, and a lack of formal processes of coordination and harmonization.

Coordination and harmonization are also required between the institutions responsible for managing other activities utilizing resources in a given area, requiring a more holistic approach to area management. Usually neither

[5] http://www.msc.org/html/content_ 458.htm

mechanisms nor institutions exist that operate at this level of complexity. Winsnes and Skjoldal (this volume) show an example of management in the Barents Sea where these complexities are addressed, based on existing institutional set-up and by creating new mechanisms to facilitate interaction and decision making.

Lack of an appropriate institutional framework can also be seen at the international level, where international organizations, including UN specialized agencies, such as FAO or UNEP, are still largely defined following a sectoral division of labour. Moving towards an ecosystem approach has triggered development towards more holistic approaches in all these agencies, with results not always consistent or coordinated with each other and reflecting a less clear, sometimes confused, understanding of respective roles.

Another aspect is related to the increasing amount of funds being channelled through NGOs and INGOs (international NGOs). In the United States of America, the amount of funds made available by philanthropic initiatives was about ten times larger in the early years of the new millennium as compared to the early 1970s (AAFRC Trust for Philanthropy/Giving USA 2001). This makes NGOs important players not only in setting the international agenda, but also in shaping and implementing policies. This can be seen as a positive development considering the major sustainability challenges to be met. However, because these processes in some cases take place largely outside established governance systems, there is a need for improved coordination and harmonization.

In addition to the slowness in incorporating principles of environmental sustainability in national policies, and the insufficient coordination between agencies, both nationally and internationally, there seems to be inertia in government institutions and a lack of synchronization with international policy processes and commitments on environmental issues. This is a slow 'institutionalization' of the high-level principles and goals such as those characterizing the ecosystem approach. By the term 'institutionalization' we denote the process of embedding the EAF concepts within government institutions and agencies and society at large. This would entail, at a minimum, the full sharing of the EAF principles and vision, the consistent implementation of these in societal and institutional activities and perhaps even a gradual reorganization of institutions and institutional arrangements to be consistent with EAF principles. For this reason, implementation may take a long time, but it can be accelerated by taking action at different levels, from promoting structural changes in the organization of relevant institutes and agencies to make them more efficient in relation to achieving EAF objectives, to more proactively sensitizing public opinion and/or including environmental concerns as part of the compulsory syllabus in primary and secondary schools. NGOs, by creating public awareness, often play an important role in promoting change and fighting institutional inertia.

Conclusions

Sustainable development, a concept that embodies both development and conservation, remains the key model for organizing human activities, despite the challenges it poses. The EAF, the framework that enables sustainable development to become operational in fisheries, has been recognized and adopted as the best framework for fisheries policy, planning and implementation by the world community and there has been good progress in putting it into practice in various parts of the world. A key message of this contribution is that it is achievable, even with limited capacity and information, and various methods and tools exist to facilitate its application. However, a change in attitudes and a different emphasis of existing societal values will be required.

References

Aqorau, T. (2003) Obligations to protect marine ecosystems under international conventions and other legal instruments. In Sinclair, M. and Valdimarsson, G. (eds), *Responsible Fisheries in the Marine Ecosystem* (pp. 25–40). CAB International, Wallingford, UK, 426 pp.

Arrow, K., Bolin, B., Costanza, R., Dasgupta, P., Folke, C., Holling, C.S., Jansson, B-O., Levin, S., Mäler, K.-G., Perrings, C. and Pimentel, D. (1995) Economic growth, carrying capacity, and the environment. *Ecological Economics*, 15, 91–95.

Bianchi, G., Cochrane, K. and Vasconcellos, M. (in press) Implementing the Ecosystem Approach to Fisheries. *Fish, Trade and Development*. The Royal Swedish Academy of Agriculture and Forestry.

Christie, P., Fluharty, D.L., White, A.T., Eisma-Osorio, L. and Jatulan, W. (2007). Assessing the feasibility of ecosystem-based fisheries management in tropical contexts. *Marine Policy*, 31, 239–250.

Christy, F.T. Jr (1992) Territorial use rights in marine fisheries: definitions and conditions. *FAO Fish. Tech. Pap.* (227), 10.

Cochrane, K.L., Augustyn, C.J., Bianchi, G., de Barros, P., Fairweather, T., Iitembu, J., Japp, D., Kanandjembo, A., Kilongo, K., Moroff, N., Nel, D., Roux, J.P., Shannon, L.J., van Zyl, B. and Vaz Velho, F. (2007) Results and conclusions of the project 'Ecosystem Approaches for Fisheries Management in the Benguela Current Large Marine Ecosystem'. FAO Fisheries Circular No. 1026. FIMF/C1026 (En)

FAO (1995a) Code of Conduct for Responsible Fisheries. FAO, Rome, Italy, 41 pp.

FAO (1995b) Precautionary approach to fisheries. Part 1: Guidelines on the precautionary approach to capture fisheries and species introductions. Elaborated by the Technical Consultation on the Precautionary Approach to Capture Fisheries (Including Species Introductions). Lysekil, Sweden, 6–13 June 1995 (A scientific meeting organized by the Government of Sweden in cooperation with FAO). FAO Fisheries Technical Paper. No. 350, Part 1. FAO, Rome, Italy, 52 pp.

FAO (1996) Precautionary approach to capture fisheries and species introductions. FAO Technical Guidelines for Responsible Fisheries 2, 54 pp.

FAO (2003) Fisheries management. The ecosystem approach to fisheries. FAO Technical Guidelines for Responsible Fisheries, 8, 68 pp.

FAO (2005a) Putting into practice the ecosystem approach to fisheries. FAO, Rome, Italy, 76 pp.

FAO (2005b) Guidelines for ecolabelling of fish and fishery products from marine capture fisheries. FAO, Rome, Italy, 90 pp.

Fletcher, W.J., Chesson, J., Fisher M., Sainsbury, K. J., Hundloe, T., Smith, A.D.M. and Whitworth, B. (2002) National ESD Reporting Framework for Australian Fisheries: The 'How To' Guide for Wild Capture Fisheries. FRDC Project 2000/145, Canberra, Australia, 120 pp.

Garcia, S.M. (1992) Ocean fishery management. The FAO programme. In Fabbri, P. (ed) *Ocean Management in Global Change* (pp. 381–418). Elsevier Applied Science, London.

Garcia, S.M. (2006) The ecosystem approach to fisheries: implementation and lessons learnt. Paper presented at the Conference organized by the Center of Law and Policy (COLP) in Dublin, Ireland, June 2006.

Garcia, S.M. and Cochrane, K.L. (2005) Ecosystem approach to fisheries. *ICES Journal of Marine Science*, 62, 311–318.

Garcia, S.M., Zerbi, A., Aliaume, C., Do Chi, T. and Lasserre, G. (2003) The ecosystem approach to fisheries. Issues, terminology, principles, institutional foundations, implementation and outlook. *FAO Fisheries Technical Paper,* 443, 71.

Kekik, L. (2007) The Economist Intelligence Unit's index of democracy. In: The World in 2007. Available at: http://www.economist.com/media/pdf/DEMOCRACY_INDEX_2007_v3.pdf

Kjærulff, U.B. and Madsen, A.L. (2006) Probabilistic networks for practitioners – A guide to construction and analysis of Bayesian networks and influence diagrams. Aalborg University, Aalborg, Denmark, 262 pp.

Pakarinen, T, Romakkaniemi, A., Kuikka, S., Uusitalo, L. and Ikonen, E. (2001) Evaluation of the quality of biological reference points of Baltic salmon management. Available at: http://www.ices.dk/products/CMdocs/2001/M/M-2001.pdf

Paterson, B., Jarre, A., Maloney, C.L., Fairweather, T.P., van der Lingen, C.D., Shannon, L.J. and Field, J.G. (2007) A fuzzy logic tool for multi-criteria decision making in fisheries: the case of the South African pelagic fishery. *Marine and Freshwater Research* 58 (11), 1056–1068.

Pikitch, E.K., Santora, C., Babcock, E.A., Bakun, A., Bonfil, R., Conover, D.O., Dayton, P., Doukakis, P., Fluharty, D., Heneman, B., Houde, E.D., Link, J., Livingston, P.A., Mangel, M., McAllister, M.K., Pope, J. and Sainsbury, K.J. (2004) Ecosystem-based fishery management. *Science* 305, 346–347.

Renard, Y. (2004) Guidelines for stakeholder identification and analysis: a manual for Caribbean natural resource managers and planners. Caribbean Natural Resource Institute, Trinidad, 30 pp.

Turrell, W.R. (2004) The policy basis of the 'Ecosystem Approach' to fisheries management. *Euro Goos Publication No. 21,* Euro Goos, Norr Köping.

Vierros, M., Douvere, F. and Arico, S. (2006) Implementing the Ecosystem Approach in Open Ocean and Deep Sea Environments. An Analysis of stakeholders, their interests and existing approaches. UNU-IAS Report, 41 pp.

3 The Ecosystem Approach of the Convention on Biological Diversity

MARJO VIERROS

Institute of Advanced Studies, United Nations University

Abstract

The Convention on Biological Diversity (CBD) ecosystem approach provides an overarching and well-developed framework that is not specific to any sector or biome, and brings together various methods and techniques that can be integrated to address current environmental, economic, social and political needs, while enabling future generations to meet their own needs. This chapter provides an introduction to the CBD ecosystem approach, with a definition of the 12 key principles and an outline of the operational guidance for its application. However, it is recognized that 'one-size-fits-all' solutions are neither feasible nor desirable, while 'learning by doing' is a priority for the broader implementation of the ecosystem approach.

Introduction to the CBD Ecosystem Approach

The ecosystem approach, as defined in the Convention on Biological Diversity (CBD) (Decision V/6), is a strategy for the integrated management of land, water and living resources that promotes conservation and sustainable use in an equitable way. Application of the ecosystem approach will help to reach a balance of the three objectives of the Convention (conservation, sustainable use and the equitable sharing of benefits). It is based on the application of appropriate scientific methodologies focused on levels of biological organization, which encompass the essential processes, functions and interactions among organisms and their environment. It recognizes that humans, with their cultural diversity, are an integral component of ecosystems.

The ecosystem approach was originally adopted as the primary framework for action under the Convention by the Conference of the Parties (COP) at its second meeting in Jakarta, in November 1995 (see Decision II/8). Subsequently, the COP acknowledged the need for a workable description and further elaboration

(eds G. Bianchi and H.R. Skjoldal)

of the ecosystem approach. This leads to the development of a description of, and principles and guidance to, the ecosystem approach under the CBD.

The CBD ecosystem approach provides an overarching and well-developed framework that is not specific to any sector or biome, although it can be applied everywhere. It brings together various methods and techniques that can be integrated to address current environmental, economic, social and political needs, while enabling future generations to meet their own needs.

In accordance with Decision V/6, the CBD ecosystem approach is underpinned by 12 principles and 5 points of operational guidance. The principles are all complementary and interlinked, and should be applied together. It is important to stress that when applying the ecosystem approach, all its principles need to be considered and appropriate weight given to each, according to the priorities of the issues being addressed. However, it is legitimate to give different weights to each principle according to the circumstances under which it is being applied.

Further work on the principles led to each one being supported by a rationale, a case study, implementation guidelines and an indicative list of tools and sources of information. The seventh meeting of the COP in Kuala Lumpur in February 2004 welcomed these implementation guidelines and annotations to rationale, which are contained in Annex 1 to Decision VII/11. At the same time, the COP also agreed that the priority at this time should be to facilitate the implementation of the ecosystem approach as the primary framework for addressing the three objectives of the Convention in a balanced way, and that a potential revision of the principles of the ecosystem approach should take place only at a later stage, when the application of the ecosystem approach has been more fully tested.

The 12 Principles of the Ecosystem Approach

Principle 1: The objectives of management of land, water and living resources are a matter of societal choices.

Different sectors of society view ecosystems in terms of their own economic, cultural and society needs. Indigenous peoples and other local communities living on the land are important stakeholders and their rights and interests should be recognized. Both cultural and biological diversities are central components of the ecosystem approach, and management should take this into account. Societal choices should be expressed as clearly as possible. Ecosystems should be managed for their intrinsic values and for the tangible or intangible benefits for humans, in a fair and equitable way.

Principle 2: Management should be decentralized to the lowest appropriate level.

Decentralized systems may lead to greater efficiency, effectiveness and equity. Management should involve all stakeholders and balance local interests with the wider public interest. The closer management is to the ecosystem, the greater is the responsibility, ownership, accountability, participation and use of local knowledge.

Principle 3: Ecosystem managers should consider the effects (actual or potential) of their activities on adjacent and other ecosystems.

Management interventions in ecosystems often have unknown or unpredictable effects on other ecosystems; therefore, possible impacts need careful consideration and analysis. This may require new arrangements or ways of organization for institutions involved in decision making to make, if necessary, appropriate compromises.

Principle 4: Recognizing potential gains from management, there is usually a need to understand and manage the ecosystem in an economic context. Any such ecosystem management programme should:

1. *Reduce those market distortions that adversely affect biological diversity.*
2. *Align incentives to promote biodiversity conservation and sustainable use.*
3. *Internalize costs and benefits in the given ecosystem to the extent feasible.*

The greatest threat to biological diversity lies in its replacement by alternative systems of land use. This often arises through market distortions, which undervalue natural systems and populations and provide perverse incentives and subsidies to favour the conversion of land to less diverse systems.

Often those who benefit from conservation do not pay the costs associated with it and, similarly, those who generate environmental costs (e.g. pollution) escape responsibility. Alignment of incentives allows those who control the resource to benefit and ensures that those who generate environmental costs will pay.

Principle 5: Conservation of ecosystem structure and functioning, in order to maintain ecosystem services, should be a priority target of the ecosystem approach.

Ecosystem functioning and resilience depends on a dynamic relationship within species, among species and between species and their abiotic environment, as well as the physical and chemical interactions within the environment. The conservation and, where appropriate, restoration of these interactions and processes is of greater significance for the long-term maintenance of biological diversity than simply protection of species.

Principle 6: Ecosystems must be managed within the limits of their functioning.

In considering the likelihood or ease of attaining the management objectives, attention should be given to the environmental conditions that limit natural productivity, ecosystem structure, functioning and diversity. The limits to ecosystem functioning may be affected to different degrees by temporary, unpredictable or artificially maintained conditions and, accordingly, management should be appropriately cautious.

Principle 7: The ecosystem approach should be undertaken at the appropriate spatial and temporal scales.

The approach should be bounded by spatial and temporal scales that are appropriate to the objectives. Boundaries for management will be defined operationally by users, managers, scientists and indigenous and local peoples. Connectivity between areas should be promoted where necessary. The ecosystem approach is based upon the hierarchical nature of biological diversity characterized by the interaction and integration of genes, species and ecosystems.

Principle 8: Recognizing the varying temporal scales and lag effects that characterize ecosystem processes, objectives for ecosystem management should be set for the long term.

Ecosystem processes are characterized by varying temporal scales and lag effects. This inherently conflicts with the tendency of humans to favour short-term gains and immediate benefits over future ones.

Principle 9: Management must recognize that change is inevitable.

Ecosystems change, including species composition and population abundance. Hence, management should adapt to the changes. Apart from their inherent dynamics of change, ecosystems are beset by a complex of uncertainties and potential 'surprises' in the human, biological and environmental realms. Traditional disturbance regimes may be important for ecosystem structure and functioning, and may need to be maintained or restored. The ecosystem approach must utilize adaptive management in order to anticipate and cater for such changes and events and should be cautious in making any decision that may foreclose options, but, at the same time, consider mitigating actions to cope with long-term changes such as climate change.

Principle 10: The ecosystem approach should seek the appropriate balance between, and integration of, conservation and use of biological diversity.

Biological diversity is critical both for its intrinsic value and because of the key role it plays in providing the ecosystem and other services upon which we all ultimately depend. There has been a tendency in the past to manage components of biological diversity either as protected or non-protected. There is a need for a shift to more flexible situations, where conservation and use are seen in context and the full range of measures is applied in a continuum from strictly protected to human-made ecosystems.

Principle 11: The ecosystem approach should consider all forms of relevant information, including scientific and indigenous and local knowledge, innovations and practices.

Information from all sources is critical to arriving at effective ecosystem management strategies. A much better knowledge of ecosystem functions and the impact of human use is desirable. All relevant information from any concerned area should be shared with all stakeholders and actors, taking into account, *inter alia*, any decision to be taken under Article 8(j)[1] of the CBD. Assumptions behind proposed management decisions should be made explicit and checked against available knowledge and views of stakeholders.

Principle 12: The ecosystem approach should involve all relevant sectors of society and scientific disciplines.

Most problems of biological diversity management are complex, with many interactions, side effects and implications, and therefore should involve the

[1] http://www.biodiv.org/convention/articles.asp?lg=0&a=cbd-08

necessary expertise and stakeholders at the local, national, regional and international level, as appropriate.

Operational Guidance for Application of the Ecosystem Approach

In applying the 12 principles of the ecosystem approach, the following five points are proposed as operational guidance.

Focus on the relationships and processes within ecosystems

The many components of biodiversity control the stores and flows of energy, water and nutrients within ecosystems, and provide resistance to major perturbations. A much better knowledge of ecosystem functions and structure, and the roles of the components of biological diversity in ecosystems, is required, especially to understand: (i) ecosystem resilience and the effects of biodiversity loss (at species and genetic levels) and habitat fragmentation; (ii) underlying causes of biodiversity loss; and (iii) determinants of local biological diversity in management decisions. Functional biodiversity in ecosystems provides many goods and services of economic and social importance. While there is a need to accelerate efforts to gain new knowledge about functional biodiversity, ecosystem management has to be carried out even in the absence of such knowledge. This implies taking risks, which can be mitigated through the use of an adaptive and precautionary approach to management (see Principle 3). The ecosystem approach can facilitate practical management by ecosystem managers (whether local communities or national policy makers).

Enhance benefit-sharing

Benefits that flow from the array of functions provided by biological diversity at the ecosystem level provide the basis of human environmental security and sustainability. The ecosystem approach seeks that the benefits derived from these functions are maintained or restored. In particular, these functions should benefit the stakeholders responsible for their production and management. This requires, *inter alia*: capacity building, especially at the level of local communities managing biological diversity in ecosystems; proper valuation of ecosystem goods and services; removal of perverse incentives that devalue ecosystem goods and services; and consistent with the provisions of the CBD, where appropriate, their replacement with local incentives for good management practices.

Use adaptive management practices

Ecosystem processes and functions are complex and variable. Their level of uncertainty is increased by the interaction with social constructs, which need to be better understood. Therefore, ecosystem management must involve a learning process, which helps to adapt methodologies and practices to the ways in which these systems are being managed and monitored. Implementation programmes should be designed to adjust to the unexpected, rather than to act on the basis of a belief in certainties. Ecosystem management needs to recognize the diversity of social and cultural factors affecting natural resource use. Similarly, there is a need for flexibility in policy making and implementation. Long-term, inflexible decisions are likely to be inadequate or even destructive. Ecosystem management should be envisaged as a long-term experiment that builds on its results as it progresses. This 'learning by doing' will also serve as an important source of information to gain knowledge of how best to monitor the results of management and evaluate whether established goals are being attained. In this respect, it would be desirable to establish or strengthen the capacities of the Parties for monitoring.

Carry out management actions at the scale appropriate for the issue being addressed, with decentralization to lowest level, as appropriate

As noted in the description of the ecosystem approach, an ecosystem is a functioning unit that can operate at any scale, depending upon the problem or issue being addressed. This understanding should define the appropriate level for management decisions and actions. Often, this approach will imply decentralization to the level of local communities. Effective decentralization requires proper empowerment, which implies that the stakeholder has both the opportunity to assume responsibility and the capacity to carry out the appropriate action, and needs to be supported by enabling policy and legislative frameworks. Where common property resources are involved, the most appropriate scale for management decisions and actions would necessarily be large enough to encompass the effects of practices by all relevant stakeholders. Appropriate institutions would be required for such decision making and, where necessary, for conflict resolution. Some problems and issues may require action at still higher levels, through, for example, transboundary cooperation, or even cooperation at global levels.

Ensure intersectoral cooperation

As the primary framework of action to be taken under the Convention, the ecosystem approach should be fully taken into account in developing and reviewing national biodiversity strategies and action plans. There is also a need to integrate the ecosystem approach into agriculture, fisheries, forestry and other production systems that have an effect on biodiversity. Management of natural resources, according to the ecosystem approach, calls for increased intersectoral communication and cooperation at a range of levels (government ministries, management agencies, etc.). This might be promoted through, for example, the formation of inter-ministerial bodies within the Government or the creation of networks for sharing information and experience.

The CBD Ecosystem Approach and Other Ecosystem Approaches

There is no single correct way to apply the ecosystem approach. Rather, the ecosystem approach provides a framework of principles and operational guidance that can be implemented in a number of different ways, and using a variety of tools, which may include, for example, marine-protected areas, fisheries management measures and species-protection measures.

The ecosystem approach as defined by the CBD provides an overarching framework for conservation and sustainable use of biological diversity, while satisfying societal and human needs for food and economic benefits. In addition, the CBD COP recognized that there exist other ecosystem approaches that are specific to certain sectors or biomes.[2] For the marine and coastal environment these include the ecosystem approach to fisheries (EAF) and integrated marine and coastal area management. In addition, many countries and some regions are attempting to integrate the management of their marine environments through ocean policies, Large Marine Ecosystem projects and other approaches, such as the 'Mountains to the Sea' concept. Such approaches can be consistent and complementary with the CBD ecosystem approach, and their application can achieve the ecosystem and societal objectives of the CBD ecosystem approach.

Both the CBD ecosystem approach and the EAF are guided by a set of principles. In the case of EAF, these principles are concepts that have been expressed in various instruments and conventions, and in particular in the Code of Conduct for Responsible Fisheries. A comparison between the principles and guidance of these two ecosystem approaches shows good consistency between them, with some differences in emphasis that are natural between a sectoral approach and an overarching approach.[3] This consistency is also explained by the fact that the CBD ecosystem approach principles were taken into account in design of the EAF.

For ocean areas, the challenge lies in integrating the various management approaches - both sectoral and cross-sectoral - into a comprehensive and cohesive plan that has the ecosystem approach as its central framework. Many countries and regions are starting to develop this type of integration for their exclusive economic zones (EEZs) through ocean policies. Consistent with the ecosystem approach, national and regional ocean policies may also need to extend their coverage into the high seas to take into account interlinkages between ecosystems.

Next Steps in the Context of the CBD

In July 2007, the 12th meeting of the CBD Subsidiary Body on Scientific, Technical and Technological Advice (SBSTTA) reviewed the implementation

[2] Decision VII/11 of the CBD Conference of the Parties.

[3] Implementing the Ecosystem Approach in Open Ocean and Deep Sea Environments: An Analysis of Stakeholders, Their Interests and Existing Approaches (2006). *UNU-IAS report.*

of the ecosystem approach. This review provided an opportunity to learn from the practical experiences of those involved in applying the ecosystem approach on the ground, including in the context of fisheries. In preparation for SBSTTA 12, the CBD Secretariat developed a web-based sourcebook, which includes a case study database.[4] A systematic analysis of case studies, including both successes and failures, has the potential to provide for eventual improvement of the implementation of the ecosystem approach.

In its review of the ecosystem approach (Recommendation XII/1), SBSTTA highlighted the importance of just this type of learning from experience, or 'learning by doing', as a priority for the broader implementation of the ecosystem approach. The discussions at SBSTTA were in agreement in that the ecosystem approach remains a useful normative framework for bringing together social, economic, cultural and environmental values. However, there is a need to translate this normative framework into methods for further application, which are tailored to the needs of specific users. It was noted that 'one-size-fits-all' solutions for the ecosystem approach are neither feasible nor desirable. Although systematic, global application of the ecosystem approach is lacking, there are many examples of successful application at the regional, national and, in particular, local scales, which should be widely promoted and communicated. Most of these examples can be considered as positive outcomes for both biodiversity and human well-being.

[4] http://www.biodiv.org/programmes/cross-cutting/ecosystem/sourcebook/search.shtml

4 The Large Marine Ecosystem Approach to Marine Resources Assessment and Management

Kenneth Sherman

National Oceanographic and Atmospheric Administration, USA

Abstract

Continued overfishing in the face of scientific warnings, fishing down food webs, destruction of habitat, and accelerated pollution loading – especially nitrogen export – have resulted in significant degradation to coastal and marine ecosystems of both rich and poor nations. Fragmentation among institutions, international agencies, and disciplines, lack of cooperation among nations sharing marine ecosystems, and weak national policies, legislation and enforcement all contribute to the need for a new imperative for adopting ecosystem-based approaches to managing human activities in these systems in order to avoid serious social and economic disruption. The global environment facility (GEF) has been approached by developing countries in overwhelming numbers for assistance in securing the futures of their shared large marine ecosystems (LMEs). This presentation describes the LME assessment approach to assist developing countries in Asia, Africa, Latin America and eastern Europe in the management of human activities affecting coastal and marine ecosystems and linked freshwater basins. At risk are renewable goods and services valued at an estimated $12.6 trillion/year. There are 110 countries involved in 16 LME projects approved by the GEF Council or under preparation. A five-module assessment and management methodology is being applied that moves the countries towards adopting practical joint governance institutions through place-based management. This LME approach engages stakeholders, fosters the participation of the science community and leads to the development of adaptive management practices that are assisting participating countries in making the transition towards recovery and sustainable development of marine resources.

Large Marine Ecosystems

Large marine ecosystems (LMEs) are natural regions of ocean space encompassing coastal waters from river basins and estuaries to the seaward boundary of continental shelves and the outer margins of coastal currents. They are relatively large regions of 200,000 km^2 or greater, the natural boundaries of which are based on four ecological criteria: bathymetry, hydrography, productivity and trophically related populations (Sherman, 1994; Sherman and Duda, 2005). The areas of the world most stressed from habitat degradation, pollution and over-exploitation of marine resources are the coastal ecosystems. Ninety per cent of the usable annual global biomass yield of marine fish and other living marine resources is produced in 64 LMEs (Fig. 4.1) identified within, and in some cases extending beyond, the boundaries of the exclusive economic zones of coastal nations located around the margins of the ocean basins (Sherman, 1994; Garibaldi and Limongelli, 2003).

Levels of primary production are persistently higher around the margins of the ocean basins, within the boundaries of the LMEs, than in the open-ocean pelagic areas (Fig. 4.2). Urban centres with high population density characterize many of these coastal ocean areas and contribute to the pollution that has its greatest impact on natural productivity cycles through eutrophication from high levels of nitrogen and phosphorus effluent from estuaries. Toxins in poorly treated sewage discharge, harmful algal blooms, and loss of wetland nursery areas to coastal development are ecosystem-level problems that also need to be addressed (GESAMP, 1990).

The theory, measurement and modelling relevant to monitoring the changing states of LMEs are embedded in reports on ecosystems with multiple steady

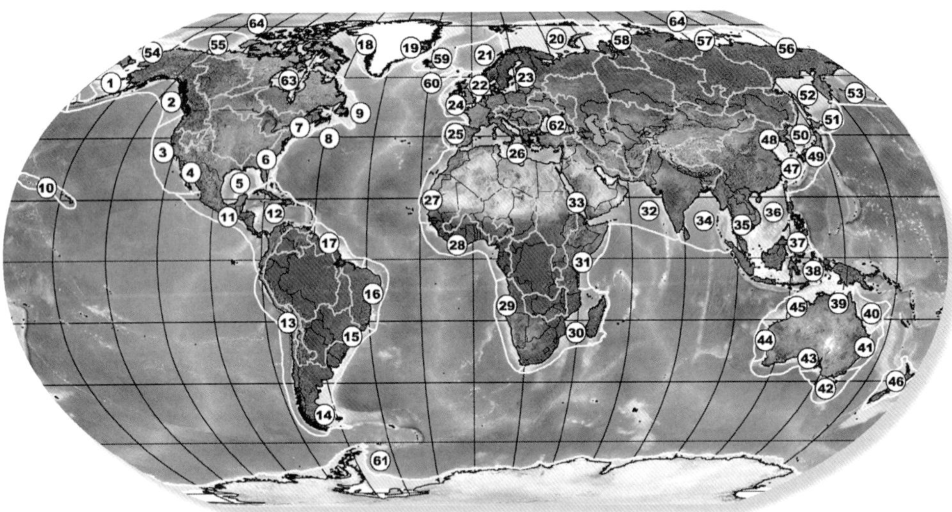

1 East Bering Sea
2 Gulf of Alaska
3 California Current
4 Gulf of California
5 Gulf of Mexico
6 Southeast US Continental Shelf
7 Northeast US Continental Shelf
8 Scotian Shelf
9 Newfoundland-Labrador Shelf
10 Insular Pacific-Hawaiian
11 Pacific Central-American Coastal
12 Caribbean Sea
13 Humboldt Current
14 Patagonian Shelf
15 South Brazil Shelf
16 East Brazil Shelf
17 North Brazil Shelf
18 West Greenland Shelf
19 East Greenland Shelf
20 Barents Sea
21 Norwegian Shelf
22 North Sea
23 Baltic Sea
24 Celtic-Biscay Shelf
25 Iberian Coastal
26 Mediterranean Sea
27 Canary Current
28 Guinea Current
29 Benguela Current
30 Agulhas Current
31 Somali Coastal Current
32 Arabian Sea
33 Red Sea
34 Bay of Bengal
35 Gulf of Thailand
36 South China Sea
37 Sulu-Celebes Sea
38 Indonesian Sea
39 North Australian Shelf
40 Northeast Australian Shelf-Great Barrier Reef
41 East-Central Australian Shelf
42 Southeast Australian Shelf
43 Southwest Australian Shelf
44 West-Central Australian Shelf
45 Northwest Australian Shelf
46 New Zealand Shelf
47 East China Sea
48 Yellow Sea
49 Kuroshio Current
50 Sea of Japan
51 Oyashio Current
52 Okhotsk Sea
53 West Bering Sea
54 Chukchi Sea
55 Beaufort Sea
56 East Siberian Sea
57 Laptev Sea
58 Kara Sea
59 Iceland Shelf
60 Faroe Plateau
61 Antarctic
62 Black Sea
63 Hudson Bay
64 Arctic Ocean

Fig. 4.1. Global map showing 64 large marine ecosystems (LMEs) and linked watersheds.

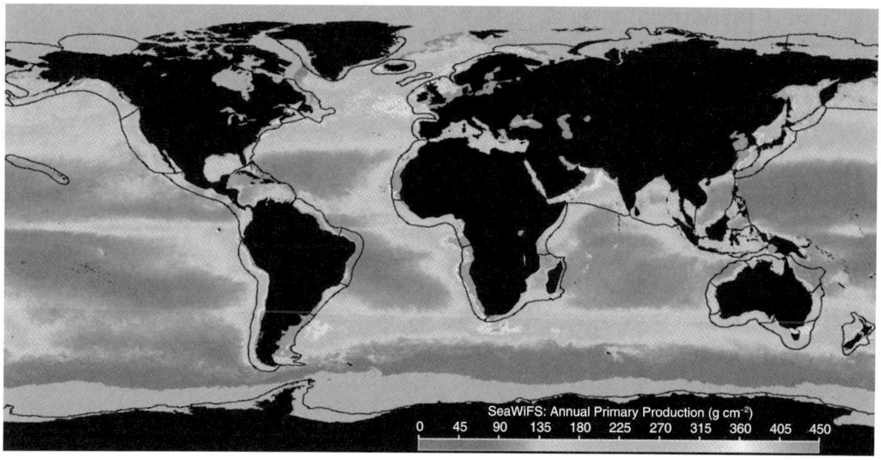

Fig. 4.2. Global map showing 64 large marine ecosystems (LMEs) and their estimated average annual productivity. Estimates are based on SeaWiFS satellite data collected between September 1998 and August 1999, and the model developed by Behrenfeld and Falkowski (1997). The original figure can be seen in colour at http://www.aslo.org/lo/pdf/vol_42/issue_1/0001.pdf from Limnology and Oceanography 42(1), 1–20, and depicts a shaded gradient of primary productivity from a high of 450g C/cm^2/year in red to <45g C/cm^2/year in purple.

states, and on the pattern formation and spatial diffusion within ecosystems (Holling, 1973; Pimm, 1984; Beddington, 1986; Holling, 1986; Sherman and Alexander, 1986; Sherman *et al.*, 1990; Mangel, 1991; Holling, 1993; Levin, 1993). Critical processes controlling the structure and function of biological communities can best be addressed on a regional basis using LMEs as the distinct units for marine resources assessment, monitoring and management (Ricklefs, 1987; Sherman, 1993a; Duda and Sherman, 2002). In turn, the concept of assessment, monitoring and management of marine resources from an LME perspective has been the topic of a series of ongoing national and international symposia, case studies and workshops initiated in 1984; in each instance, the geographic extent of the LME has been defined on the basis of bathymetry, hydrography, productivity and trophodynamics. A list of peer-reviewed published volumes of LME case studies is given in Table 4.1.

Within the geographic limits of LMEs, domains or subsystems can be defined. For example, the Adriatic Sea is a subsystem of the Mediterranean Sea LME. In other LMEs, geographic limits are defined by the character of continental shelves. Among these are the US Northeast Continental Shelf and its four subsystems - Gulf of Maine, Georges Bank, Southern New England and Mid-Atlantic Bight (Sherman 1988; Sherman *et al.,* 1998). Other examples of Continental Shelf LMEs are the Icelandic Shelf, Yellow Sea, East Bering Sea, North Sea and Barents Sea. LMEs with narrow shelf areas and well-defined currents are bounded by the outer margins of the major coastal currents. The Humboldt Current, California Current, Canary Current, Kuroshio Current and Benguela Current are examples of coastal current LMEs.

Table 4.1. Published large marine ecosystem (LME) case studies and volumes.

LME	Vol.	Author(s)	LME	Vol.	Author(s) cont.
Barents Sea	2	Skjoldal & Rey		13	Edwards *et al.*
	4	Borisov		13	Cho *et al.*
	5	Skjoldal		13	Grigalunas *et al.*
	10	Dalpadado *et al.*	Scotian Shelf	10	Zwanenburg *et al.*
	12	Matishov	Caribbean Sea	3	Richards & Bohnsack
Norwegian Shelf	3	Ellertsen *et al.*			
	5	Blindheim & Skjoldal	Patagonian Shelf	5	Bakun
			South Brazil Shelf	12	Ekau & Knoppers
North Sea	1	Daan	East Brazil Shelf	12	Ekau & Knoppers
	9	Reid	North Brazil Shelf	12	Ekau & Knoppers
	10	McGlade	Baltic Sea	1	Kullenberg
	12	Hempel		12	Jansson
Iceland Shelf	10	Astthorsson & Vilhjálmsson	Celtic-Biscay Shelf	10	Lavin
Faroe Plateau	10	Gaard *et al.*	Iberian Coastal	2	Perez-Gandaras
Antarctic	1	Scully *et al.*		10	Wyatt & Porteiro
	3	Hempel	Mediterranean Sea	5	Caddy
	5	Scully *et al.*			
California Current	1	MacCall	Canary Current	5	Bas
	4	Mullin		12	Roy & Cury
	5	Bottom	Guinea Current	5	Binet & Marchal
	12	Lluch-Belda *et al.*		11	Koranteng & McGlade
Pacific American Coastal	8	Bakun		11	Mensah & Quaatey
Humboldt Current	5	Bernal		11	Lovell & McGlade
	12	Wolff *et al.*		11	Cury & Roy
Gulf of Thailand	5	Piyakarnchana		11	Koranteng
	11	Pauly & Chuenpagdee	Benguela Current	2	Crawford *et al.*
				12	Shannon & O'Toole
South China Sea	5	Christensen		14	Ahanhanzo
Indonesian Sea	3	Zijlstra & Baars		14	Shillington *et al.*
Northeast Australian Shelf	2	Bradbury & Mundy		14	Monteiro & van der Plas
	5	Kelleher		14	Hutchings *et al.*
	8,12	Brodie		14	Pitcher & Weeks
Gulf of Mexico	9	Shipp		14	van der Lingen *et al.*
	9	Gracia & Vasquez		14	Fréon *et al.*
		Baden		14	Reason *et al.*
Southeast US Shelf	4	Yoder		14	Jarre *et al.*
				14	Bernard *et al.*
Northeast US Shelf	1	Sissenwine		14	Monteiro *et al.*
	4	Falkowski		14	Gründlingh *et al.*
	6	Anthony		14	Brundrit *et al.*
	10,12	Sherman	Black Sea	5	Caddy
	13	Dyer & Poggie		12	Daskalov

Continued

Table 4.1. Continued

Volume no.	Volume description
1	1986. *Variability and Management of Large Marine Ecosystems.* Sherman and Alexander, eds. AAAS Symposium 99. Westview Press, Boulder, Colorado, 319 p
2	1989. *Biomass Yields and Geography of Large Marine Ecosystems.* Sherman and Alexander, eds. AAAS Symposium 111. Westview Press, Boulder, Colorado, 493 p
3	1990. *Large Marine Ecosystems: Patterns, Processes and Yields.* Sherman, Alexander and Gold, eds. AAAS Symposium. AAAS Press, Washington, DC., 242 p
4	1991. *Food Chains, Yields, Models and Management of Large Marine Ecosystems.* Sherman, Alexander and Gold, eds. AAAS Symposium. Westview Press, Boulder, Colorado, 320 p
5	1992. *Large Marine Ecosystems: Stress, Mitigation and Sustainability.* Sherman, Alexander and Gold, eds. AAAS Press, Washington, DC., 376 p
6	1996. *The North-east Shelf Ecosystem: Assessment, Sustainability and Management.* Sherman, Jaworski and Smayda, eds. Blackwell Science, Cambridge, Massachusetts, 564 p
7	1998. *Large Marine Ecosystems of the Indian Ocean: Assessment, Sustainability and Management.* Sherman, Okemwa and Ntiba, eds. Blackwell Science, Malden, Massachusetts, 394 p
8	1999. *Large Marine Ecosystems of the Pacific Rim: Assessment, Sustainability and Management.* Sherman and Tang, eds. Blackwell Science, Malden, Massachusetts, 455 p
9	1999. *The Gulf of Mexico Large Marine Ecosystem: Assessment, Sustainability and Management.* Kumpf, Steidinger and Sherman, eds Blackwell Science, Malden, Massachusetts, 736 p
10	2002. *Large Marine Ecosystems of the North Atlantic: Changing States and Sustainability.* Sherman and Skjoldal, eds. Elsevier Science, New York and Amsterdam, 449 p
11	2002. *Gulf of Guinea Large Marine Ecosystem: Environmental Forcing and Sustainable Development of Marine Resources.* McGlade, Cury, Koranteng, Hardman-Mountford, eds. Elsevier Science, Amsterdam and New York, 392 p
12	2003. *Large Marine Ecosystems of the World: Trends in Exploitation, Protection and Research.* Hempel and Sherman, eds. Elsevier Science, New York and Amsterdam, 423 p.
13	2005. *Sustaining Large Marine Ecosystems: The Human Dimension.* Hennessey and Sutinen, eds. Elsevier Science, New York and Amsterdam, 368 p
14	2006. *Benguela: Predicting a Large Marine Ecosystem.* Shannon, Hempel, Malanotte-Rizzoli, Moloney, Woods, eds. Elsevier Science, New York and Amsterdam, 401 p

Monitoring and Assessment of LMEs

Temporal and spatial scales influencing biological production and changing ecological states in marine ecosystems have been the topic of a number of theoretical and empirical studies. The selection of scale in any study is related to the processes under investigation. An excellent treatment of this topic can be found

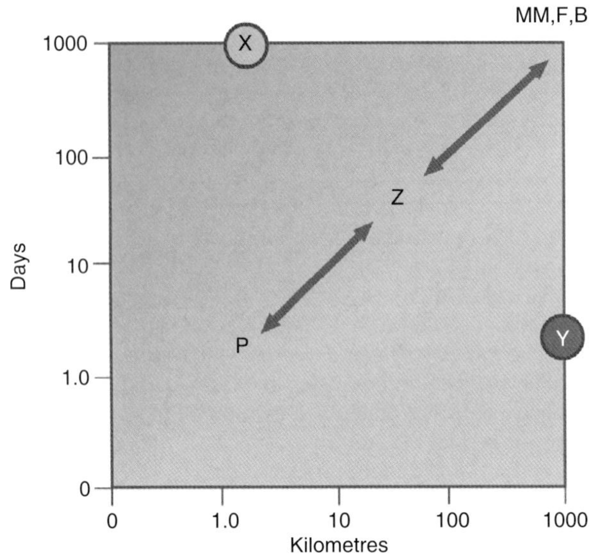

Fig. 4.3. A simple set of scale relations for the pelagic food web (P = phytoplankton, Z = zooplankton, F = fish, MM = marine mammals, B = birds, X = predictable fronts with small cross-front dimensions, and Y = weather events occurring over relatively large scales. (Adapted from Steele, 1988.)

in Steele (1988). Steele indicates that in relation to the general ecology of the sea, the best-known models in marine population dynamics include those by Schaefer (1954), Beverton and Holt (1957) and following the earlier pioneering approach of Lindemann (1942). However, as noted by Steele, this array of models is unsuitable for dealing with temporal or spatial variability in the ocean. A heuristic projection was produced by Steele to illustrate scales and ecosystem indicators of importance in monitoring pelagic components of the ecosystem, including phytoplankton, zooplankton, fish, frontal processes and short-term but large-area episodic effects (Fig. 4.3).

Integrated Ecosystem Assessments

A key factor in reaching a determination on the status of ecosystem condition is the quantitative output from five modules of spatial and temporal indicators of ecosystem: (i) productivity; (ii) fish and fisheries; (iii) pollution and ecosystem health; (iv) socio-economics; and (v) governance (Fig. 4.4). Advances in technology now allow for cost-effective measuring of the changing states of LMEs using these suites of indicators. The five-module indicator approach to the integrated assessment and management of LMEs has proven useful in ecosystem-based projects in the USA and elsewhere. The modules are customized for each LME through a transboundary diagnostic analysis (TDA) process and a strategic action plan (SAP) development process for the groups of nations or states sharing an LME. These processes are critical for integrating science into management

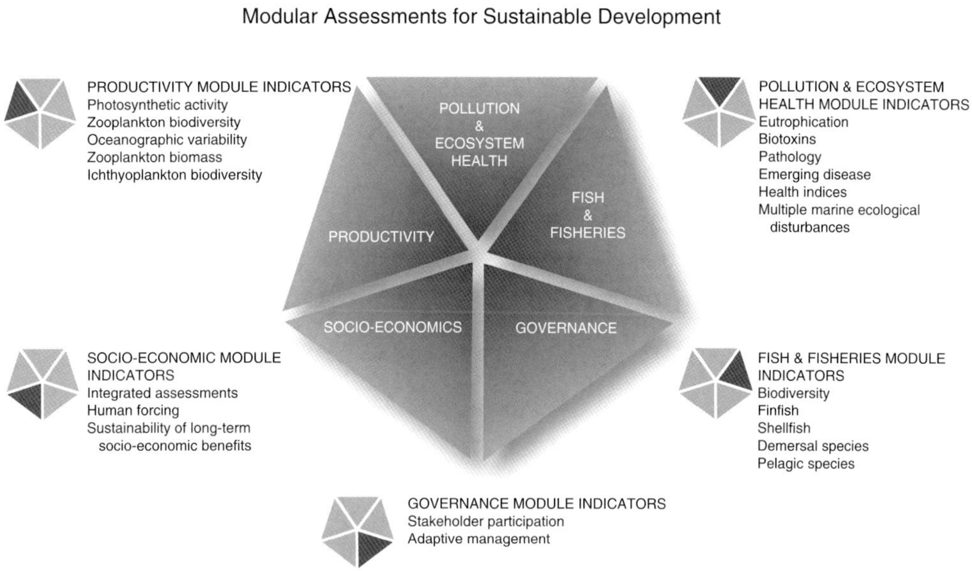

Fig. 4.4. Large marine ecosystem (LME) modules as suites of condition indicators for inputs to integrated ecosystem assessments.

in a practical way, and for establishing appropriate governance regimes (Duda and Sherman, 2002; Olsen *et al.,* 2006). Of the five modules, three are science-based indicators that focus on productivity, fish/fisheries and pollution/ecosystem health. The other two modules, socio-economics and governance, support the development of indicators that improve measures of economic benefits to be derived from a more sustainable resource use, as well as advance legal and administrative support for ecosystem-based management practices (Fig. 4.4). The first four modules support the TDA process, while the governance module is associated with periodic updating of the SAP development process. Adaptive management regimes are encouraged through periodic assessment processes (i.e. TDA updates) and through updating the action plans as gaps are filled (Wang, 2004).

Productivity module indicators

Primary productivity can be related to the carrying capacity of an ecosystem for supporting fish resources (Pauly and Christensen, 1995). It has been reported that the maximum global level of primary productivity for supporting the average annual world catch of fisheries has been reached, and that further large-scale unmanaged increases in fisheries yields from marine ecosystems are likely to be at trophic levels below fish in the marine food web (Beddington, 1995). Measurements of ecosystem productivity can be useful indicators of the growing problem of coastal eutrophication. In several LMEs, excessive nutrient loadings of coastal waters have been related to algal blooms implicated in mass

Fig. 4.5. NuShuttle, an undulating oceanographic sampling platform, carries sensors for temperature, salinity, chlorophyll, primary productivity and dissolved oxygen. The shuttle also contains a Continuous Plankton Recorder (CPR) mechanism.

mortalities of living resources, emergence of pathogens (e.g. cholera, vibrios, red tides, and paralytic shellfish toxins), and explosive growth of non-indigenous species (Epstein, 1993).

The ecosystem parameters measured and used as indicators of changing conditions in the productivity module are zooplankton biodiversity and species composition, zooplankton biomass, water-column structure, photosynthetically active radiation, transparency, chlorophyll-*a*, nitrite, nitrate and primary production. Plankton can be measured over decadal timescales by deploying continuous plankton recorder systems (Fig. 4.5) monthly across ecosystems from commercial vessels of opportunity. Advanced plankton recorders can be fitted with sensors for temperature, salinity, chlorophyll, nitrate/nitrite, petroleum hydrocarbons, light, bioluminescence and primary productivity, providing the means for *in situ* monitoring and for calibrating satellite-derived oceanographic data. Properly calibrated satellite data can provide information on ecosystem conditions including physical state (i.e. surface temperature), nutrient characteristics, primary productivity and phytoplankton species composition (Aiken *et al.*, 1999; Berman and Sherman, 2001; Melrose *et al.*, 2006).

Fish and fisheries module indicators

Changes in biodiversity and species dominance within fish communities of LMEs have resulted from excessive exploitation, naturally occurring environmental shifts due to climate change and coastal pollution. Changes in biodiversity and species dominance in a fish community can move up the food web to apex predators and cascade down the food web to plankton components of the ecosystem. The fish and fisheries module includes both fisheries-independent bottom-trawl surveys and pelagic-species acoustic surveys to obtain time-series information on changes in fish biodiversity and abundance levels.[1] Standardized sampling procedures, when employed from small calibrated trawlers, can provide important information on changes in fish species (Sherman, 1993b). Fish catch provides biological samples for stock identification, stomach content analyses, age-growth relationships, fecundity and coastal pollution monitoring for possibly associated pathological conditions, as well as data for preparing stock assessments and for clarifying and quantifying multi-species trophic relationships. The survey vessels can also be used as platforms for obtaining water, sediment and benthic samples for monitoring harmful algal blooms, diseases, anoxia and changes in benthic communities.

Pollution and ecosystem health module indicators

In several LMEs, pollution and eutrophication have been important driving forces of change in biomass yields. Assessing the changing status of pollution and health of an entire LME is scientifically challenging. Ecosystem health is a concept of wide interest for which a single precise scientific definition is difficult. The health paradigm is based on multiple-state comparisons of ecosystem resilience and stability, and is an evolving concept that has been the subject of a number of meetings (Sherman, 1993b). To be healthy and sustainable, an ecosystem must maintain its metabolic activity level and its internal structure and organization, and must resist external stress over time and space scales relevant to the ecosystem (Costanza, 1992).

The pollution and ecosystem health module measures pollution effects on the ecosystem through the bivalve mollusc monitoring strategy of the US Environmental Protection Agency's (EPA) Mussel-Watch Program, through the pathobiological examination of fish and fish tissue, through the estuarine and nearshore monitoring of contaminants and contaminant effects in the water column, substrate and selected groups of organisms and through similar efforts. Where possible, bioaccumulation and trophic transfer of contaminants are assessed, and critical life-history stages and selected food web organisms are examined for indicators of exposure to, and effects from, contaminants. Effects of impaired reproductive capacity, organ disease and impaired growth from contaminants are measured. Assessments are made of contaminant impacts at both species and population levels. Implementation of protocols to assess the frequency and effect of harmful algal blooms, emergent diseases and multiple marine ecological disturbances (Sherman, 2000) are included in the pollution module.

In the USA, the EPA has developed a suite of five coastal condition indices - water quality, sediment quality, benthic communities, coastal habitat and fish tissue contaminants - as part of an ongoing collaborative effort with NOAA, the US Fish and Wildlife Service, the US Geological Survey and other agencies

[1] Northeast Fisheries Science Center research vessel groundfish surveys overviews are available online at http://www.nefsc.noaa.gov/sos/vesurv/vesurv.html. A description of the Alaskan Fisheries Science Center 2006 Aleutian Islands Cooperative Acoustic Surveys is available at http://www.afsc.noaa.gov/quarterly/amj2006/divrptsREFM7.htm

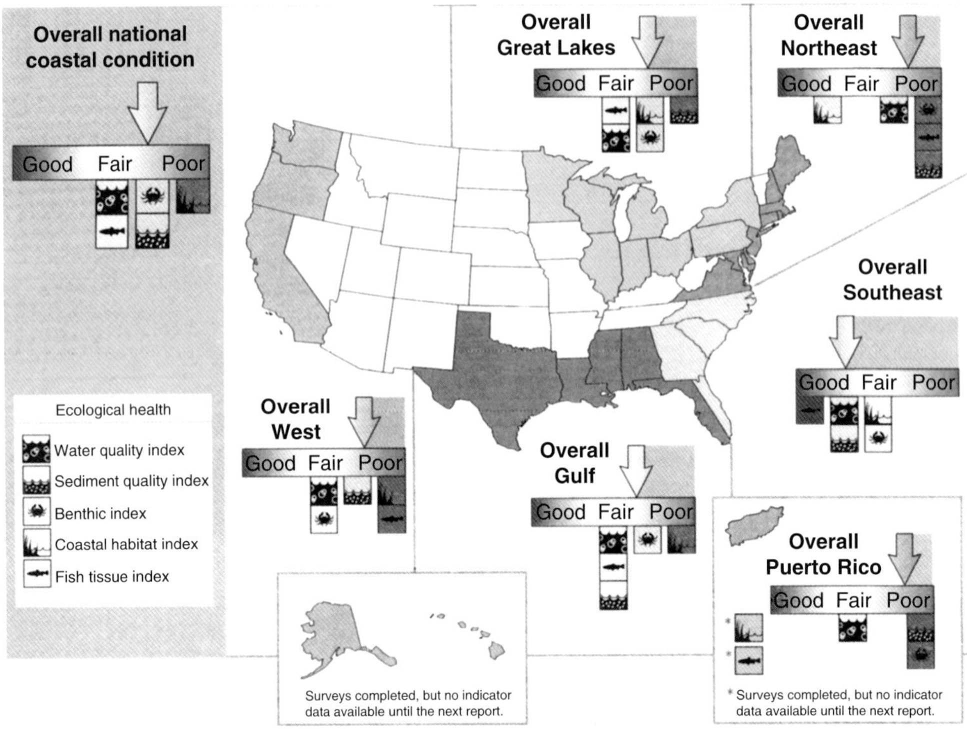

Fig. 4.6. Indicators from the Environmental Protection Agency's (EPA) National Coastal Condition Report. (USEPA, 2004.)

representing states and tribes (Fig. 4.6). The 2004 report, 'National Coastal Condition Report II', includes results from EPA's analyses of coastal condition indicators and NOAA's fish stock assessments by LMEs aligned with EPA's national coastal assessment regions (USEPA, 2001, 2004).

Socio-economic module indicators

The LMEs annually contribute US$12.6 trillion to the global economy (Costanza *et al.*, 1997).The socio-economic module emphasizes the practical application of scientific findings to managing LMEs, and the explicit integration of social and economic indicators and analyses with all other scientific assessments, to assure that prospective management measures are cost-effective. Economists and policy analysts work closely with ecologists and other scientists to identify and evaluate management options that are both scientifically credible and economically practical with regard to the use of ecosystem goods and services. In order to respond adaptively to enhanced scientific information, socio-economic considerations must be closely integrated with science.This component of the LME approach to marine resources management has recently been described as the human dimensions of LMEs. A framework has been developed by the Department of Natural

Resource Economics at the University of Rhode Island for monitoring and assessment of the human dimensions of LMEs, and for incorporating socio-economic considerations into an adaptive management approach for LMEs (Sutinen, 2000). One of the more critical considerations, a method for economic valuations of LME goods and services, has been developed using framework matrices for ecological states and economic consequences of change (Hoagland *et al.*, 2005; Hoagland and Jin, 2006; Olsen *et al.*, 2006).

Governance module indicators

The governance module is evolving, based on demonstration projects now underway in several ecosystems, such that ecosystems will be managed more holistically than in the past. In LME assessment and management projects supported by the global environment facility for the Guinea Current, and Benguela Current LMEs, agreements have been reached among the Environmental, Fisheries, Energy and Tourism ministers of the countries bordering these LMEs to enter into joint transboundary, international resource assessment and management Commissions.[2] Elsewhere, the Great Barrier Reef and Antarctic LMEs are also being managed from an ecosystem perspective, the latter under the Commission for the Conservation of Antarctic Marine Living Resources. Governance profiles of LMEs are being explored to determine their utility in promoting long-term sustainability of ecosystem resources (Juda and Hennessey, 2001). In each of the LMEs, governance jurisdiction can be scaled to ensure conformance with existing legislated mandates and authorities (Olsen *et al.*, 2006). An example of multiple governance-related jurisdictions is shown in Fig. 4.7.

Application of Indicator Modules to Integrated LME Assessment and Management

Continued overfishing in the face of scientific warnings, fishing down food webs, destruction of habitat, and accelerated pollution loading, especially nitrogen in the world's LMEs, are actions underway in both rich and poor nations engaged in encouraging economic growth. Fragmentation among institutions, international agencies and disciplines, lack of cooperation among nations sharing marine ecosystems and weak national policies, legislation and enforcement all contribute to the need for a new imperative for adopting ecosystem-based approaches to managing human activities in these systems in order to avoid serious social and economic disruptions. Indicator data derived from spatial and temporal applications of the five modules are being applied in projects underway by a growing number of nations in the assessment and management of LMEs with the financial assistance of the GEF. The GEF allocated $3.2 billion in grant financing, supplemented

[2] Benguela Commission 2006 is announced at http://www.bclme.org/news/mediaflash_bcc.asp; Guinea Current Commission 2006 Ministerial signing of the Abuja Declaration is announced at http://www.unep.org/GC/GC24/download.asp?ID=88 p.12.

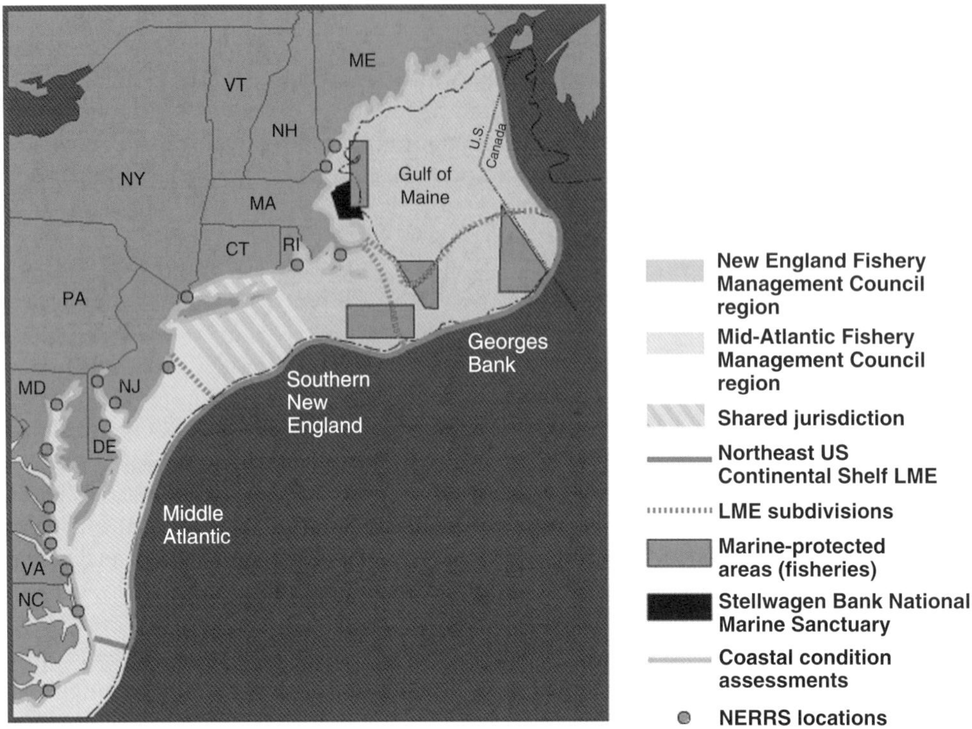

Fig. 4.7. Example of multijurisdictional large marine ecosystem (LME) governance. Included are: (i) jurisdictions covered by the New England and Mid-Atlantic Fishery Management Councils; (ii) LME subareas; (iii) marine-protected areas and the boundaries of the Stellwagen Bank National Marine Sanctuary; (iv) near-coastal areas assessed for 'condition' determinations by the EPA; and (v) locations of National Estuarine Research Reserve Sites (NERRS).

by more than $8 billion in additional financing, for 800 projects in 156 developing countries and those in economic transition. All six thematic areas of GEF, including the land degradation cross-cutting theme, have implications for coastal and marine ecosystems. Priorities have been established by the GEF Council in its Operational Strategy adopted in 1995 (GEF, 1995). The international waters focal area was designed to be consistent with both Chapters 17 and 18 of Agenda 21 of UNCED, the United Nations Conference on Environment and Development (UNCED, 1992).[3] In 1995, the GEF Council included the concept of LMEs in its Operational Strategy as a vehicle for promoting ecosystem-based management of coastal and marine resources in the international waters focal area within a framework of sustainable development. The Report of the Second Meeting of the UN Informal, Open-ended Consultative Process on Ocean Affairs (UNGA, 2001), which was related to the UN Convention on the Law of the Sea, recognized the contribution of GEF in addressing LMEs through its ecosystem-based approach.

[3] UNCED Report available at http://www.un.org/documents/ga/conf151/aconf15126–1annex1.htm

Among the stressors affecting the sustainability of LMEs are the growing problem of the depletion of fish and fishery resources and biomass yields and nitrogen overloading in coastal waters contributing to eutrophication.

Fisheries depletion and recovery

The growing awareness that biomass yields are being influenced by multiple driving forces has broadened monitoring strategies from fish stock assessment surveys to encompass food chain dynamics and the effects of environmental perturbations and pollution on living marine resources from an ecosystem perspective. Evidence for species biomass recovery following significant reduction in fishing effort through mandated fishery management actions is encouraging. In the USA, the reduction of fisheries effort within the boundaries of the Northeast Shelf LME (NESLME) initiated recoveries for several important fisheries populations including herring, mackerel, sea scallop, striped bass, haddock and yellowtail flounder. Two management decisions led to the recovery trend: (i) the control and elimination of foreign fishing effort in 1975 through passage of national legislation (Magnuson, 1976); and (ii) further reduction in domestic fishing effort in 1994 as a result of action taken by the New England and North Atlantic Fishery Management Councils (NEFMC, 2004) (see final Amendment 13 to the NE Multispecies Fishery Management Plan, FMP, accessible from http://www.nefmc.org[4]). The declining trend in the NESLME fisheries biomass and catch during the 1960s through to the early 1990s is depicted in Fig. 4.8. Following the exclusion of foreign fisheries and in the absence of any significant US fishing of herring and mackerel, the spawning stock biomass (SSB) increased from a level of less than 500,000 t for each species in 1982 to an estimated combined total of 3.5 million metric tonnes (mmt) in 1994 (Fig. 4.9). During this period the multidecadal trend

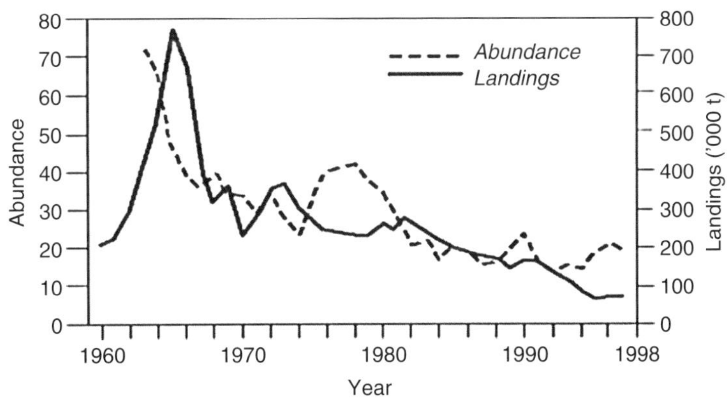

Fig. 4.8. Landings in tonnes and abundance index of principal groundfish and flounders, 1960–2000.

[4] Amendment 13 final rule document is recorded in the Federal Register of Tuesday, April 27, 2004 as 50 CFR Part 648; Docket No. 040112010 – I.D.122203A; RIN 0648–AN17.

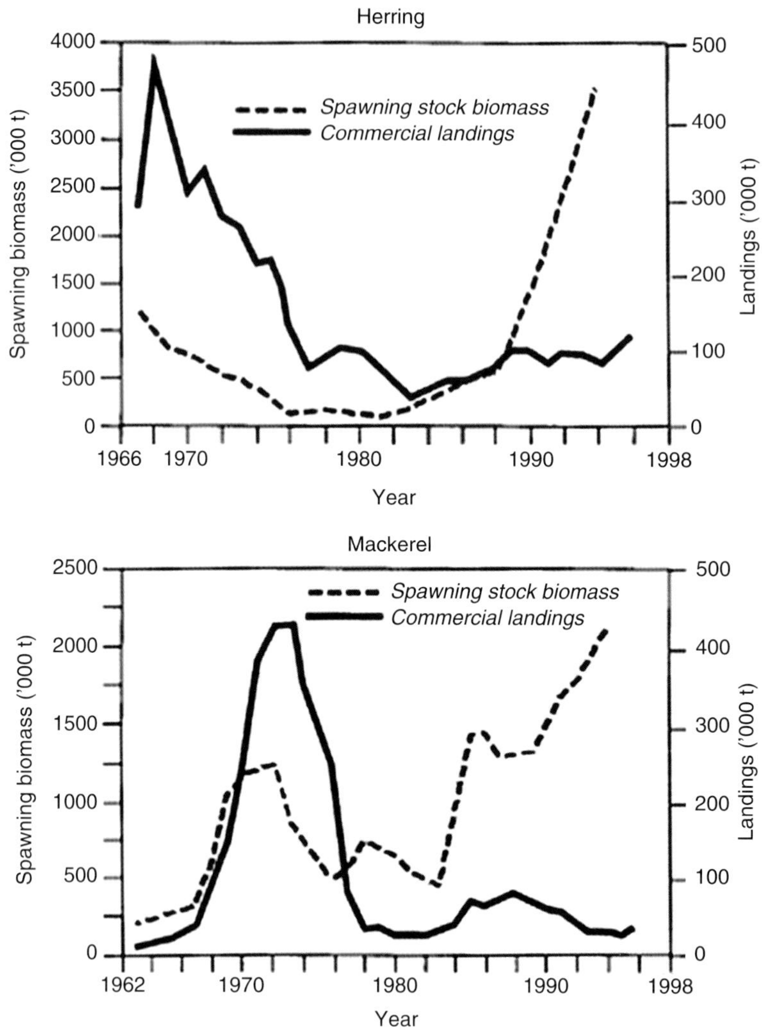

Fig. 4.9. Top: Atlantic herring commercial landings and spawning stock biomass (SSB), 1967 to 1996. Bottom: Atlantic mackerel landings and SSB, 1963 to 1996.

in the ecology of the NESLME was relatively stable. Evidence is shown in a limited interannual departure from a relatively stable long-term annual trend, but there is no evidence of either persistent ascending or descending temperature (Fig. 4.10). This relative ecosystem stability was found in zooplankton biomass (Fig. 4.11), in the abundance of the three dominant zooplankton species and combined abundance levels of dominant copepod species (Fig. 4.12).

At the base of the ecosystem food web, the long-term trend of primary productivity showed little consecutive upward or downward trend from the mean annual *in situ* value of 350 g/cm^2/year of the 1977–1987 period (O'Reilly *et al.*, 1987) and the recent value of 337 g/cm^2/year for the period 1998–2005 (Fig. 4.13) based on SeaWiFS data (O'Reilly and Belkin, 2006).

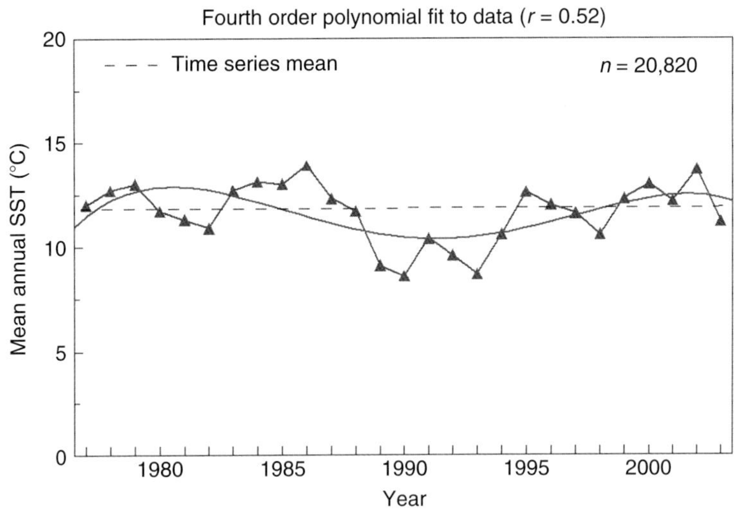

Fig. 4.10. Mean annual surface temperature pattern based on 20,820 measurements taken simultaneously with each of the zooplankton samples, 1977–2003. The trend line is a fitted polynomial with an *r* value of 0.52. The long-term mean is represented by a dashed line.

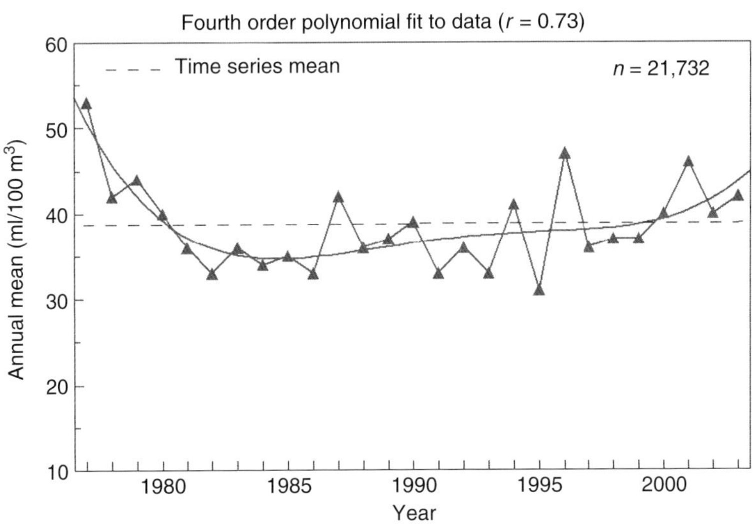

Fig. 4.11. Mean annual values of zooplankton biomass (solid line) and long-term mean (dashed line) for 21,732 zooplankton samples from four subareas of the Northeast Shelf ecosystem, 1977–2003.

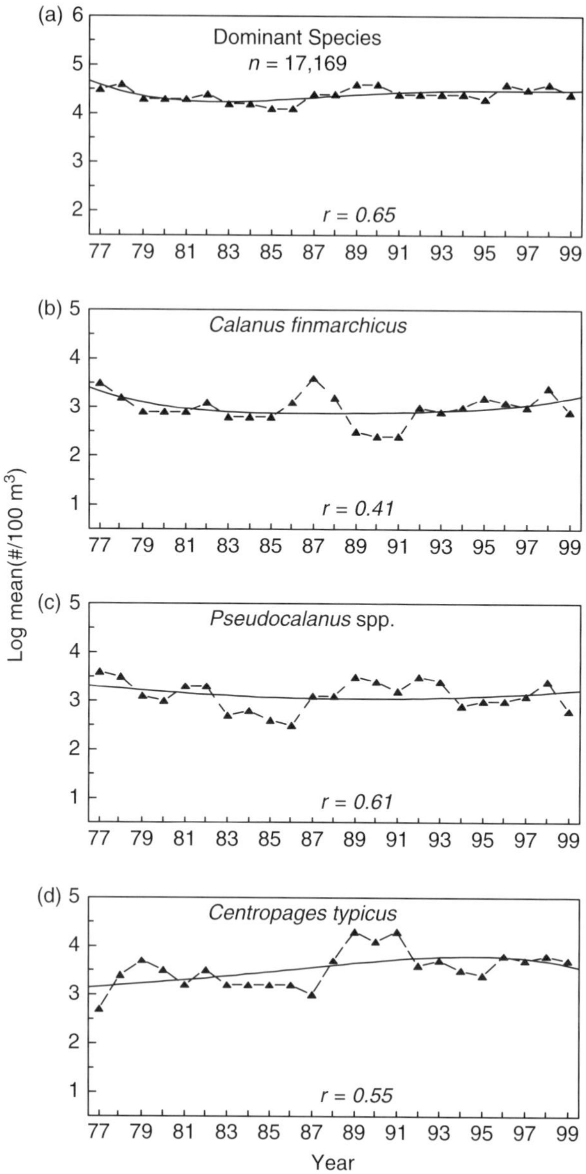

Fig. 4.12. Trends in the abundance of the three dominant zooplankton species inhabiting the Northeast Shelf ecosystem, 1977–1999. (a) mean annual abundance of the combined three species; (b) *Calanus finmarchicus*; (c) *Pseudocalanus* spp.; (d) *Centropages typicus*.

The subsequent reduction of fishing effort as the principal management action for controlling excessive fishing effort of the US fishing vessels within the Northeast Shelf ecosystem, resulted in an increase in spawning biomass and recruitment for haddock, yellowtail flounder (Figs. 4.14 and 4.15) and other species (NMFS, 2004).

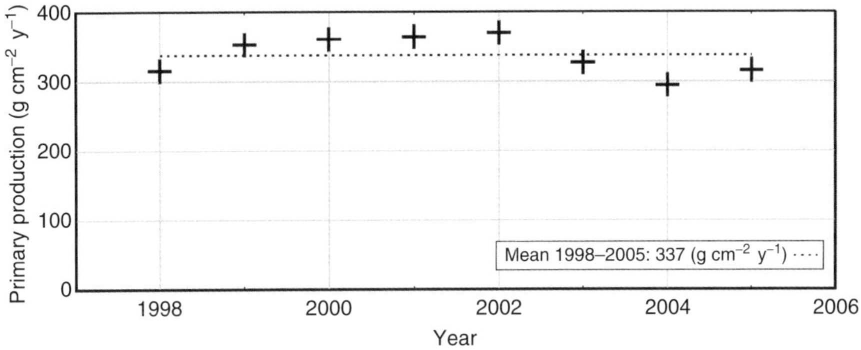

Fig. 4.13. Mean annual primary production in the Northeast Shelf LME (NESLME), 1998–2005. (From O'Reilly and Belkin, 2006.)

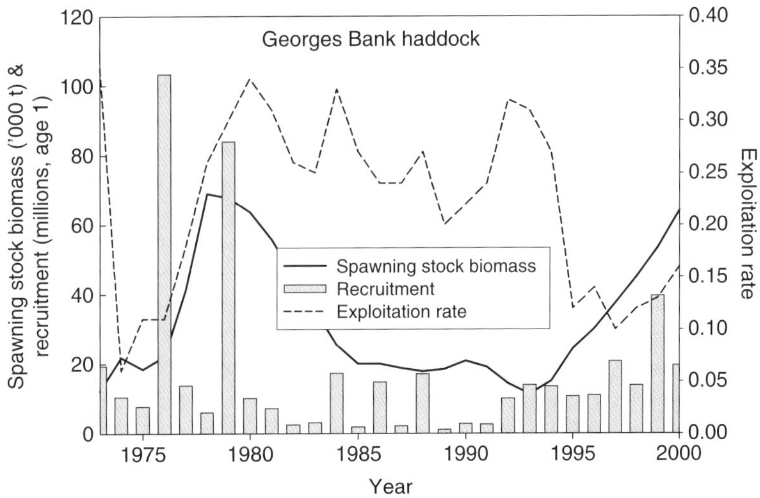

Fig. 4.14. Trends in spawning stock biomass (SSB) and recruitment in relation to reductions in exploitation rate (fishing effort) for the commercially important species haddock inhabiting the Georges Bank subarea of the Northeast Shelf ecosystem.

The Northeast Shelf ecosystem is presently undergoing a significant trend towards biomass recovery of pelagic and demersal fish species important to the fisheries of the adjacent northeast states from Maine to North Carolina under a fisheries stock rebuilding programme under the authority of the New England and Mid-Atlantic Fishery Management Council.[5] Although the recovery has not as yet been fully achieved, the corner has been turned from declining overharvested fish stocks towards a condition wherein the stocks can be managed to

[5] Councils in Action, a 2006 NOAA press release on 2005 Fishery Management Council actions on stock rebuilding, available at http://www.publicaffairs.noaa.gov/releases2006/jun06/noaa06–061A.html

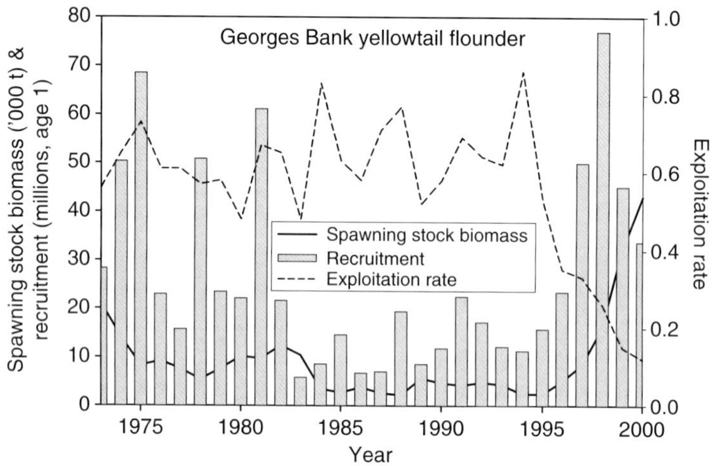

Fig. 4.15. Trends in spawning stock biomass (SSB) and recruitment in relation to reductions in exploitation rate (fishing effort) for the commercially important species yellowtail flounder inhabiting the Georges Bank subarea of the Northeast Shelf ecosystem.

sustain their long-term potential yield levels. The management decisions taken to reduce fishing effort to recover lost biomass were based on assessment of, and supported by science-based monitoring and assessment information from, the productivity, fish and fisheries, pollution and ecosystem health, socioeconomics and governance modules that have been operational by NOAA's Northeast Fisheries Science Center for several decades in collaboration with state, federal and private stakeholders from the region. This case study can serve to underscore the utility of the modular approach to ecosystem-based management of marine fish species. In an effort to stem the loss of fisheries biomass in other parts of the world, applications of this modular approach to LME management are presently underway by countries bordering the Yellow Sea, Benguela Current, Baltic Sea, and Guinea Current LMEs with financial assistance of the GEF, collaborating UN agencies, and the technical and scientific assistance of other governmental and non-governmental agencies and institutions.

The observation of Pauly and Christensen (1995) that excessive fishing effort can alter the structure of the ecosystem, resulting in a shift from relatively high-priced, large-sized, long-lived, demersal species, down the food chain towards lower-valued, smaller-sized, shorter-lived, pelagic species, is supported by the LME data on species biomass yields. Evidence from the East China Sea, Yellow Sea and Gulf of Thailand suggests that these three LMEs are approaching a critical state of change, wherein recovery to a previous ratio of demersal-to-pelagic species may become problematic. In all three cases, the fisheries are now being directed towards fish protein being provided by catches of smaller-sized species of low value (Chen and Shen, 1999; Pauly and Chuenpagdee, 2003; Tang, 2003). The species change in biomass yields of the Yellow Sea represents an extreme case wherein the annual demersal species biomass yield was reduced from 200,000 t in 1955 to less than 25,000 t in 1980.The fisheries then targeted the pelagic anchovy, and between 1990 and 1995, landings of anchovy reached an historic high of 500,000 t (Tang, 2006).

The GEF-LME projects presently funded or in the pipeline for funding in Africa, Asia, Latin America and Eastern Europe represent a growing network of marine scientists, marine managers and ministerial leaders who are pursuing ecosystem and fishery recovery goals. The annual fisheries biomass yields from the ecosystems in the network are significant at 44.8% of the global total (Table 4.2), and are a firm basis for moving towards the goals of the 2002 World Summit on Sustainable Development (WSSD) for introducing an ecosystem-based assessment and management approach to global fisheries by 2010, and for achieving fishing at maximum sustainable yield (MSY) levels by 2015.

The FAO Code of Conduct for Responsible Fishery Practice (FAO, 1995) is supported by most coastal nations, and has immediate applicability to reaching the WSSD fishery goals. The code argues for moving forward with a precautionary approach to fisheries sustainability, using available information in a more conservative approach to total allowable catch levels than has been the general practice in past decades. Based on Garibaldi and Limongelli (2003), it appears that the biomass and yields of 11 species groups in six LMEs have been relatively stable or have shown marginal increases over the 1990–1999 period. The yield for these six LMEs – the Arabian Sea, Bay of Bengal, Indonesian Sea, North Brazil Shelf, Mediterranean Sea and the Sulu-Celebes Sea – was 8.1 million t, or 9.5% of the global marine fisheries yield in 1999 (Fig. 4.16). The countries bordering

Table 4.2. Reported 1999 annual fisheries biomass yields of LMEs where stewardship ministries were implementing or planning GEF-LME Projects.

LME	Reported 1999 annual biomass yield
South China Sea	13.9
Humboldt Current	12.0
Bay of Bengal	2.3
Patagonian Shelf	1.7
Canary Current	1.6
Benguela Current	1.1
Guinea Current	1.0
Mediterranean Sea	1.0
Gulf of Mexico	1.0
Baltic Sea	0.9
Yellow Sea[a]	0.6
Black Sea	0.5
Caribbean Sea	0.35
Red Sea	0.08
Agulhas/Somali Currents	0.07
Total	38.10 mmt
Percentage of global marine yield	44.8

[a]Biomass yield data for 1995 from Tang (2003).

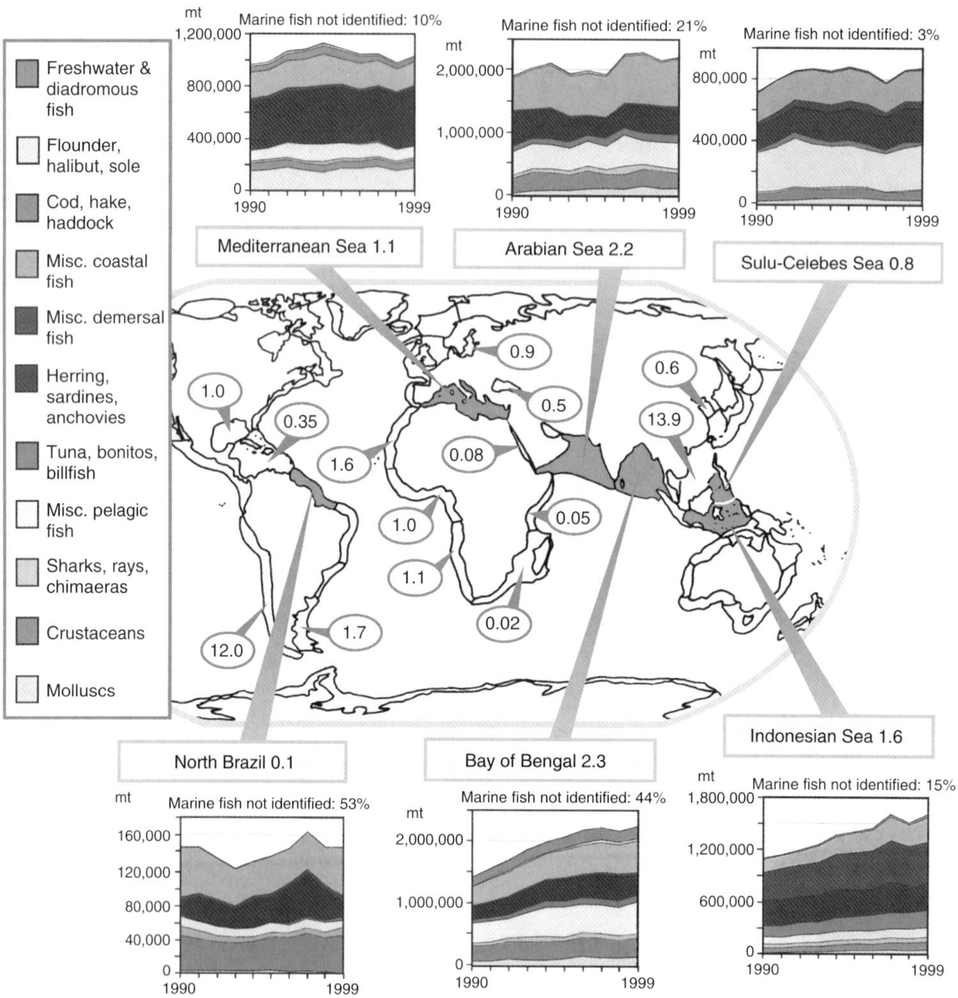

Fig. 4.16. Decadal trends (1990–1999) in biomass yields (mmt) of the six candidate large marine ecosystems (LMEs) for precautionary approach actions to preclude total fish biomass reductions. Value after LME name represents 1999 biomass yield level. Data are based on FAO statistics, as reported to the FAO by official national sources, in Garibaldi and Limongelli (2003). Unfortunately, fisheries effort data are not available for trend analyses. (Reproduced with permission from FAO.)

these six LMEs are among the world's most populous, representing approximately one-quarter of the total human population. These LME border countries increasingly depend on marine fisheries for food security and for national and international trade. In the absence of national reporting of effort data for catches in these six LMEs, and given the risks of fishing down the food chain, it would appear opportune for the stewardship agencies responsible for the fisheries of the LME-bordering countries to mandate precautionary total allowable catch levels (FAO, 1995).

Eutrophication and nitrogen over-enrichment

Nitrogen over-enrichment has been reported as a coastal problem for two decades, from the southeast coast of the USA (Duda, 1982) to the Baltic Sea and other systems (HELCOM, 2001). More recent estimates of nitrogen export to LMEs from linked freshwater basins are summarized by Jaworski (Jaworski and Howarth, 1996; Jaworski, 1999). These recent human-induced increases in nitrogen flux range from fourfold to eightfold in the USA from the Gulf of Mexico to the New England coast, while no increase was documented in areas with little agricultural or few population sources in Canada (Howarth *et al.*, 2000). In European LMEs, recent nitrogen flux increases have been recorded ranging from threefold in Spain, to fourfold in the Baltic Sea and to 11-fold in the Rhine River basin draining to the North Sea LME (Howarth *et al.*, 2000). Duda and El-Ashry (2000) described the origin of this disruption of the nitrogen cycle from the Green Revolution of the 1970s as the world community converted wetlands to agriculture, utilized more chemical inputs, and expanded irrigation to feed the world. As noted by Duda (1982) for the southeast estuaries of the USA and by Rabalais *et al.* (1999) for the Gulf of Mexico, much of the large increase in nitrogen export to LMEs is from agricultural inputs, both from the increased delivery of fertilizer nitrogen as wetlands were converted to agriculture and from concentrations of livestock (Duda and Finan, 1983) for eastern North Carolina, where the increase in nitrogen export over the forested areas ranged from 20- to 500-fold in the late 1970s. Industrialized livestock production during the last two decades has increased the flux, the eutrophication and the oxygen depletion even more as reported by the National Research Council (NRC, 2000). GESAMP (2001) also identifies as significant contributors to eutrophication both sewage from drainages from large cities and atmospheric deposition from automobiles and agricultural activities, with the amounts depending on proximity of sources.

GEF is being asked more frequently by countries to help support the agreed-upon incremental cost of actions to reduce such nitrogen flux. Actions range from assisting in: (i) development of joint institutions for ecosystem-based approaches for adaptive management described in this chapter; to (ii) on-the-ground implementation of nitrogen abatement measures in the agricultural, industrial and municipal sectors; to (iii) breaching of floodplain dikes so that wetlands recently converted to agriculture may be reconverted to promote nitrogen assimilation. The excessive levels of nitrogen contributing to coastal eutrophication constitute a new global environment problem that is cross-sectoral in nature. Excessive nitrogen loadings and oxygen depletion events causing significant mortalities among marine resource species have been identified as problems in the following LMEs that are receiving GEF assistance: Baltic Sea, Black Sea, Adriatic portion of the Mediterranean Sea, Yellow Sea, South China Sea, Bay of Bengal, Gulf of Mexico (Fig. 4.17) and Plata Maritime Front/Patagonia Shelf.

Preliminary global estimates of nitrogen export from freshwater basins to coastal waters were assembled by Seitzinger and Kroeze (Kroeze and Seitzinger,

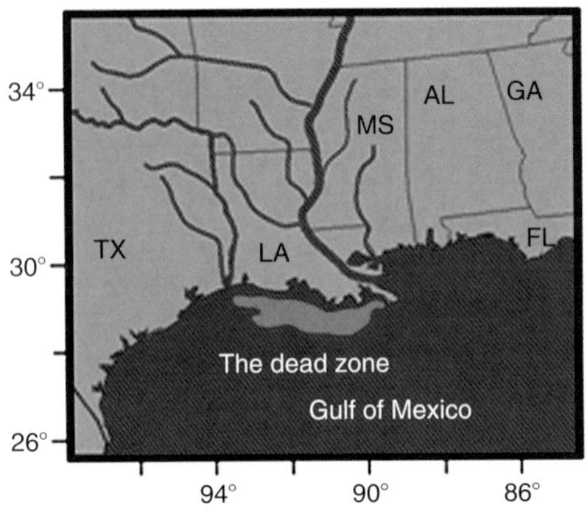

Fig. 4.17. The Gulf of Mexico Dead Zone reached 20,000 km^2 as reported by Dr Nancy Rabalais, Chief Scientist for Northern Gulf of Mexico Hypoxia Studies in a 28 July 2007 press release from the Louisiana Universities Marine Consortium (LUMCON).

1998; Seitzinger and Kroeze, 1998). Their model predicts a doubling of nitrogen in coastal waters by 2050. Included as Fig. 4.18 and adapted from an image provided courtesy of S.P. Seitzinger, these preliminary estimates of global freshwater basin nitrogen export are alarming for the future sustainability of LMEs. Given the expected future increases in population and fertilizer use, without significant nitrogen mitigation efforts, LMEs will be subjected to a future of increasing

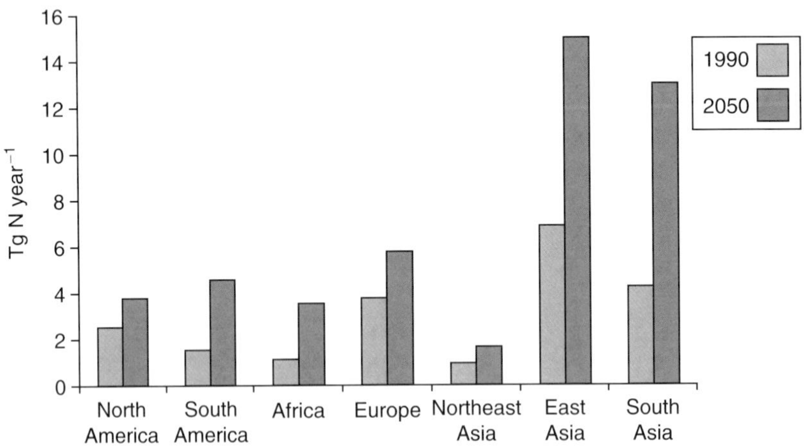

Fig. 4.18. Model-predicted dissolved inorganic nitrogen (DIN) export by rivers to coastal systems in 1990 and 2050. (T_gN = teragrams of nitrogen). Predictions are based on a business-as-usual (BAU) scenario. (Adapted from an image provided courtesy of S.P. Seitzinger; see further Kroeze and Seitzinger, 1998.)

harmful algal bloom events, reduced fisheries and hypoxia that further degrades marine biomass and biological diversity.

Contributions of LME Modelling to Policy-making

The sequence for improving the understanding of the possible mechanisms underlying observed patterns in LMEs is described by Levin (1990) as: (i) examination of statistical analyses of observed distributional patterns of physical and biological variables; (ii) construction of competing models of variability and patchiness based on statistical analyses and natural scales of variability of critical processes; (iii) evaluation of competing models through experimental and theoretical studies of component systems; and (iv) integration of validated component models to provide predictive models for population dynamics and redistribution. The approach suggested by Levin (1990) is consistent with the observation by Mangel (1991) that empirical support for the currently used models of LMEs is relatively weak, and that a new generation of models is needed that serves to enhance the linkage between theory and empirical results. Three models of ecosystem structure and function are being applied to LMEs with financial assistance from GEF through one mid-sized project, 'Promoting Ecosystem-based Approaches to Fisheries Conservation and LMEs' (http://www.gefonline.org/projectList.cfm). Estimates of carrying capacity using ECOPATH/ECOSIM food web approaches for the world's 64 LMEs are being prepared collaboratively by scientists from the University of British Columbia and marine specialists from developing countries. Similarly, a 24-month training project is being implemented by scientists from Rutgers University in collaboration with the IOC to estimate expected nitrogen loadings for each LME over the next 50 years. Scientists from Princeton University are examining particle spectra pattern formation within LMEs. Additionally, the American Fisheries Society and the World Council of World Fisheries Societies are collaborating to create an electronic network to expedite information access and communication among marine specialists participating in GEF-supported LME projects. There is a growing awareness among marine scientists, geographers, economists, government representatives and lawyers of the utility of a more holistic ecosystem approach to resource management (Byrne, 1986; Christy, 1986; Alexander, 1989; Belsky, 1989; Crawford *et al.*, 1989; Morgan, 1989; Prescott, 1989). On a global scale, the loss of sustained biomass yields from LMEs from mismanagement and overexploitation has not been fully investigated, but is likely very large. Effective management strategies for LMEs will be contingent on identification of major driving forces causing large-scale changes in biomass yields. Management of species responding to strong environmental signals will be enhanced by improving the understanding of the physical factors forcing biological change, thereby enhancing forecasts of El Niño-type events. In other LMEs, where the prime driving force is overfishing, options can be explored for reductions of fishing effort and implementing adaptive management strategies (Collie, 1991). Further, remedial actions are required to ensure that the pollution of the coastal zone of LMEs is reduced and does not become a principal driving force in an LME. Recent reports explore the application of ecosystem-based research and modelling that are focused on management (Browman and Stergiou, 2005) and on macroecology (Belgrano, 2004).

World Summit Targets

Since 1993, the NOAA Fisheries Service has been cooperating with GEF, IUCN, IOC and several other UN agencies (i.e. Industrial Development Organization, UNDP, UNEP and FAO) to assist developing countries in planning and implementing ecosystem-based management focused on LMEs as the principal assessment and management unit for coastal ocean resources. NOAA contributes scientific and technical assistance and expertise to aid developing countries in reaching the targets of the 2002 WSSD (Duda and Sherman, 2002). The WSSD targets, agreed on by officials of more than 100 countries, call for the achievement of 'substantial' reductions in land-based sources of pollution by 2006, introduction of the ecosystems approach to marine resource assessment and management by 2010, designation of a network of marine-protected areas by 2012 and the maintenance and restoration of fish stocks to MSY levels by 2015. The GEF-LME strategy supports the WSSD targets for addressing coastal and marine issues by jointly analysing scientific information on transboundary problems and their root causes, and setting priorities for action on these problems. The TDA process noted earlier provides a useful mechanism to foster participation at all levels in this information analysis and priority-setting effort. Countries then determine the national and regional policy, legal and institutional reforms and investments needed to address the priorities in a country-driven SAP. Project goals and milestones of the SAP promote vertical integration across the LME indicator modules on an annual basis, leading to an adaptive, ultimately self-financing, management regime (Fig. 4.19).

Reforms are taking place among the participating countries in operationalizing the integrated ecosystem-based approach to managing human activities in the different economic sectors that contribute to place-specific degradation of the LME and adjacent waters. The WSSD target for introducing ecosystem-based assessment and management practices by 2010 is likely to be met by most of the 121 countries constituting the existing LME network (Duda and Sherman, 2002). It is unlikely that the WSSD target for maintaining and restoring fishery resources to MSY levels by 2015 will be met. However, some initial progress is being made in recovery of depleted fish stocks through mandated reductions in fishing effort within the US NESLME (Sherman *et al.*, 2002).

With regard to the 2006 target for significant control and reduction of land-based sources of pollution, considerable additional effort will be required to achieve substantial reductions in land-based sources of pollution, whereas good progress has been made in designating marine-protected areas within the GEF-LME project network before 2012. The 'US Ocean Action Plan' published on 17 December 2004 by the Office of the President, Washington, DC, in response to the US Commission on Ocean Policy's Final Report (USCOP, 2004), supports the LME concept and strategy for ecosystem-based management and actions that

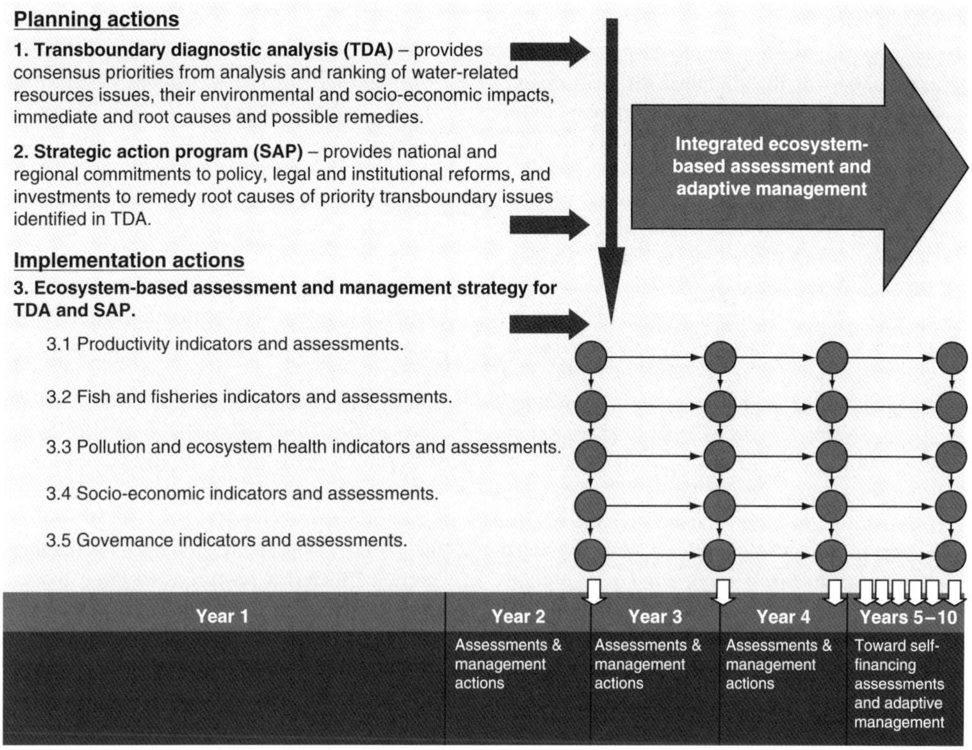

Fig. 4.19. Integrated Large Marine Ecosystem assessments based on Transboundary Diagnostics and Strategic Action Plans (SAPs) framework in support of annual adaptive management actions.

support and assist developing nations recover and move towards the WSSD targets within the UN regional seas programmes and international fisheries bodies (EOPUS, 2004a):

Advancing International Oceans Science

> Advance the Use of Large Marine Ecosystems. The United States will promote, within the UN Environment Program's regional seas programs and by international fisheries bodies, the use of the Large Marine Ecosystems (LME) concept as a tool for enabling ecosystem-based management to provide a collaborative approach to management of resources within ecologically bounded transnational areas. This will be done in an international context and consistent with customary international law as reflected in 1982 UN Convention on the Law of the Sea.
>
> (EOPUS, 2004a, b; Executive Office of the President of the United States)

Additional information on contributions to the global LME movement towards ecosystem-based management and resource sustainability is available from the LME Program Office, Northeast Fisheries Science Center, Narragansett Laboratory, Narragansett, RI and from the GEF web site, http://www.thegef.org and the LME web site: http://www.lme.noaa.gov

References

Aiken, J., Pollard, R., Williams, R., Griffiths, G. and Bellan, I. (1999) Measurements of the upper ocean structure using towed profiling systems. In: Sherman, K. and Tang, Q. (eds) *Large Marine Ecosystems of the Pacific Rim:Assessment, Sustainability, and Management*. Blackwell Science, Malden, Massachusetts, pp. 346-362.

Alexander, L.M. (1989) Large marine ecosystems as global management units. In: Sherman, K. and Alexander, L.M. (eds) *Biomass Yields and Geography of Large Marine Ecosystems. AAAS Selected Symposium 111*. Westview Press, Boulder, Colorado, pp. 339-344.

Beddington, J.R. (1986) Shifts in resource populations in large marine ecosystems. In: Sherman, K and Alexander, L.M. (eds) *Variability and Management of Large Marine Ecosystems. AAAS Selected Symposium 99*. Westview Press, Boulder, Colorado, pp. 9-18.

Beddington, J.R. (1995) The primary requirements. *Nature* 374, 213-214.

Behrenfeld, M. and Falkowski, P.G. (1997) Photosynthetic rates derived from satellite-based chlorophyll concentration. *Limnology and Oceangraphy* 42, 1-20.

Belgrano, A., Coord. (2004) Theme section: emergent properties of complex marine systems: a macroecological perspective. *Marine Ecology Progress Series* 273, 227-302.

Belsky, M.H. (1989) The ecosystem model mandate for a comprehensive United States ocean policy and Law of the Sea. *San Diego Law Review* 26, 417-495.

Berman, M.S. and Sherman, K. (2001) A towed body sampler for monitoring marine ecosystems. *Sea Technology* 42, 48-52.

Beverton, J.R.J. and Holt, S.J. (1957) On the dynamics of exploited fish populations. *Fish investigations*. Ministry of Agriculture Fisheries and Food. Series II, 1-533.

Browman, H.I. and Stergiou, K.I., Coord. (2005) Theme section: perspectives on ecosystem-based approaches to the management of marine resources. *Marine Ecology Progress Series*, 274, 269-298.

Byrne, J. (1986) Large marine ecosystems and the future of ocean studies. In: Sherman, K. and Alexander, L.M. (eds) *Variability and Management of Large Marine Ecosystems. AAAS Selected Symposium 99*. Westview Press, Boulder, Colorado, pp. 299-308.

Chen, Y.Q. and Shen, X.Q. (1999) Changes in the biomass of the East China Sea ecosystem. In: Sherman, K. and Tang, Q. (eds) *Large Marine Ecosystems of the Pacific Rim: Assessment, Sustainability and Management*. Blackwell Science, Malden, Massachusetts, pp. 221-239.

Christy, F.T. (1986) Can large marine ecosystems be managed for optimum yields? In: Sherman, K. and Alexander, L.M. (eds) *Variability and Management of Large Marine Ecosystems. AAAS Selected Symposium 99*, Westview Press, Boulder, Colorado, pp. 263-267.

Collie, J.S. (1991) Adaptive strategies for management of fisheries resources in large marine ecosystems. In: Sherman, K. Alexander, L.M. and Gold, B.D. (eds) *Food Chains, Yields, Models, and Management of Large Marine Ecosystems*. Westview Press, Boulder, Colorado, pp. 225-242.

Costanza, R., (1992) Toward an operational definition of ecosystem health. In: Costanza, R., Norton, B.G. and Haskell, B.D. (eds) *Ecosystem Health: New Goals for Environmental Management*. Island Press, Washington, DC, pp. 239-256.

Costanza, R., d'Arge, R., de Groot, R., Farber, S., Grasso, M., Hannon, B., Limburg, K., Naeem, S., O'Neil, R.V., Paruelo, J., Raskin, R.G., Sutton, P. and van den Belt, M. (1997) The value of the world's ecosystem services and natural capital. *Nature* 387(15 May 1997), 253-260.

Crawford, R.J.M., Shannon, L.V. and Shelton, P.A. (1989) Characteristics and management of the Benguela as a large marine ecosystem. In: Sherman, K. and Alexander, L.M. (eds) *Biomass Yields and Geography of Large Marine Ecosystems. AAAS Selected Symposium 111*. Westview Press, Boulder, Colorado, pp. 169-219.

Duda, A.M. (1982) Municipal point sources and agricultural non-point source contributions to coastal eutrophication. *Water Resources Bulletin* 18, 397-407.

Duda, A.M. and El-Ashry, M.T. (2000) Addressing the global water and environmental crises through integrated approaches to the management of land, water, and ecological resources. *Water International* 25, 115–126.

Duda, A.M. and Finan, D.S. (1983) Influence of livestock on nonpoint source nutrient levels of streams. *Transactions of American Society of Agricultural Engineers* 26, 1710–1726.

Duda, A.M. and Sherman, K. (2002) A new imperative for improving management of large marine ecosystems. *Ocean and Coastal Management* 45, 797–833.

EOPUS (2004a) Executive Order 121704—Committee on Ocean Policy. Office of the President of the United States.

EOPUS (2004b) US Ocean Action Plan. Executive Office of the President of the United States.

Epstein, P.R. (1993) Algal blooms and public health. *World Resource Review* 5, 190–206.

FAO (1995) Code of Conduct for Responsible Fisheries: Precautionary Principle. In: FAO (ed) FAO Document Repository.

Garibaldi, L. and Limongelli, L. (2003) Trends in oceanic captures and clustering of large marine ecosystems: two studies based on the FAO capture database, as reported to the FAO by official national sources. *FAO Fisheries Technical Paper* 435, 71.

GEF (1995) GEF Operational Strategy, Global Environment Facility, Washington, DC.

GESAMP (1990) The state of the marine environment, UNEP Regional Seas Reports and Studies No. 115, Nairobi.

GESAMP (2001) Protecting the oceans from land-based activities - Land-based sources and activities affecting the quality and uses of marine, coastal, and associated freshwater environment. 71, Joint Group of Experts on the Scientific Aspects of Marine Environmental Protection and Advisory Committee on Protection of the Sea.

HELCOM (2001) Environment of the Baltic Sea area, Baltic Sea Environment Proceedings No. 82A. Helsinki Commission, pp. 23.

Hoagland, P. and Jin, D. (2006) *Accounting for Economic Activities in Large Marine Ecosystems and Regional Seas*, UNEP Regional Seas Report and Studies No. 181. DEPI Regional Seas Programme, 2006.

Hoagland, P., Jin, D., Thunberg, E. and Steinback, S. (2005) Economic activity associated with the Northeast Shelf large marine ecosystem: application of an input-output approach. In: Hennessey, T. and Sutinen, J. (eds) *Sustaining Large Marine Ecosystems: The Human Dimension*. Elsevier, Amsterdam, The Netherlands, pp. 159–181.

Holling, C.S. (1973) *Resilience and Stability of Ecological Systems*, Institute of Resource Ecology, University of British Columbia, Vancouver, Canada.

Holling, C.S. (1986) The resilience of terrestrial ecosystems, local surprise and global change. In: Clark, W.C. and Munn, R.E. (eds) *Sustainable Development of the Biosphere. Cambridge University Press*, London, pp. 292–317.

Holling, C.S. (1993) Investing in research for sustainability. *Ecological Applications* 3, 552–555.

Howarth, R., Anderson, D., Cloern, J., Elfring, C., Hopkinson, C., Lapointe, B., Malone, T., Marcus, N., McGlathery, K., Sharpley, A. and Walker, D. (2000) Nutrient pollution of coastal rivers, bays and seas. *ESA Issues in Ecology* 7, 1–15.

Jaworski, N.A. (1999) Comparison of nutrient loadings and fluxes into the US Northeast Shelf LME with the Gulf of Mexico and other LMEs. In: Kumpf, H. Steidinger, K. and Sherman, K. (eds) *The Gulf of Mexico Large Marine Ecosystem: Assessment, Sustainability, and Management.* Blackwell Science, Malden, Massachusetts, pp. 360–371.

Jaworski, N.A. and Howarth, R. (1996) Preliminary estimates of the pollutant loads and fluxes into the Northeast Shelf ecosystem In: Sherman, K., Jaworski, N.A. and Smayda, T.J. (eds) *The Northeast Shelf Ecosystem: Assessment, Sustainability and Management*. Blackwell Science, Cambridge, Massachusetts.

Juda, L. and Hennessey, T. (2001) Governance profiles and the management of the uses of large marine ecosystems. *Ocean Development and International Law* 32, 41–67.

Kroeze, C. and Seitzinger, S.P. (1998) Nitrogen inputs to rivers, estuaries and continental shelves and related nitrous oxide emissions in 1990 and 2050: a global model. *Nutrient Cycling in Agroecosystems* 52, 195–212.

Levin, S.A. (1990) Physical and biological scales and the modelling of predator-prey interactions in large marine ecosystems. In: Sherman, K., Alexander, L.M. and Gold, B.D. (eds) *Large Marine Ecosystems: Patterns, Processes and Yields.* American Association for the Advancement of Science, Washington, DC, pp. 179–187.

Levin, S.A. (1993) Approaches to forecasting biomass yields in large marine ecosystems. In: Sherman, K., Alexander, L.M. and Gold, B.D. (eds) *Large Marine Ecosystems: Stress, Mitigation and Sustainability.* AAAS Press, Washington, DC, pp. 36–39.

Lindemann, R.L. (1942) The trophic dynamic aspect of ecology. *Ecology* 23, 399–418.

Magnuson (1976) Magnuson Act. In: US Congress (ed).

Mangel, M. (1991) Empirical and theoretical aspects of fisheries yield models for large marine ecosystems. In: Sherman, K., Alexander, L.M. and Gold, B.D. (eds) *Food Chains, Yields, Models, and Management of Large Marine Ecosystems.* Westview Press, Boulder, Colorado, pp. 243–261.

Melrose, D.C., Oviatt, C.A., O'Reilly, J.E. and Berman, M.S. (2006) Comparisons of fast repetition rate fluorescence estimated primary production and ^{14}C uptake by phytoplankton. *Marine Ecology Progress Series* 311, 37–46.

Morgan, J.R. (1989) Large marine ecosystems in the Pacific Ocean. In: Sherman, K. and Alexander, L.M. (eds) *Biomass Yields and Geography of Large Marine Ecosystems. AAAS Selected Symposium 111.* Westview Press, Boulder, Colorado, pp. 377–394.

NEFMC (2004) Magnuson-Stevens Fishery Conservation and Management Act Provisions; Fisheries of the Northeastern United States; Northeast (NE) Multi-species Fishery; Amendment 13; Final Rule. In: N.R.a.R. Federal Register/Vol. 69 (ed). National Archives and Records Administration.

NMFS (2004) Our Living Oceans: Report on the Status of US living Marine Resources, 1999, US DOC, NOAA, NMFS.

NRC (2000) Clean coastal waters: understanding and reducing the effects of nutrient pollution. National Research Council, National Academy Press, Washington, DC, 405 pp.

O'Reilly, J.E. and Belkin, I. (2006) Seasonal and Annual Variability in Primary Production, Chlorophyll and SST in the Northeast US Large Marine Ecosystem. *Eos Trans. AGU*, 87(36), Ocean Sci. Meet. Suppl., Abstract OS25G-21.

O'Reilly, J.E., Evans-Zetlin, C. and Busch, D.A. (1987) Primary production. In: Backus, R.H. (ed) *Georges Bank.* MIT Press, Cambridge, Massachusetts, pp. 220–233.

Olsen, S.B., Sutinen, J.G., Juda, L., Hennessey, T.M. and Grigalunas, T.A. (2006) *A Handbook on Governance and Socio economics of Large Marine Ecosystems*, Coastal Resources Center, University of Rhode Island, Kingston, Rhode Island.

Pauly, D. and Christensen, V. (1995) Primary production required to sustain global fisheries. *Nature* 374, 255–257.

Pauly, D. and Chuenpagdee, R. (2003) Development of fisheries in the Gulf of Thailand large marine ecosystem: analysis of an unplanned experiment. In: Hempel, G. and Sherman, K. (eds) *Large Marine Ecosystems of the World: Trends in Exploitation, Protection and Research.* Elsevier Science, Amsterdam, The Netherlands, pp. 337–354.

Pimm, S.L. (1984) The complexity and stability of ecosystems. *Nature* 307, 321–326.

Prescott, J.R.V. (1989) The political division of large marine ecosystems in the Atlantic Ocean and some associated seas. In: Sherman, K. and Alexander, L.M. (eds) *Biomass Yields and Geography of Large Marine Ecosystems. AAAS Selected Symposium* 111. Westview Press, Boulder, Colorado, pp. 395–442.

Rabalais, N.N., Turner, R.E. and Wiseman, W.J. Jr (1999) Hypoxia in the northern Gulf of Mexico: linkages with the Mississippi River. In: Kumpf, H. Steidinger, K. and Sherman, K. (eds) *The Gulf of Mexico Large Marine Ecosystem: Assessment, Sustainability and Management.* Blackwell Science, Malden, Massachusetts, pp. 297–322.

Ricklefs, R.E. (1987) Community diversity: relative roles of local and regional processes. *Science* 235(4785), 161–171.

Schaefer, M.B. (1954) Some aspects of the dynamics of populations important to the management of the commercial marine fisheries. *Bulletin Inter-American Tropical Tuna Commission* 1, 27–56.

Seitzinger, S.P. and Kroeze, C. (1998) Global distribution of nitrous oxide production and

N inputs to freshwater and coastal marine ecosystems. *Global Biogeochemical Cycles* 12, 93–113.

Sherman, B. (2000) Marine ecosystem health as an expression of morbidity, mortality, and disease events. *Marine Pollution Bulletin* 41, 232–254.

Sherman, K. (1988) Large marine ecosystems as global units for recruitment experiments. In: Rothschild, B.J. (ed) *Towards a theory on biological-physical interactions in the world ocean*. Kluwer Academic Publishers, Dordrecht, The Netherlands.

Sherman, K. (1993a) Large marine ecosystems as global units for management: an ecological perspective. In: Sherman, K., Alexander, L.M. and Gold, B.D. (eds) *Stress, Mitigation and Sustainability of Large Marine Ecosystems. Proceedings of Symposium on Large Marine Ecosystems, 1–6 October 1990, Monaco*. AAAS Press, Washington, DC, pp. 3–14.

Sherman, K. (ed) (1993b) Emerging theoretical basis for monitoring changing states (health) of large marine ecosystems. Summary reports of two workshops: Narragansett, RI, April 1992; Cornell University, July 1992.

Sherman, K. (1994) Sustainability, biomass yields, and health of coastal ecosystems: an ecological perspective. *Marine Ecology Progress Series* 112, 277–301.

Sherman, K. and Alexander, L.M. (eds) (1986) *Variability and Management of Large Marine Ecosystems.* AAAS Symposium 99. Westview Press, Boulder, Colorado, 319 pp.

Sherman, K. and Duda, A.M. (2005) Applications of the large marine ecosystem approach toward World Summit targets. In: Hennessey, T. and Sutinen, J. (eds) *Sustaining Large Marine Ecosystems: The Human Dimension*. Elsevier, Amsterdam, The Netherlands, pp. 297–318.

Sherman, K., Alexander, L.M. and Gold, B.D. (eds) (1990) *Large Marine Ecosystems: Patterns, Processes and Yields*. AAAS Symposium. Washington, DC, 242 pp.

Sherman, K., Solow, A., Jossi, J. and Kane, J. (1998) Biodiversity and abundance of the zooplankton of the northeast shelf ecosystem. *Journal of Marine Science* 55, 730–738.

Sherman, K., Kane, J., Murawski, S., Overholtz, W. and Solow, A. (2002) The US northeast shelf large marine ecosystem: zooplankton trends in fish biomass recovery. In: Sherman, K. and Skjoldal, H.R. (eds) *Large marine ecosystems of the north atlantic: changing states and sustainability.* Elsevier, Amsterdam, pp. 195–215.

Steele, J.H. (1988) Scale selection for biodynamic theories. In: Rothschild, B.J. (ed) *Toward a Theory on Biological-Physical Interactions in the World Ocean*. Kluwer Academic Publishers, Dordrecht, The Netherlands, pp. 513–526.

Sutinen, J. (ed) (2000) A framework for monitoring and assessing socio economics and governance of large marine ecosystems. NOAA Technical Memorandum NMFS-NE-158, Northeast Fisheries Science Center, Woods Hole, Massachusetts, 32 pp.

Tang, Q. (2003) The Yellow Sea and mitigation action. In: Hempel, G. and Sherman, K. (eds) *Large Marine Ecosystems of the World: Trends in Exploitation, Protection and Research*. Elsevier Science, Amsterdam, The Netherlands, pp. 423.

Tang, Q. (2006) Changes in biomass yield in the central to southern part of Yellow Sea. In: Sherman, K., Bernal, P. and Lundin, C.J. (eds) IOC 8th LME Consultative Meeting. UNESCO, Paris. Available online at http://unesco.org

UNCED (1992) Report of the United Nations Conference on Environment and Development; Rio de Janeiro, Brazil, 3–14 June 1992.

UNGA (2001) Report on the work of the United Nations Open-ended Informal Consultative Process established by the General Assembly in its resolution 54/33 in order to facilitate the annual review by the Assembly of developments in ocean affairs at its second meeting. Rep. A/56/121 (22 June 2001); 62 pp., United Nations General Assembly.

USCOP (2004) US Commission on Ocean Policy Report, US Commission on Ocean Policy.

USEPA (2001) National Coastal Condition Report, United States Environmental Protection Agency, Washington, DC.

USEPA (2004) National Coastal Condition Report, United States Environmental Protection Agency, Washington, DC.

Wang, H. (2004) An evaluation of the modular approach to the assessment and management of large marine ecosystems. *Ocean Development and International Law* 35, 267–286.

5 Ecosystem-based Management of Marine Capture Fisheries: Not a Theoretical Concept but Useful Operational Reality!

KATHERINE SHORT,[a] ALISTAIR GRAHAM[a] AND CHRIS GRIEVE[b]

[a]*WWF International, Surrey, UK;* [b]*Meridian Prime*

Abstract

Ecosystem-based management (EBM) is now an accepted approach to managing humanity's needs of natural systems. WWF has a holistic, conservation-based EBM framework supporting the restoration and recovery of marine capture fisheries. Internationally, there are many initiatives to apply elements of EBM to the marine environment with varying objectives and methodologies. Other maritime sectors need support to apply EBM, thus enabling multiple objectives for use of the marine environment to be reconciled. Further definition of terminology or debate over methodology needs to give way to focusing on the commonalities, analysing applications and determining the most efficient pathways to accommodate multiple objectives and facilitate improved ocean health.

Introduction

The growing acceptance about the urgent need to adopt an ecosystem-based management approach (EBM) to manage human activities in the oceans is encouraging. That the 2006 meetings on the United Nations Informal Consultative Process on Oceans and the Law of the Sea (UNICPOLOS) and UN Fish Stocks Agreement (UNFSA) focused on this only reinforced it. They grounded EBM within the broader oceans governance reform agenda and established the practicality and applicability of the approach to fisheries within exclusive economic zones (EEZs) and on the high seas. Fishing is an important use of the world's oceans, but EBM has not been clearly accepted nor fully operationalized within fisheries management or other maritime agencies, let alone by user groups themselves. The Code of Conduct for Responsible Fisheries and the FAO Guidelines for an Ecosystem Approach to Fisheries (EAF) are very useful tools but they are voluntary, largely production-oriented rather than conservation-oriented, and their application varies greatly (Pitcher *et al.*, 2006, 2008) – to the potential continued detriment of marine ecosystems. WWF has a comprehensive conservation-based EBM Policy Framework (Ward *et al.*,

2002) that is informing global debate and providing a workable approach for managers of fishing activities.

This chapter seeks to clarify some of the critical bottom lines of EBM for fisheries while briefly surveying international efforts to implement EBM through Oceans' approaches. These international, 'ocean-based' agendas have largely arisen in response to the need to mitigate the effects of fishing, but are not restricted to managing fishing impacts alone. However, this chapter is not an exhaustive review of the full range and scope of the use and application of the many, extensive and varying EBM initiatives mentioned.

Despite limited *comprehensive* implementation of EBM in fisheries, there are many EBM elements already operating internationally. This chapter presents some of these, demonstrating that EBM is no longer an abstract, theoretical concept, but a burgeoning, useful, operational reality. Some elements have been underway for many years such as setting catch limits. However, it is accepted that these individual elements are insufficient and marine ecosystems and human wants of it are not that simple. A comprehensive, integrated approach to the management of all human uses of the oceans is needed if ecosystems are to be recovered, resources sustained and values restored or maintained.

EBM Frameworks

WWF was the first international conservation NGO to describe EBM for fisheries with a document containing policy proposals and operational guidance (Ward *et al.*, 2002). It was produced as a specific conservation and ecologically based contribution to the then growing international discourse on applying an ecosystem-based approach to fisheries management.

There are many concepts and terminology related to the ecosystem approach (EA) to ocean management and fisheries. The UN Convention on Biological Diversity (CBD) has adopted the EA as a core element in its work. The EA is explained as 'a strategy for the integrated management of land, water and living resources that promotes conservation and sustainable use in an equitable way', and is elaborated with 12 principles and 5 points of operational guidance to its application (Vierros, this volume). The key element of the CBD's EA concept is the need for an 'integrated' approach where all sectors need to take into account their own uses and values, the interactions between the sectors, and the full range of stakeholders involved. Individual sectors, such as fisheries, can take an 'ecosystem approach' in order to be fully prepared to become an effective part of an integrated approach. However, it is conceptually not possible for sectors, acting alone, and in the absence of overarching integrative arrangements, to fully implement EBM.

In addition to the CBD EA, there are a number of other concepts and terms that are all in regular use in various, mostly fisheries-related, policy, science, industry and NGO fora (Box 5.1).

In the early years of the 21st century, published literature reflected the vigorous academic debate about the subtle differences between the frameworks.

Box 5.1. Concepts and terms.

Ecosystem-based management – *management* of the uses and values of ecosystems in conjunction with *stakeholders* to ensure *ecological integrity* is maintained, and recognizing that ecosystems are dynamic and inherently uncertain (Ward *et al.*, 2002).

Ecosystem management – a synonym for EBM; often interpreted incorrectly to imply management of ecosystems, but more correctly interpreted to mean management of human activities that affect ecosystems, often detrimentally (Ward *et al.*, 2002).

Ecosystem-based fisheries management – a new direction for fishery management, essentially reversing the order of management priorities to start with the ecosystem rather than the target species (Pikitch *et al.*, 2004).

Ecosystem approach to management – an *ecosystem* is a *geographically specified* system of organisms (including humans), the environment and the processes that control its dynamics. Characteristics of EAM are: adaptive; incremental; takes account of ecosystem knowledge and uncertainties; considers multiple external influences; strives to balance diverse social objectives and geographically specified (NOAA working definition in Murawksi, 2005).

Ecosystem approach to fisheries – strives to balance diverse societal objectives, by taking account of the knowledge and uncertainties of biotic, abiotic and human components of ecosystems and their interactions and applying an integrated approach to fisheries within ecologically meaningful boundaries (FAO, 2002).

Marine ecosystem-based management – EBM is an integrated approach to management that considers the entire ecosystem, including humans. The goal of EBM is to maintain an ecosystem in a healthy, productive and resilient condition so that it can provide the services humans want and need. EBM differs from current approaches that usually focus on a single species, sector, activity or concern; it considers the cumulative impacts of different sectors.

(*Compass Consensus Statement*, 2005. http://www.compassonline.org/pdf_files/EBM_Consensus_Statement_v12.pdf)

The debate has cooled somewhat now that more EBM professionals are focused on the resolution of the challenges to ocean health. And while it is more productive to focus on the commonalities, there are sometimes glaring inadequacies in interpretation and use that must be consciously addressed when developing initiatives. Competing values can also arise from different cultures 'clashing' with different objectives, i.e. some within communities wishing to manage fisheries for export, while others recognizing the need for food security and domestic consumption.

While there are some subtle differences between the main frameworks internationally, the key commonalities are: (i) managing fisheries within a more spatially based or 'whole ocean' view, which considers cumulative impacts of all human activities; (ii) mitigating the ecological effects of fishing activity; and (iii) restoring degraded marine ecosystems including fished populations. It is clearly more important and useful to focus on these commonalities than to continue to debate the subtle, and often culturally based, differences.

Even though the overarching commonalities are becoming clearer to the scientific and policy communities applying EBM, considerable need remains to clarify the subtleties and communicate and translate the policy and science jargon for practitioners, government managers, industry operators and the conservation NGO community at large. There also remains a significant risk that the conservation NGO community, while trying to promote the use of EBM, could diminish its conservation impact by failing to use the different terms appropriately, misunderstanding the distinctions between, and the subtleties of, the various frameworks. In 2006, SeaWeb, the communications-based non-profit organization that uses social marketing techniques to advance ocean conservation, began a process to further understand and resolve some of the issues in communicating EBM (SeaWeb, 2008). The risk that varying terminology and the various understandings can play in delaying meaningful progress towards restoring marine ecosystems cannot be understated and this new guide will be very helpful.

Oceans-type Initiatives

Overfishing of targeted stocks has led to widespread marine ecosystem decline. Doing EBM of fisheries 'right' should result in the recovery of populations for production purposes *and* the restoration and recovery of the associated ecosystem, what some might suggest to be a '*revolution*' in thinking, and hence practice. Additionally, taking a multi-sectoral, oceans policy approach can address other marine environment user needs and impacts if relevant objectives are set.

The area needing the greatest change and restructuring is that of governance structures, including management agencies, to ensure their more efficient integration to deliver the challenges of multi-sectoral, multi-jurisdictional management of human activities. Experience gained during oceans' process discussions in the last 10 years has highlighted several examples; there are an estimated 27 marine enforcement entities in Australia including the navy, coastguard and various fisheries and marine park enforcement agencies across Federal and State waters. In California there are over 40 agencies responsible for various elements of marine management. There may be hundreds of pieces of often conflicting legislation for managing marine space in some countries. This could make even the most responsible and committed operator despair when trying to navigate an efficient and sustainable option or solution for a new business opportunity or in addressing the impacts of existing operations.

At the time of writing there were an estimated 12 'oceans-type' agendas running internationally, as listed below. Oceans-type refers to marine spatial planning agendas designed to improve the overall health of an area of marine 'space', involving multiple stakeholders. This is by no means an exhaustive or comprehensive review; to do so would have required criteria and a meaningful, i.e. ecosystem-based analytical framework. It does, however, reflect anecdotal evidence gathered from practitioners and information sources within each one. The good news is that these initiatives demonstrate the momentum and interest, across different cultures and political and economic landscapes, in grappling with the multi-sectoral nature of ocean use management internationally. Local and regional contexts are crucial to the success of EBM oceans-type initiatives, and it is exciting to ensure lessons are shared and best practice promoted.

1. The US Joint Ocean Commission Initiative (JOCI) – The Joint Ocean Commission Initiative is a collaborative effort of the members of the Pew Oceans Commission and the US Commission on Ocean Policy. Its purpose is to advance and catalyse meaningful ocean policy reform consistent with the recommendations of the two commissions. http://www.jointoceancommission.org/
2. The Canadian Oceans Act – The only country to enact legislation for an oceans agenda. Very gradual implementation within agencies, spatial management agenda very slow to gain traction and no significant institutional restructuring. http://www.pac.dfo-mpo.gc.ca/oceans/default_e.htm
3. The New Zealand Oceans Taskforce – Elements of marine research to support better spatial planning underway. http://www.mfe.govt.nz/issues/oceans/
4. Pacific Islands Regional Oceans Policy – After excellent energy initially, it appears to have stalled.
5. UK Marine Act – Designed to overcome the complicated mishmash of laws and manage the growing pressures of activities in UK seas. Marine life not well protected to date, risk of delay in Marine Act entry into parliament. http://www.wwf.org.uk/marineact/main.asp
6. Norwegian Barents Sea Management Plan – An excellent plan on paper at present. http://www.regjeringen.no/en/dep/md/Selected-topics/Integrated-Management-of-the-Barents-Sea.html?id=457531
7. EU Maritime Policy – A single EU Maritime Policy is a real possibility to break the deadlock by bringing together all the existing management tools that could protect Europe's seas. http://ec.europa.eu/maritimeaffairs/
8. Australian Oceans Policy and the Northwest Regional Marine Plan – Australia followed hot on the heels of Canada in bringing in its policy for Oceans Management and the Southeast Regional Marine Plan was the first to be tackled. The National Oceans Ministerial Board has been disbanded and the National Oceans Office brought under the marine Division of the Department of Environment and Water. Marine Bioregional Plans, including a system of marine-protected areas (MPAs), will be established over Australia's 14 million square kilometre ocean jurisdictions. http://www.environment.gov.au/coasts/mpa/index.html
9. Japan Oceans Foundation – Ocean Declaration. Sign of the non-government community with shipping origins starting to consider these challenges. http://www.sof.or.jp/
10. Global Forum on Oceans, Coasts and Islands – A portal, projects, facility and community advance oceans agenda. http://www.globaloceans.org/nippon/index.html
11. Asia Pacific Economic Cooperation Forum – Ministerial Conference. http://www.dfo-mpo.gc.ca/media/infocus/2005/20050914_e.htm. Follow up through specific fisheries underway with recent Coral Triangle mentions (APEC, 2007). http://www.apec2007.org/apec.aspx?inc=lw/lw_syd_dec
12. Benguela Current Commission – Signed into being at the time of the Bergen EBM conference, end of 2006. Ministerial Conference of the Benguela Current Commission took place in Namibia in July 2007. http://www.bclme.org/bcc/

To our knowledge, although considerable time and resources have been spent on advancing Ocean Agendas or Ocean Policies in countries such as Australia, Canada and the USA, very little integrating change has occurred in national or international governance, let alone 'in the water', between sectors. Indeed in New Zealand, although a Parliamentary Commissioner for the Environment report in 1999 (PCE, 1999) recommended an Oceans Taskforce, and one was established in 2000, the completion of the proposed Oceans Policy options package was delayed in June 2003 to take account of government decisions on public access and customary rights to the foreshore and seabed. Now, a more technical approach is being taken, (http://www.mfe.govt.nz/issues/oceans/current-work/index.html), leading the national environmental group, Environment and Conservation Organizations of New Zealand (ECO), to call for an 'Ocean's Policy with teeth' (http://www.eco.org.nz/news_item.asp?sID=113). While these efforts within national EEZs have so far failed to produce integrated domestic inter-agency structures, the Benguela Current Commission, to the authors' knowledge, appears to be the first attempt to create a multi-sectoral, multi-agency and multinational approach to managing human activities in a body of water (Sherman, this volume). The appointment of the Namibian Fisheries Minister, Dr Abraham Iyambo as the Commission Chairman, and Namibia hosting the Commission are excellent signs of its intention to meaningfully deliver EBM.

The WWF Framework

The WWF framework was deliberately structured to draw on the collaborative work of the FAO Code of Conduct for Responsible Fisheries. WWF's framework, identified alongside others as useful implementation guidance for EBM (Garcia and Cochrane, 2005) was prepared as a contribution to the 2001 international Reykjavík Conference on Responsible Fisheries and the resulting FAO EAF technical guidelines (FAO, 2003). The development of WWF's framework was also partly prompted by concerns within the environmental and conservation NGO community in the late 1990s that, although EA is the terminology under the CBD, it was apparent at many international fisheries meetings that, as a term, it was emerging as a new code among some fisheries managers and fishing industry lobbyists to argue that ecosystems could and should be manipulated through culling and other measures to meet production ambitions.

As described in Ward *et al.* (2002), WWF's EBM principles are:

1. Maintaining the natural structure and function of ecosystems, including the biodiversity and productivity of natural systems and identification of important species, as the focus for management.
2. Human use and values of ecosystems are central to establishing objectives for use and management of natural resources.
3. Ecosystems are dynamic; their attributes and boundaries are constantly changing and consequently, interactions with human uses also are dynamic.
4. Natural resources are best managed within a management system that is based on a shared vision and a set of objectives developed among stakeholders.

5. Successful management is adaptive and based on scientific knowledge, continual learning and embedded monitoring processes.

Although comprehensive Oceans Approaches are an 'ideal', they are rare, and thus it is useful to highlight specific elements of implementation for further application worldwide. Following the principles in WWF's EBM Framework, there are *12 operational steps*, which have been described in a set of case studies (Grieve and Short, 2007a). All over the world, working with the FAO, the leading Norwegian fishing gear company, Mustad,The Nature Conservancy, local NGOs, the shipping company Wallenius Wilhelmsen and coastal fishermen's associations, such as the circle hook work with the mahi-mahi and tuna fishermen in the Eastern Pacific, WWF is a partner with private and public sector organizations advancing EBM. These partnerships tackle the challenges of climate change, shipping and oceans governance, species protection, improving fishing techniques and management, illegal fishing and protecting marine ecosystems and biodiversity.

The steps and relevant case studies include:

Step	Case study
1. Identify stakeholder community	Yellow Sea Marine Ecoregion
2. Prepare a map of ecoregions and habitats	East Africa Marine Ecoregion
3. Identify partners and their interests/ responsibilities	Baltic Sea Marine Ecoregion
4. Establish ecosystem values	Fiji Islands Marine Ecoregion
5. Determine major factors influencing ecosystem values	Northwest Atlantic Marine Ecoregion – Grand Banks, Canada
6. Conduct ecological risk assessment (ERA)	Benguela Current Marine Ecoregion
7. Establish objectives and targets	Heard and McDonald Island/ Prince Edward Islands/ Kerguelen and Crozet Islands Initiatives
8. Establish strategies for achieving targets	Southwest Atlantic/Patagonian Shelf Marine Ecoregion – San Matias Gulf, Argentina
9. Design information systems, including monitoring	Gulf of California Marine Ecoregion – Mexico
10. Establish research and information needs and priorities	Bismarck Solomon Seas Marine Ecoregion – Bird's Head Peninsula Seascape, Indonesia
11. Design performance assessment and review processes	Southern Ocean – Antarctic Krill
12. Prepare education and training packages for fishers	West Africa Marine Ecoregion and New Zealand Marine Ecoregion

WWF and its partners internationally have experience and evidence showing that, if applied, EBM can produce significant wins for stakeholders, marine ecosystems and their surrounding environment. Much of this experience was gained through developing ecoregional conservation in the early 1990s, a methodology designed to bring a larger-scale, multi-stakeholder and science-based approach to conservation (Hails, 2006). The case studies in Grieve and Short (2007a) were specifically selected to illustrate the range of the various tools being used and, most importantly, the critical involvement of stakeholders. It is an important aspect of EBM that no one organization or approach can solve complex resource management challenges.

While the policy framework suggests the operational steps should be guided by, and nested within, EBM principles, the research revealed that following them rigidly or sequentially is not essential. Fisheries stakeholders rarely have the luxury of beginning with a blank canvas, nor following a neat 12-step process. While the pace of change needed is revolutionary, making EBM operational is best characterized as evolutionary, negotiated incrementally through existing political, economic and socio-cultural realities, with the right elements in place for some EBM steps and more work to do on others. Thus, the EBM framework was adapted uniquely and activity was determined by the reality confronting people working on the ground. The case studies show how stakeholders implemented some operational elements, trying to create an enabling environment for EBM activity. Others demonstrate the use of a more holistic approach. Three case studies, for example, highlight aspirations and achievements relevant to the European Union (Grieve and Short in Anzuelo, 2007b).

The Baltic study focuses on the challenging socio-political context within which stakeholder relationships are evolving via the Baltic Sea Regional Advisory Council. While the jury is still out on whether the forum will be a positive force in the management of Baltic Sea fisheries, some stakeholders sound a note of optimism. If cross-sectoral stakeholders can form productive relationships, gradually positive action may achieve sustainable outcomes for the Baltic Sea Ecoregion.

Meanwhile, in 2006, the Northwest Atlantic Fisheries Organization (NAFO), of which the EU are members, confronted severely overfished stocks, with high levels of by-catch for several moratoria stocks. WWF-Canada's independent scientific report, *High Seas Reform: Actions to Reduce By-catch and Implement Ecosystem-Based Management for the North-west Atlantic Fisheries Organization*, provides a scientific basis for its work in this forum. WWF-Canada became the first environmental NGO-granted observer status at NAFO, which enables staff to strengthen relations with decision makers, engaging them on the need to reduce cod by-catch and implement EBM.

Finally, the Commission for the Conservation of Antarctic Marine Living Resources (CCAMLR) is pioneering and leading the way for EBM in marine capture fisheries internationally, especially when it comes to assessing fishery performance and reviewing fishery management outcomes against ecosystem-based objectives. Many players, including delegation members, environmental NGOs and industry and scientific advisers, agree that work is needed to refine the management system, especially when it comes to krill. However, EBM is nothing if it is not adaptive and based on scientific knowledge, embedded monitoring processes and continual learning, which CCAMLR showcases in action.

The Challenge of Balancing Use and Conservation

The objectives of the UN CDB are the conservation of biological diversity, the sustainable use of its components, as well as the fair and equitable sharing of benefits arising from utilization of genetic resources. The EA of CBD is seen as a way to reach a balance between conservation and sustainable use. Where this balance is to be placed is a major issue to settle by societal choice when implementing the EA (CDB EA Principles 1 and 10; see Vierros, this volume).

Recovering Ocean Ecosystems

WWF argues that a more cautious approach to fisheries management must be taken. The balance needs to shift more to conservation and recovery to secure the long-term sustainability of marine ecosystems and their delivery of ecosystem goods and services. Too many commercially fished populations are depleted and many marine ecosystems are degraded such that precautionary management to allow recovery is necessary, forthwith. The problems are broad and various – from the impacts of marine pollution on the lowest trophic levels, to targeted, indiscriminate or wasteful removal of species throughout the food chain. Our society, as users of global marine resources, cannot afford to fail in applying EBM or compromise the future of the wondrous marine life and the livelihoods and food security of millions.

WWF promotes recovery, pragmatic evolution and challenging revolution and the integration of marine conservation and fisheries management, use and protection; EBM provides a mechanism to integrate these seemingly competing demands. In particular, while evolution is necessary at the human cultural level, i.e. gradually educating, promoting champions and success and enabling innovation, revolution is needed in the cultures of both governments and the users, i.e. fishery managers, ocean-related agencies and the fishing industry, owners, skippers and crew. Some would argue that the pace of adoption by retailers of the sustainable seafood agenda and explosive growth in the variety of Marine Stewardship Council (MSC)-certified seafood products now available is a revolution underway and there is a growing pool of industry champions, largely around reducing the amount of illegal product on the market and the promotion of technical by-catch mitigation measures that also herald enormous potential to tackle the harder challenges of overcapacity and overfishing. Clearly, the MSC is becoming a mechanism supporting the delivery of EBM in fisheries.

The presentation to the Bergen conference that this chapter summarizes recommended three points in particular:

1. Ecological health is the primary goal of EBM.
2. An international EBM 'information clearing house' be established.
3. The revolution in thinking, where not underway already, must begin.

Some progress is clearly underway with the arising EBM communications momentum, via SeaWeb and the need for a revolution reinforced in the conference report (Bianchi *et al.*, this volume). Continued broad support for the EAF was also reinforced at the FAO Committee on Fisheries Meeting in March 2007 (FAO, 2007).The many case studies and oceans initiatives demonstrate that EBM is clearly working.The key focus of its application should be to ensure improvements in the quality and health of marine ecosystems and the livelihoods of those that depend upon them.

References

APEC (2007) Sydney APEC Leaders' Declaration on Climate Change, Energy Security and Clean Development. Strengthening our Community, Building a Sustainable Future Sydney, Australia, 9 September 2007. Available at: http://www.apec2007.org/apec.aspx?inc=lw/lw_syd_dec

COMPASS (2005) Scientific Consensus Statement on Marine Ecosystem-Based Management. Prepared by scientists and policy experts to provide information about coasts and oceans to US policy-makers.

FAO (2002) Report of the Reykjavík Conference on Responsible Fisheries in the Marine Ecosystem. Reykjavík, Iceland, 1–4 October 2001. FAO Fisheries Report No. 658. FAO, Rome, Italy, 128 pp.

FAO (2003) The ecosystem approach to fisheries. FAOTechnical Guidelines for Responsible Fisheries. No. 4, Suppl. 2. FAO, Rome, Italy, 112 pp.

FAO (2007) Report of the Twenty-Seventh Session of the Committee on Fisheries. FAO, Rome, Italy, 5–9 March 2007.

Garcia, S.M. and Cochrane, K.L. (2005) Ecosystem approach to fisheries: a review of implementation guidelines. *ICES Journal of Marine Science* 62, 311–318.

Grieve, C. and Short, K. (2007a) Implementation of Ecosystem-Based Management in Marine Capture Fisheries. Case Studies from WWF's Marine Ecoregions. WWF International. Available at http://assets.panda.org/downloads/wwf_ebm_toolkit_2007.pdf

Grieve, C. and Short, K. (2007b) WWF International. *Anzuelo*, Issue 19. Institute for European Environmental Policy, 8 pp.

Hails, C. (2006) Conservation 1961-2006. Available at: http://www.panda.org/about_wwf/how_we_work/conservation/wwf_conservation_1961_2006/index.cfm

Murawksi, S. (2005) Implementing Marine Ecosystem-Based Management: Integrating Perspectives from Science and Management. NOAA Fisheries.

Pitcher, T.J., Kalikoski, D. and Pramod, G. (eds) (2006) Evaluations of Compliance with the FAO (UN) Code of Conduct for Responsible Fisheries. *Fisheries Centre Research Reports 2006*, Vol. 14(2), 1142 pp.

Pitcher, T., Kalikoski, D., Short, K., Varkey, D. and Ganapathiraju, P. (2008) An evaluation of progress in implementing ecosystem-based management in fisheries for 33 countries. Marine Policy, 142 pp.

Parliamentary Commissioner for the Environment (PCE) (1999) Setting Course for a Sustainable Future. Wellington: Parliamentary Commissioner for the Environment. New Zealand, 5 pp.

Pikitch, E.K., Santora, C., Babcock, E.A., Bakun, A., Bonfil, R., Conover, D.O., Dayton, P., Doukakis, P., Fluharty, D., Heneman, B., Houde, E.D., Link, J., Livingston, P.A., Mangel, M., McAllister, M.K., Pope, J. and Sainsbury, K.J. (2004) Ecosystem-based fishery management. *Science* 305(5682), 346–347.

SeaWeb (2008) An Ecosystem-based Management Messaging Guide for NGO Strategies for Using Research to Shape Communications (in press).

Ward, T., Tarte, D., Hegerl, E. and Short, K. (2002) Ecosystem-based management of marine capture fisheries. *World Wide Fund for Nature*, Australia, 80 pp.

6 The Human Side of the Ecosystem Approach to Fisheries Management: Preliminary Results of an FAO Expert Consultation

Cassandra De Young[1]

FAO, Rome, Italy

Abstract

This chapter will discuss the preliminary outcomes of the UN FAO Expert Consultation on the Social, Economic and Institutional (SEI) Considerations of Applying the Ecosystem Approach to Fisheries Management (6–9 June 2006). This meeting was designed to provide an understanding of the roles played by these human activity-focused perspectives within the ecosystem approach to fisheries (EAF) process as presented through the meeting's background document, including: (i) as a driving force for EAF; (ii) as a means for valuation of potential costs and benefits; (iii) as instruments in the application of EAF; and (iv) as institutions supporting or constraining the EAF. In addition, the meeting provided guidance on proposed FAO Technical Guidelines focusing on the human dimension of the EAF.

The Need for the Meeting

In 2003, the FAO published its *Technical Guidelines on the Ecosystem Approach to Fisheries* (FAO, 2003) in accordance with a request in the Reykjavík Declaration on Responsible Fisheries in the Marine Ecosystem (FAO, 2001), focusing on fisheries management. Recognizing the wide range of interpretations of the approach, the FAO proposed the following definition, which is aligned with the more general ecosystem approach (EA) (UNEP, 2000); however,

[1] Listed as sole author; however, the chapter represents the results of the work of the 15 participants and the FAO Secretariat to the workshop. The official meeting report may be found on the FAO Internet as the 'Report of the Expert Consultation on the Economic, Social and Institutional Considerations of Applying the Ecosystem Approach to Fisheries Management, Rome, Italy from 6 to 9 June 2006'. *FAO Fisheries Report*. No. 799. FAO, Rome, 2006, 16 pp. However, any errors and omissions are the personal responsibility of the listed author.

taking a pragmatic approach in that the ecosystem approach to fisheries (EAF) remains mainly bounded by the ability of fisheries management to implement the EA, but not downplaying the fisheries sector's responsibility in collaborating in a broader multi-sectoral application of the EA:

> [A]n ecosystem approach to fisheries (EAF) strives to balance diverse societal objectives, by taking account of the knowledge and uncertainties of biotic, abiotic and human components of ecosystems and their interactions and applying an integrated approach to fisheries within ecologically meaningful boundaries.
>
> (FAO, 2003)

The call for the EAF has continued to increase in volume, as seen through an increasing number of conferences, projects and policy statements broaching the subject explicitly and an increase in awareness with regards to fisheries–ecosystems interactions has been the fruit of these discussions. However, there remains a sense of frustration at the management and policy levels due to a lack of full understanding of how the EAF should be applied in practice. Questions remain such as: 'What are the entry points into the EAF?' and 'What are the variables we can control in order to implement the EAF?' Figure 6.1 provides an oversimplified interpretation of the variables available to implement the EAF: the right-hand side representing the ecosystem goods and services and the left-hand side the human activities. The former are variables that are given to us (i.e. we can affect but not control them); while the latter are those variables that we can control, for the most part, and, therefore, must be considered the entry point for the application of the EAF. The arrows in between represent the linkages between the two activities (i.e. impacts of fishing activities on the ecosystem and impacts of changes in the ecosystem on fishing activities), some of which are known to us, while others are currently unknown but need to be recognized as uncertainties in the management process.

More specifically, this 'human side', comprising a wide range of social, economic and institutional (SEI) considerations, is relevant to the implementation of an EAF as:

1. Social, economic and institutional objectives and factors may be driving forces behind the need for EAF management (e.g. macroeconomic context, societal values and economic efficiency).

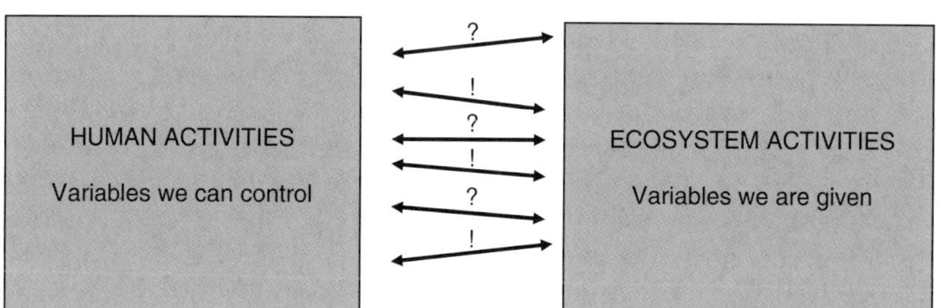

Fig. 6.1. Entry points for the application of the ecosystem approach to fisheries management.

2. The costs and benefits to individuals and to society of applying the EAF have social, economic and institutional aspects (e.g. distribution, scope and temporal elements).
3. Social, economic and institutional instruments are all crucial in the application of the EAF (e.g. incentives for adoption).
4. Social, economic and institutional factors can play supporting or constraining roles in EAF (e.g. governance, local context and buy-in).

Basically, the EAF must take place in the context of societal and/or community objectives, which inherently reflect human aspirations and values, and, as the implementation of the EAF is a human pursuit, there are actions to be undertaken in terms of the governance-institutional arrangements that are needed, the social and economic forces at play to be understood and the carrots (incentives) and sticks (disincentives) to be investigated that can induce actions compatible with societal objectives.

To gain a deeper understanding of the role that SEI should play in the implementation of the EAF, the decision to hold an Expert Consultation focusing on the human side of the EAF was made and the meeting was held at the FAO headquarters in Rome from 6 to 9 June 2006. The following two sections will provide a description of the process followed during the meeting and of the results emanating from this process.

The Meeting's Approach

The participants

In the spirit of the EAF, 15 individuals were invited to participate in the workshop representing:

- A global distribution of experiences.
- A multidisciplinary team of biologists, ecologists, economists, sociologists and institutional experts.
- A wide range of interests (small-scale and large-scale fisheries, government, non-governmental organizations (NGOs), donor agencies, academia, research centres and government).

What EAF meant to the group

During the group introductions, at least two participants identified themselves as 'non-believers' in the EAF. Reasons for incredulity included: (i) the belief that EAF was merely the latest fad or sexy term of the moment and that it did not represent anything new; (ii) given the very high data requirements involved in the approach, only the very rich countries could afford to implement the EAF; and (iii) that the term EAF contained the word 'ecosystem' for a reason: it was of concern only to ecologists and biologists; therefore, there was no role for social scientists in the approach.

The participants took advantage of these criticisms to help define what the approach meant to them; not only as a starting point for the meeting, but also as a means of solidifying their own understanding of the approach.

Why is the EAF needed?

The participants recognized several factors that have led to the call for the EAF at the international level: the inadequacies of management approaches focused solely on target species, the promotion of conservation-oriented policies, the need for more participatory/co-management approaches, and the fact that EAF-type management approaches have already been used implicitly in many local-level and/or community management schemes.

The nature and contents of the EAF

The group affirmed that the EAF represents a holistic, participatory and integrated approach to fisheries management, as opposed to a strictly biological/ecological approach. However, there were concerns about the high data/information needs that the EAF seems to imply, especially when those required for single stock management have often proved unsuccessful. In this regard, the participants discussed the notion of using the 'best available (scientific) information' that, for example, in small-scale fisheries could, in some cases, be confined to traditional knowledge. The participants agreed that data inadequacy in itself should not hinder the application of EAF. The group also acknowledged that there was often an imbalance in available data and information across disciplines with a bias towards natural science data and information. This was both a consequence of how management objectives are being formulated and stated (often in biological terms) and the shortcomings in the allocation of research funding and staffing to social sciences.

The participants agreed that the move towards EAF would, in many instances, be accomplished on an incremental and adaptive management basis in view of the much greater uncertainties and risks, the time needed to learn and acquire new knowledge, and the need to carefully assess the distributional implications of EAF interventions. In many developing countries, EAF would have to be applied in a 'low-cost' manner to be feasible and become widely adopted.

The background document: the SEI of the EAF

In recognition of the wide range of backgrounds of the participants, a substantial background report on the SEI components of EAF was prepared (De Young *et al.*, 2008) so that the participants would have a broad overview of the concepts, vocabulary and methodologies used by the various disciplines. The hope behind the background document was to facilitate communication within a multidisciplinary group, to provide a better understanding of the role of the economic, institutional and socio-cultural components within the EAF process, and to examine some potential methods and approaches that may facilitate the adoption of EAF management.

The document was divided into four parts. The first part discussed concepts and issues relating to the EAF, with emphasis on social, economic and institutional aspects, as well as interactions with complementary approaches. The second part highlighted the many ways in which aquatic ecosystem goods and services are valued, socially, culturally and economically, as well as the various non-market and market valuation techniques for assessing those values. The third part of the document covered the key issues of implementing the EAF: (i) the various benefits and costs involved, from social, economic, ecological and management perspectives, and how these are measured; (ii) intra-fisheries incentives (economic, social and institutional) that can be created and utilized for promoting, facilitating and funding the adoption of EAF management; (iii) extra-fisheries approaches for financing EAF implementation; and (iv) social considerations to be taken into account in implementing EAF management, including equity issues. The fourth part of the background document examined some aspects of the policy and institutional frameworks – within fisheries and more broadly – that relate to EAF implementation.

Testing the usefulness of the background document

Angel Alcala, Patrick McConney and John Ward had been asked by the FAO Secretariat to initiate the discussion on the background document through a discussant panel. These thoughtful reviews were then followed by an open discussion in which the whole group reviewed the background document, section by section, providing comments and discussion points to be incorporated into a final document to be published as an FAO Technical Paper.

As another aid towards testing the background document for its usefulness in the application of the EAF, three case profiles (the Mesoamerican reef spiny lobster fishery, the Tanzanian coastal mixed-species fisheries and the Norwegian Barents Sea cod fishery[2]) were examined by three multidisciplinary break-out groups. The idea was not to analyse the case profiles themselves, as the limited time available would not have done justice to the case examples, but to place the multidisciplinary groups into various situations providing for different EAF contexts, complete with their own specific needs in terms of applying the EAF.

Through these varied mechanisms, many useful comments were received from the participants, including the need to strengthen the discussions on: (i) the understanding and application of decision making theory to address situations of limited information and uncertainty; (ii) more attention to the notion of nested institutional arrangements at various scales, from local to national to international; and (iii) additional issues to be included such as power and political considerations, demographics, local indigenous knowledge, change management, issues of legitimacy and transparency and greater attention to gender and family planning. Although an annotated outline was defined by the group, only the concise version is reproduced in Table 6.1.

[2] Prepared by Juan Carlos Seijo, Cassandra De Young/Magnus Ngoile and Björn Hersoug, respectively.

Table 6.1. Suggested outline for the technical paper: Human Dimensions of the EAF – An Overview of Context, Concepts, Tools and Methods.

Part I. Setting the economic, social and institutional context for the EAF
- Introduction and background
- Human values of ecosystem services
- Policies, legal and institutional frameworks relevant to the EAF
- Social and economic considerations of applying the EAF

Part II. Facilitating the implementation of the EAF
- Introduction
- General principles
- Approaches and decision making tools
- Mechanisms for applying/achieving the EAF
- Sustaining an EAF

Moving Towards an FAO Code of Conduct Technical Guidelines on the Human Dimensions of the Ecosystem Approach to Fisheries

With the background document providing guidance and technical information, the participants felt it necessary to produce a separate document targeting, *inter alia*,'decision makers, managers, researchers, leaders of fishing communities and industry; who would be part of, or drive, the development and implementation of EAF'.To this aim, the participants recommended the development of FAO Technical Guidelines, which would be 'generic guidelines, not a step-by-step guide, and would offer a suite of approaches, mechanisms and tools which could be applied and adopted to specific contexts and situations'.

As the use of smaller, multidisciplinary break-out groups had proved successful during the analysis of the background document, the participants decided to employ this technique to draft three separate outlines for the proposed Technical Guidelines (FAO, in preparation) (Table 6.2).Although risky at the outset as three completely different outlines would be difficult to combine, there was, in fact, very quick convergence of ideas among the three groups.This was due, in part, to the efforts of the previous days: by the time the group began discussing the Technical Guidelines, there was already a strong consensus and mutual understanding of the SEI needs with respect to implementing the EAF. The resulting draft outline for the Technical Guidelines is as below and reflects a four-step logic: (i) understanding where we are and what the entry point to the EAF is for the given situation; (ii) understanding where we want to go given step (i); (iii) how do we get there?; and (iv) how do we sustain and continue the process once we have attained our objectives?

Next steps

The authors of the background document will incorporate the comments received from the Expert Consultation, draft the relevant Technical Paper and

Table 6.2. Suggested outline for the FAO technical guidelines: Human Dimensions of the Ecosystem Approach to Fisheries Management.

Chapter 1. Background and Introduction

- Nature, scope target group of this guideline
- Objectives of this guideline
- Rationale of EAF
- Setting the scene, historical development of EAF
- Explanation of principles (e.g. multi/interdisciplinary, participatory, holistic, integrated, transparent, legitimate, accountability, fairness, equity) and definitions
- Starting/entry points for engaging with EAF

Chapter 2. Understanding the SEI Context for EAF

- Participatory situation analysis (clarify your entry point, looking at the social, economic and institutional context)
- SEI issues including:
  - Risks and uncertainties; vulnerabilities (references to tools how to do it)
  - National differences in terms of fishery type
  - Political realities
  - Power
  - Assess current governance relative to good governance principles (accountability, transparency, etc.)
  - Know societal goals and values (get consensus on what these are)
  - Relevant policy frameworks
  - External influences (e.g. climate change, trade and global market forces, regime shifts, natural disasters)
  - Special considerations (GMOs, alien species, etc.)

Chapter 3. Formulating Objectives

- Identifying and agreeing on EAF priorities (at different levels)
- Policy and management – decide on entry point
- Policy and strategy considerations
- Take into account aspects of:
  - Gender, poverty, livelihoods, equity
  - Human realities of interaction between politics, etc., and EAF implementation
  - Inter-generational and intra-generational equity issues (i.e. serving needs and interests of current and future generations)
  - Management and access regimes

Chapter 4. SEI Aspects of Developing and Implementing an EAF Plan of Action

- Clarify context:
  - Wide variety of contexts (e.g. coastal community, middle-scale, industrial, freshwater)
- The plan of action:
  - Identifying and evaluating options for action
  - Resources and resource mobilization (i.e. human and financial resources, capacities and capabilities)
  - Capacity building
  - Institutional requirements (legislative change, local organization development, encouraging ownership, cooperation and support)
  - Incentive mechanisms (rights, economic incentives, correcting market failures, encouraging social organization)
  - Livelihood diversification, economic diversification and other non-fishery approaches to fishery issues
  - Enabling appropriate access regimes, such as rights-based management, where appropriate (implement with high degree of participation; add value to the asset, implications

Continued

Table 6.2. Continued

for who does the research; take into account all use rights to ecosystem services including fishery resources)

- Conflict management and cooperation (more skills and institutional arrangements for conflict management/resolution, negotiation, mediation, arbitration)
- Considerations to take into account:
  - Adaptive process
  - Inter-sectoral linkages
  - Sustainability of the plan of action
- Analytical approaches:
  - Assessment and evaluation framework (giving attention to cumulative, secondary and induced impacts)
  - Identifying information requirements
  - Decision making under uncertainty
  - Assessing economic, social and institutional implications and trade-offs, using various approaches and tools (SLA, valuation methods, SIA, CBA, decision making techniques and tools)

Chapter 5. Monitoring and Evaluation

- Iterative, adaptive process needed throughout
- Agreeing on criteria for M&E
- Identification of M&E indicators, using baseline information, reference points, etc.
- Mechanisms and methods for M&E
- Implications of increased scale and scope of EAF for MCS
- Learning lessons and revision of policy, strategy and plan of action

Chapter 6. Research, Data and Information

- What SEI info is needed? Need a balance of SEI and biophysical information
- Participatory process of data collection and research, in every relevant field
- Members of community and fishermen can be involved in all aspects
- Provide mechanisms for fishery-related SEI research (e.g. universities)
- Appropriate capacity building and acquisition (SEI) – in broader context (IM)
- Education for society; getting into schools
- GIS with relevant socio-economic data of all ecosystem users
- Application and use of traditional ecological knowledge (TEK)
- Utilize simplified low-cost approaches where possible

Chapter 7. Sustaining an EAF

- Political commitment and buy-in at relevant levels
- Awareness raising; education/training
- Sustainable financing:
  - Differs depending on the fishery context (e.g. inshore versus offshore)
  - Broaden perspective on who is receiving benefits, who pays costs, self-reliant financing to ensure long-term sustainability (through, e.g. cost-recovery, resource fees and beneficiary-pays principle)
  - Adaptive management and institutional learning

provide the document to the participants for their comments before publishing the final version by the FAO.

The Technical Guidelines on the SEI of implementing the EAF will be drafted by a multidisciplinary subset of the Expert Consultation participants to provide the FAO Committee on Fisheries (COFI) with a draft version of the document during its 2009 session.

In addition, an ongoing effort to collect examples of applications of the EAF and to provide the lessons learned from these applications would further assist in the wider application of the EAF around the world. International projects, such as the ECOST and the various GEF-funded large marine ecosystem projects[3] are examples of ongoing efforts to explicitly implement the EAF. Other examples to investigate include local-scale and nationally initiated projects that are implementing the EAF either implicitly or explicitly.

References

De Young, C., Charles, A., and Hjort, A. (2008) Human dimensions of the ecosystem approach to fisheries: an overview of context, concepts, tools and methods. *FAO Fisheries Technical Paper*. No. 489. FAO, Rome, Italy, 152 pp.

FAO (2001) The Reykjavík Conference on Responsible Fisheries in the Marine Ecosystem, Reykjavík, 1–4 October 2001, *FAO Conference Report* C 2001/INF/25. FAO, Rome, Italy, 426 pp.

FAO (2003) The ecosystem approach to fisheries, *FAO Technical Guidelines for Responsible Fisheries* No. 4 (Suppl. 2). FAO, Rome, Italy, 112 pp.

FAO (in preparation) The Human Dimensions of the Ecosystem Approach to Fisheries, *FAO Technical Guidelines for Responsible Fisheries* No. 4 (Suppl. 3). FAO, Rome, Italy.

UNEP (2000) *Report of the Fifth Meeting of the Conference of the Parties to the Convention on Biological Diversity*, Nairobi, 15–26 May 2000. UNEP/CBD/COP/5/23. 206 pp.

[3] See http://www.port.ac.uk/research/cemare/ and http://www.gefweb.org

7 Research Requirements of an Ecosystem Approach to Fisheries

Robert O'Boyle, Michael Sinclair and Tana Worcester

Bedford Institute of Oceanography, Dartmouth, NS, Canada

Abstract

In recent years, there has been considerable discussion within the international scientific and ocean policy community on an ecosystem approach to the integrated management (IM) of multiple ocean uses, and specifically its application to fisheries. Much of this discussion has focused on the planning hierarchy composed of high-level conceptual objectives and ocean industry-level operational objectives with associated indicators and reference points. There has also been discussion on governance systems including planning area considerations and the engagement of stakeholders in the management process. While a lot has been learned from these efforts, it is timely to consider the scientific research requirements of effective implementation. This chapter presents a methodology for defining the research needs of an ecosystem approach to fisheries (EAF) based upon experience on Canada's East Coast. The methodology is applied to the IM of the Eastern Scotian Shelf (ESS), which is a regional pilot initiative in Canada. The research issues identified focus on interactions between fisheries and the environment, as well as those required to evaluate cumulative impacts. Attention is also given to the large-scale influences of climatic and oceanographic systems on ecosystems and their dependent fisheries.

Introduction

Since the principles of an ecosystem approach to management and more specifically to fisheries management (ecosystem approach to fisheries (EAF)) were first introduced in the 1990s, there have been numerous papers exploring the policy, institutional structures and management planning hierarchies required to implement the approach (see O'Boyle and Jamieson, 2006, for a recent perspective). Scientific work on required indicators and reference points has also progressed, with the 2004 Paris Symposium on ecosystem indicators noting the strong linkage needed between scientific expertise and managers to ensure that the knowledge developed is of practical use (Cury and Christensen, 2005). Overall, though, there has been less focus on the specific research issues that require attention in relation to the management questions that arise from an EAF.

 The Ecosystem Approach to Fisheries (eds G. Bianchi and H.R. Skjoldal)

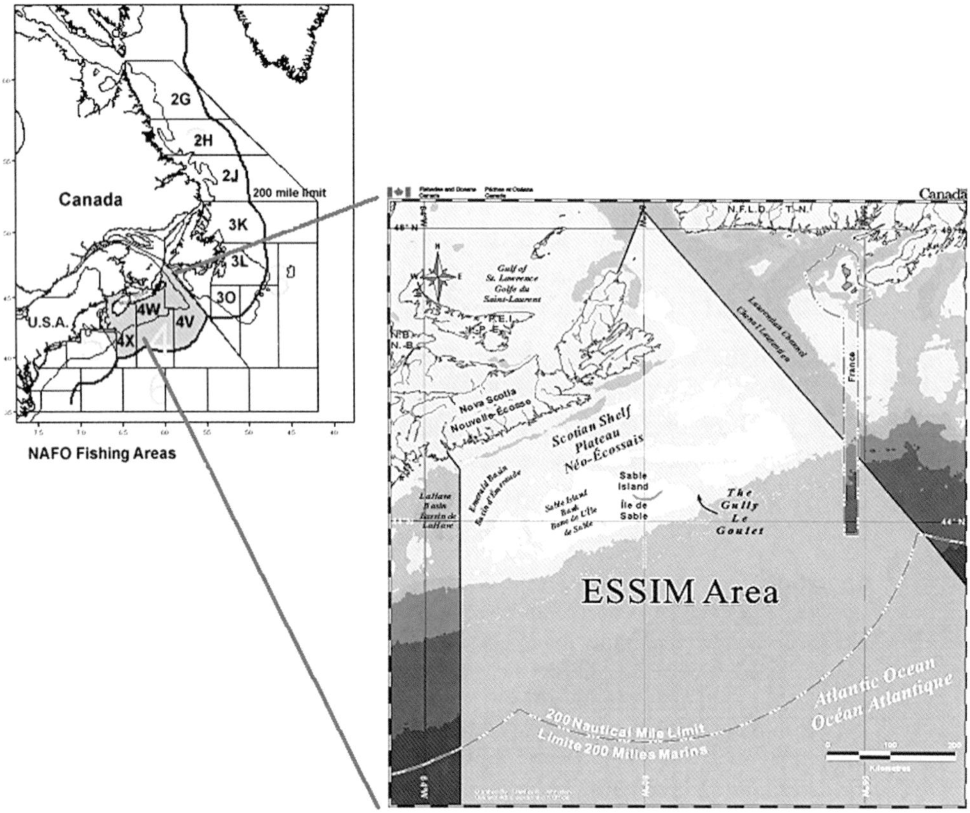

Fig. 7.1. Eastern Scotian Shelf Integrated Management (ESSIM) Area.

This chapter outlines a structured approach to identifying the research issues associated with emerging management questions under an EAF, based upon experience developed in the Eastern Scotian Shelf Integrated Management (ESSIM) initiative.The ESSIM initiative (Fig. 7.1) was established in 1998 as a pilot to test how ecosystem-based management could be implemented on Canada's East Coast (Rutherford *et al.*, 2005). Since 1998, much has been learnt not only about governance and planning structures, but also about the ecosystem issues that need to be addressed (Zwanenburg *et al.*, 2006).

We present an overview of the research needs of ESSIM over the next 3–5 years, based upon the management questions of an EAF. We then consider, based upon this experience, the broader implications for research.

Approach to Consider Research Needs of an EAF

There are many management issues to consider when defining research needs for an EAF. We consider these in three categories: (i) the impact of a fishery on the ecosystem (e.g. bottom impacts of trawling); (ii) the impact of the ecosystem on a fishery (e.g. cooling water temperatures changing target species distributions and seal predation on depleted cod populations); and (iii) the manipulation of ecosystems through management and habitat mitigation efforts (such as lowering capelin removals to enhance cod productivity, the cull of seals, and artificial habitat enhancement for lobster). For issues under the first category, we referred to the ESSIM Plan (DFO, 2005) and a DFO Science/Fisheries Management Review of the region's fisheries management plans to evaluate their compliance with an EAF (Gavaris, personal communication). We identified issues under the second and third categories from the Ecosystem Status Report for the ESSIM area (DFO, 2003), O'Boyle *et al.* (2004, 2005a), discussions at the Scotia-Fundy Fishing Industry Roundtable (a fishing industry-wide body established to encourage discussion on broad and inter-fleet issues), and our own experience overseeing the science programs of the Maritimes region of the Department of Fisheries and Oceans (DFO).

Since the issues cover a broad range of ecosystem processes, it is necessary to take a structured approach. Early in its ecosystem-based management efforts, DFO realized that it would be useful to provide regional integrated management (IM) projects with a suite of nationally articulated overarching conservation objectives, which provide a link between ratified international agreements (e.g. the Convention of Biological Diversity) and the national ocean management policy and regionally established IM priorities. This suite, established in 2001, states that an ecosystem's biodiversity, productivity and habitat should be conserved, with sub-objectives dealing with different components of the ecosystem (Fig. 7.2; O'Boyle and Jamieson, 2006). We have found these national conservation objectives effective in providing guidance to our IM efforts, both at the ecosystem and fishery planning level. We thus sorted the management issues according to the relevant conservation objective. This brought all issues relating to species biodiversity together, those on population productivity together and so on. We then phrased each of the issues identified as a question that might be posed by a fisheries manager. In our interactions with managers, we have found it useful to ask them to articulate a question that science could respond to. We then considered what research activities would be required to answer these management questions. At this stage, we noted significant overlap between issues identified under each of the initial three categories and thus the research issues were aggregated by management questions and conservation objectives.

We initially thought that it would be possible to prioritize the issues, but soon determined that all could be considered medium to high priority. This is not surprising as the initial selection process tended to 'weed out' low priority issues. We found it more productive to evaluate how best these issues could be addressed and over what time frame. First, based on our own judgement, we considered the scope of the required research effort. Could it be undertaken at our institute or was it necessary to engage other institutes and organizations on the Atlantic coast? Is the issue of such a nature that a more global effort needs to be mounted? In judging this, we recognized that some issues could be addressed at a local scale, while others were better suited to the collaborative efforts of many scientists from different research communities. It is then

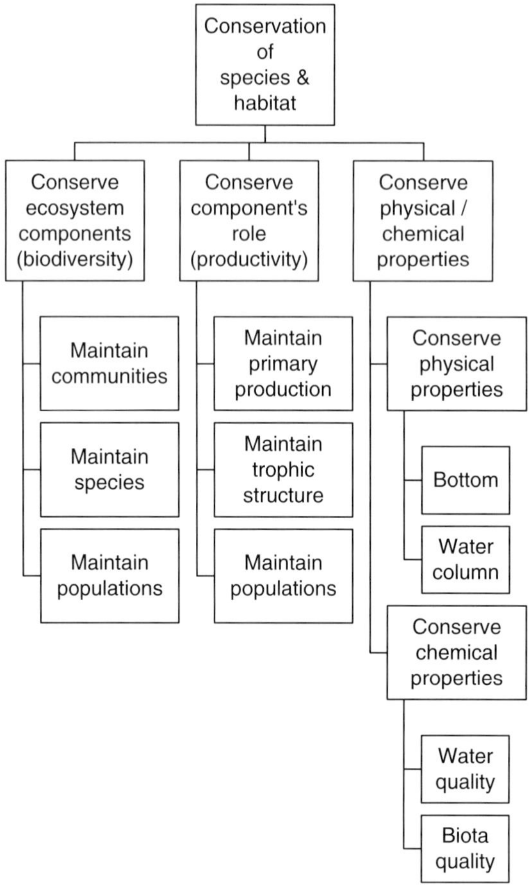

Fig. 7.2. Conceptual objectives tree for the environmental dimension of integrated management in Canada. (From O'Boyle and Jamieson, 2006.)

important to ask over what time period could research be expected to provide products of use to managers, again based upon our experience with research programmes both here and elsewhere. In asking this, we do not intend to imply that the intractable issues should not be studied; rather, given the current state of ecological theory and ecosystem understanding, we felt that products of use to managers would not be available in a relatively short (3–5 year) time frame.

This approach to defining issues of importance to an EAF might not capture some overarching considerations – those that implicate a number of conservation objectives. Some thoughts on overarching research needs are thus provided at the end of the chapter.

The sections below provide an overview of the issues identified for ESSIM, organized by the national conservation objectives structure. The 12 management questions and associated research issues are listed in Table 7.1 and briefly discussed below (grouped by the conceptual conservation objectives).

Table 7.1. Synopsis of management questions and related research issues along with their tractability and scale of required research effort.

Fisheries management question	Research activity	Probability research can be resolved in 3–5 years	Scale of research (local,[a] northwest Atlantic, global)
How should fisheries management and industry prepare for and respond to large-scale ecosystem community changes?	Investigate large-scale biogeographic patterns of fish and invertebrate communities, and how these respond to circulation and mixing	High	Northwest Atlantic
What are the implications to fisheries management and industry of changes to large-scale species distributional patterns?	Determine if snow crab and shrimp on the Eastern Scotian Shelf (ESS) are extension of distribution from Gulf or separate self-sustaining populations	Low	Local
	Study factors controlling rate of invasion of *Codium*	Low	Local
What can be done to recover species at risk (with the detailed questions outlined in the diverse recovery plans for leatherback turtle, porbeagle shark, bottlenose whale, etc.)?	Determine survivorship of turtles (leatherback and loggerhead) following release from fishing gear	High	Local
	Determine impact of porbeagle by-catch on recovery	Medium	Northwest Atlantic
	Determine productivity characteristics of data-poor species from data-rich species for use in modeling recovery (e.g. white hake, skates, cusk)	Medium	Local
How should temporal trends of the abundance of species at risk on the ESS be monitored?	Investigate reliability of indicator trends based upon current monitoring	Medium	Northwest Atlantic

Continued

Table 7.1. Continued

Fisheries management question	Research activity	Probability research can be resolved in 3–5 years	Scale of research (local,[a] northwest Atlantic, global)
Do present management practices allow the accurate estimation of discards and by-catch?	Estimate by-catch and discards of the fisheries in ESS	Medium	Local
	Determine impact of discarding and by-catch on sustainability of non-commercial species	Low	Local
Should fisheries management be concerned about population substructure for ESS stocks?	Develop spatially realistic models to examine recolonization processes of recovering populations (e.g. cod and herring)	Medium	Local
What are management implications of systematic removal of large fish on ecosystem functioning?	Continue investigation of trophic regulation of ESS and other ecosystems	Low	Global
	Continue investigation of source and impact of high natural mortality in cod and other depleted species (e.g. winter skate)	Low	Northwest Atlantic
	Continue investigation of population processes (e.g. density-dependent) that regulate grey seals	Medium	Northwest Atlantic
	Determine food requirements of apex predators (e.g. marine mammals, sea birds, large fish)	Medium	Local

What is minimum spawning stock biomass (SSB_{min}) for Scotian Shelf stocks?	Estimate minimum spawning stock biomass for a number of commercially exploited stocks	Medium	Northwest Atlantic
Can fisheries management incorporate the impact of climate change in the management of harvested stocks?	Study link between climate change and trends in recruitment of commercially exploited species	Low	Northwest Atlantic
Have current harvest practices caused growth reduction in gadoids?	Determine relative roles of environment and genetic selection on growth changes in gadoid (e.g. cod and haddock)	Low	Local
Are current fishery closures and gear restrictions adequate to protect benthic habitat?	Compare predicted spatial patterns of benthic communities on ESS from geological and oceanographic parameters with observations	Medium	Global
	Determine proportion of benthic habitat types (e.g. coral, sand dollar) needed to be protected to ensure sustainability (includes temporal processes)	Medium	Global
	Investigate relationship between size and location of refugia and benthic community sustainability	Low	Global
Can the impacts of climate change on habitat be predicted?	Investigate impacts of climate change on the oceanography of the ESS (e.g. NAO impact on bottom water temp)	Low	Northwest Atlantic

[a]Local research is research that could be done at the Bedford Institute of Oceanography and Saint Andrews Biological Station, Nova Scotia, Canada.

Biodiversity Issues

At ecosystem level

Fisheries managers and the fishing industry are concerned about the potential consequences of climate change on the mix of species inhabiting the Eastern Scotian Shelf (ESS). The question that they might pose is: how should fisheries management and industry prepare for and respond to large-scale ecosystem community changes?

Adaptation to the impact of ecosystem-level change due to climate variability requires an understanding of the predicted changes in spatial patterns at the community and 'seascape' level. An investigation of large-scale biogeographic patterns of fish and invertebrate communities off the North American east coast and how these respond to circulation and mixing would be useful. Similar work was conducted in the early 1990s in which changes in the large-scale distributions of over 100 species on the North American east coast were examined. This earlier study identified transition zones within which the demersal fish communities changed in composition (Mahon *et al.*, 1998), consistent with circulation and mixing patterns. Understanding how the location of these zones might change with coast-level atmospheric and ocean processes would greatly assist management efforts.

At species level

While there have been large-scale changes in the distribution of communities, so too have there been changes in the distribution of individual species of interest to managers. Fisheries managers might pose of science a question similar to the one above: what are the implications to fisheries management and industry of changes to large-scale species distributional patterns?

Target species have traditionally been the focus of fishery managers' attention. For instance, snow crab and shrimp are cold water species that have increased in abundance and productivity on the ESS since the mid- to late 1990s in association with declines in bottom water temperatures (DFO, 2003). Managers would like to know if these species are ESS residents or colonizers from the Gulf of St. Lawrence where large populations of these species exist and whether or not these stocks will disappear in response to warmer bottom water conditions, which have subsequently occurred.

Increasingly, non-target species are demanding the attention of managers. *Codium* is a non-native species of kelp with no natural predators, which is spreading in its distribution from New England onto the Scotian Shelf. Managers and industry would like to know what is controlling the spread of this species in order to undertake appropriate mitigation measures. Research is required on the rate of change in the distribution of this species and its impact on the habitat of commercial species such as sea urchin and lobster.

In Canada, legislation to facilitate conservation of species diversity has been enacted in the Species at Risk Act (SARA), which enables the identification, evaluation and protection of species at risk. In the ESSIM area, this includes species such as bottlenose whale, leatherback turtle and porbeagle shark, among others. The intent of the objective related to species diversity is to ensure that no one species is impacted to the extent that it becomes at risk of extinction. A related management question is: what can be done to recover species at risk (with the detailed questions outlined in the diverse recovery plans for leatherback turtle, porbeagle shark, bottlenose whale, etc.)?

A number of species-specific research initiatives are required, a small sampling of which is provided here. For leatherback turtles, one of the key research questions is survivorship estimates of individuals that have been hooked and released; these are only currently available for loggerhead turtles (O'Boyle, 2001). For porbeagle sharks, there is a need to quantify sources of juvenile mortality and the location of the North Atlantic pupping grounds.

More generally, data and information on the life-history traits of many species at risk are very limited, yet there is a regulatory requirement to develop recovery schedules for all listed species. For many of these data-poor species, it may be possible to infer growth and mortality from related data-rich species using meta-analyses and other like approaches (Myers *et al.*, 1999).

To properly determine whether or not a species has recovered, there must be reliable indicators of abundance. Managers are asking science: how should temporal trends of the abundance of species at risk on the ESS be monitored?

One of the primary monitoring tools has been the DFO multi-species trawl survey. Given that the survey catchability of many species at risk is very low, it is important to determine if these surveys provide reliable indicators of temporal trends in abundance. For example, the preferred habitat of cusk is rocky bottom that is inaccessible to the survey (Harris *et al.*, 2002). It is possible, if not likely, that cusk distribution conforms to the 'basin' habitat model (MacCall, 1990). When abundance is high, cusk can be found outside their preferred habitat and are thus accessible to the survey, but when abundance declines, they 'contract' into their preferred habitat and thus become inaccessible to the survey. If this is the case, survey catch rate and cusk abundance are not linearly related. Other species (e.g. deepwater species inhabiting the Scotian Slope) might be exhibiting similar dynamics. In the case of species such as barndoor skate and Atlantic halibut, immature individuals appear to be more accessible to the survey than mature individuals (Simon *et al.*, 2002). Thus, it is not possible to use the survey to monitor trends in the abundance of mature fish of these species.

The accounting of discards and by-catch has become a priority within fisheries management planning (for both biodiversity and productivity conservation objectives). Managers are asking: do present management practices allow accurate estimation of discards and by-catch?

Research is required to estimate the level of discarding and by-catch of all species (commercial and non-commercial) in fisheries, as well as the impact of discarding and by-catch on the productivity of non-commercial species. These are not insignificant tasks. A May 2006 workshop on by-catch estimation for fisheries on the Scotian Shelf and in the Gulf of Maine area identified a number of significant data collection and processing issues that need to be addressed before adequate by-catch estimates can be calculated and impacts evaluated.

At population level

The importance of population substructure to its viability is recognized as being critical to the sustainability of these resources. Managers are asking: should fisheries management be concerned about population substructure for ESS stocks?

A number of stocks inhabiting the Scotian Shelf exhibit the characteristics of metapopulations that are important to consider not only in their management, but also in their recovery (Mohn, 1996). For instance, the cod stock on the ESS historically consisted of two bank spawning components (spring and autumn), as well as coastal spawning aggregations. During the period when the fishery was open (it has been closed since September 1993), there was no spawning closure in the spring and the fishing fleets were attracted to the spring spawning grounds by the high catch rates, which may have led to the loss of the spring spawning component. The implications of this loss for the recovery of this stock have been examined by Frank and Brickman (2001) and corroborate the findings of Smedbol and Wrobleski (2000). Recolonization depends upon the spatial relationship among these subpopulations and will take considerable time to occur. However, the detailed understanding of these processes and their genetic consequences are rudimentary. The development of spatially realistic models to examine recolonization processes in these populations would greatly aid the understanding and prediction of recovery events, and would be broadly applicable to endangered and threatened species being considered under SARA.

Productivity Issues

Trophic processes

Large fish predators have been removed from the Scotian Shelf ecosystem by both fishing activities and grey seals, the abundance of which has been increasing for several decades. Managers would like to know the implications of these removals: what are the management implications of the systematic removal of large fish on ecosystem functioning?

It has been hypothesized (DFO, 2003; Choi *et al.*, 2005; Frank *et al.*, 2005) that removal of large fish was responsible for a trophic cascade. Under this interpretation, declines in groundfish in the 1970s and 1980s allowed pelagic biomass to increase, which drove down zooplankton abundance, which in turn was responsible for an increase in phytoplankton. An alternate hypothesis is based upon physical oceanographic forcing (C. Greene, personal communication) and 'bottom-up' processes.

A related issue is the source of the high natural mortality observed on ESS cod and other groundfish species prior to and since the closure of the fisheries in 1993. Based upon fishery-independent surveys, total mortality appears to have remained high after the fishery was closed in 1993. It is probable that natural mortality had increased prior to the collapse of the groundfish stocks in this area, but existing methodologies cannot partition the total mortality between natural and fisheries causes during this earlier period. A possible source of higher natural mortality is predation by grey seals, the population of which has increased at a rate of about 13% per year (approximate doubling of the population every 6 years) since the 1950s (Bowen *et al.*, 2003). The natural processes that control seal abundance are of interest to both managers and harvesters. In addition to understanding the role of predators in food chains, there is a need to quantify predator consumption by other apex predators. While much work has been done on grey seals, other predators such as sharks and sea birds also require attention. However, even in the relatively data-rich cod - grey seal situation, different assumptions on the interaction dynamics produce dramatically different estimates of cod consumption (Mohn and Bowen, 1996).

Fisheries management has traditionally focused upon conserving the productivity of individual commercially exploited populations. Estimates of minimum spawning biomass reference points are still required for many exploited stocks: what is the minimum spawning stock biomass (SSB_{min}) for Scotian Shelf stocks?

There is a need to consider both historical and current (1950–present) trends when evaluating options for SSB_{min}. Pauly (1995) refers to the 'shifting baselines syndrome' in which modern datasets based upon heavily fished resources do not provide an adequate view of reference points that should be based upon a virgin population. Rosenberg *et al.* (2005), in their analysis of historical New England fleet fishing logs, dramatically illustrate this effect for Scotian Shelf cod. They show that estimates of virgin biomass are several times the abundance observed in modern datasets.

A broader question related to these and already developed indicators and reference points of stock productivity is the linkage between life-history parameters and ocean processes in the setting of management strategies. On the Scotian Shelf, managers are increasingly asking how climate change may influence the productivity of commercially exploited populations: can fisheries management incorporate the impact of climate change in the management of harvested stocks?

The Hjort–Cushing hypothesis of food availability during the early life history of fish has long been proposed as a mechanism responsible for recruitment variation. Platt *et al.* (2003) recently reported that ESS haddock recruitment is strongly influenced by the timing of the spring plankton bloom. Haddock stocks in the northwest Atlantic have been observed to exhibit exceptional recruitment (e.g. 1963 year-class on the Scotian Shelf and Georges Bank; 2003 year-class on Georges Bank) followed by periods of weak-to-moderate recruitment. It has been argued (Brander, 2005) that such recruitment events are not as important to the dynamics of pristine as compared to heavily exploited, reduced, populations. Notwithstanding this, the linkage of these to broader oceanographic events would be influenced by climate change-induced changes in circulation and mixing.

Managers are also interested in determining whether or not harvest practices have caused the significant growth rate declines that have been observed in a number of northwest Atlantic gadoid populations since the early 1980s: have current harvest practices caused growth reduction in gadoids?

There have been suggestions that the selection of the fast-growing fish has led to genetically 'stunted' populations (Kenchington, 2003). Alternatively, changes in the food chain may be impacting growth. Which processes are responsible for the decrease in growth would have very different management consequences.

Habitat Issues

Spatial processes

There are a number of area closures in place on the Scotian Shelf to control fishing activity (Fig. 7.3a and b). These have been implemented gradually over time to meet emergent fishery issues and have not benefited from a more synoptic evaluation of their collective utility in preserving benthic habitat. Managers are asking science: are current fishery closures and gear restrictions adequate to protect benthic habitat?

To answer this, it is necessary to determine the sensitivity of the various benthic community types to human impacts and to map these sensitive communities. Kostylev (2002) and Kostylev *et al.* (2005) used a theory developed by

(a)

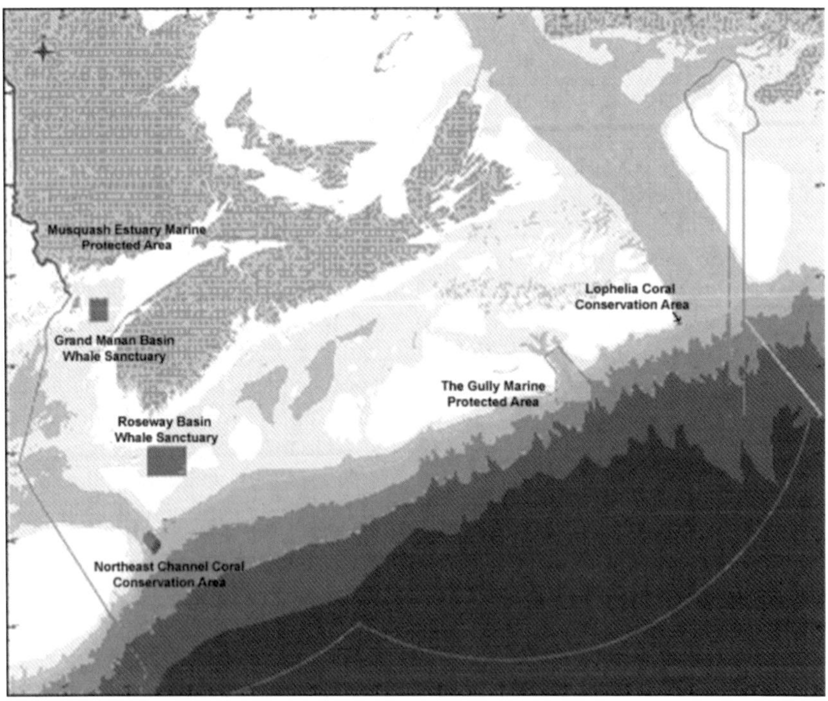

Fig. 7.3a. Current area closures to regulate human impacts on benthic community biodiversity on the Scotian Shelf.

(b)

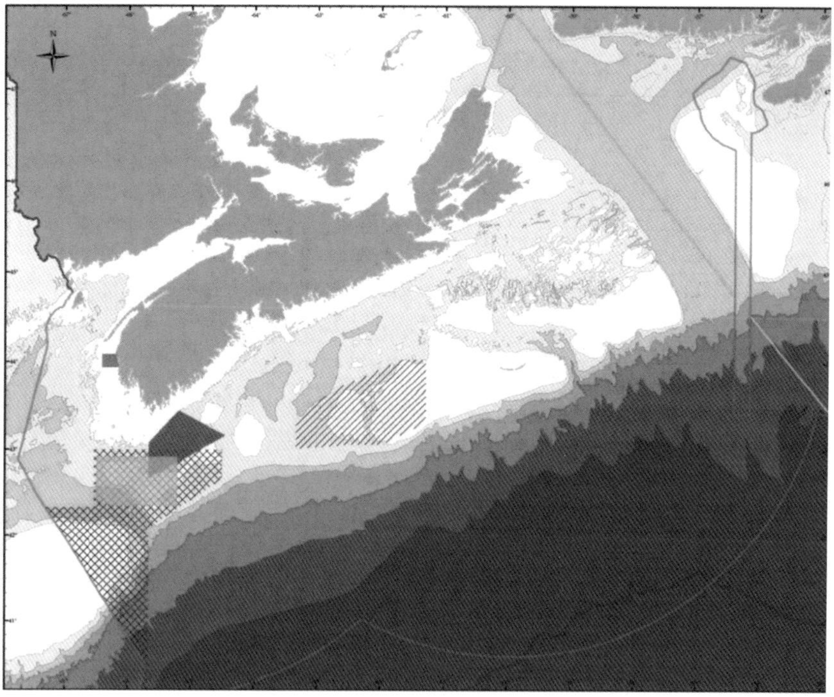

Fig. 7.3b. Current area closures to regulate fishery impacts on specified groundfish stocks on the Scotian Shelf.

Southwood (1977, 1988) to predict the distribution of benthic community types based on physical, chemical and biological features. This modelling approach was also used to identify areas of potential sensitive benthic communities (O'Boyle *et al.*, 2005b). Research is required to rigorously compare these predictions to empirical observations.

There is increasing interest by managers to use regulatory tools such as marine protected areas (MPAs) to conserve benthic community diversity. Managers are concerned with the placement of these. Should MPAs be few and large, or many and small, and in what circumstances? These are important questions that we are only now starting to consider.

Climate change

The influence of climate change on ocean resources and management will become increasingly important over the next two to three decades and managers are asking: can the impacts of climate change on habitat be predicted?

Determining the linkages between large-scale atmospheric systems (e.g. ENSO, PDO) and ocean circulation is critical to understanding the long-term implications of climate change on ocean resources. The Scotian Shelf is an area of considerable oceanographic complexity with strong influences from the

Labrador Current, Gulf Stream and outflow from the Gulf of St. Lawrence. Recent work by B. Petrie (personal communication) on the influence of the North Atlantic Oscillation (NAO) on coastal ocean circulation is starting to unravel some of these linkages. An analysis of winter-season anomalies in the NAO during 1970–2003, a period during which these anomalies have been increasing from high negative values early in the time series to positive values more recently, indicated that the NAO impacts the eastern and western Scotian Shelf quite differently; the ESS exhibits a spatial pattern in surface water temperature anomalies more similar to areas to the north rather than the south. It is of considerable interest to determine whether or not this pattern is periodic and predictable, and how it might be influenced by climate change. The research on the influence of the NAO on fish community distributions mentioned earlier (under the community-level biodiversity conservation objective) would complement this work and provide insight on the potential consequences of climate change.

Synopsis and Emergent Research Issues

Twenty one research activities associated with the 12 fisheries management questions have been identified (Table 7.1). They span all elements of the conservation objectives structure (Fig. 7.2), and deal with issues in each of the three categories of an EAF mentioned at the beginning of this chapter. However, only two of the questions we felt could be resolved in a 3–5 year period, with 9 and 10 considered respectively as being of medium and low tractability during this time frame. This has consequences in the short term for an EAF on the ESS. Scientific understanding will be of weak or limited scope to assist many management decisions.

Regarding the scope of the research effort required, many questions can be resolved at either the institute or northwest Atlantic scale. This has implications for the local management of these research programmes. Properly resourced and managed, much can be done locally and regionally (e.g. northwest Atlantic partnerships). However, a number would require international collaboration. Three research areas are notable in this regard. The first is how ecosystems are regulated – bottom-up, top-down or wasp waist control of food chain processes? Much has been learned in recent years through the regional study of ecosystems throughout the world (Cury *et al.*, 2003). What is needed is more comparative analyses of why one system works one way and another in a different way. Developments in this area would lead to more predictive power on how ecosystem food chains react to human impacts. The next two areas worthy of attention are the sensitivity of different benthic community types and the spatial scales of connectivity between benthic communities. To be useful to management, we need to do more than just classify habitat. An understanding of the characteristics of different habitats, including their sensitivity to human impacts, is required to better inform managers of the consequences of their actions. The work of Kostylev *et al.* (2005) is exemplary in exploring new applications of existing theory to provide managers with the predictive capacity to assist regulation, but much more is required in this area.

While the Canadian DFO conservation objectives' structure is useful in organizing the analysis of research needs for an EAF, it has shortcomings. The approach does not readily allow the identification of issues across objectives. One such issue is the link between biodiversity and productivity, and the implications for ecosystem resilience. While much research has focused on how productivity changes might be reflected by biodiversity, there is recent evidence (Duffy and Stachowicz, 2006) to suggest that ecosystems with high biodiversity tend to be able to both better withstand impacts as well as recover from these. Hubbell (2001) made a start at providing a theoretical basis for patterns in biodiversity and his efforts have stimulated much thinking in this area. Expansion of this theoretical work and others that include productivity might provide a critical link to an ecosystem's response to impacts and perhaps a predictive basis for use by managers. Such work will require international efforts over the medium to long term and should not be restricted to marine ecosystems.

Another issue is the need for ecological risk analysis. The move towards an EAF (and ocean management more generally) will require decisions on which issues are priorities to address and which are not. Fletcher (2005) provides a qualitative risk assessment approach that has been employed in fisheries management in Western Australia. A. Smith (personal communication) provides a risk assessment structure in which qualitative analyses are used to filter out low-risk issues, semi-quantitative analyses used to identify moderate risk issues and quantitative analyses used to assess high-risk issues. While these have been applied to fisheries situations, the requirement for risk assessment at the ecosystem level across competing ocean industries is also necessary.

The last high-level issue is the need for contextual ecosystem modelling. At the level of the impact (e.g. fishing), there will always be a need for models that synthesize knowledge and understanding of the cause–effect relationship between an impact and the population's response. There is an additional need for models that not only describe what we know about the cumulative impacts on identified ecosystem components across ocean industries, but also provide context to the ocean industry-specific impacts. Models have been used to explore and describe ecosystem behaviour for quite some time (e.g. Ecopath with Ecosim 'EwE', Walters *et al.*, 1997). These efforts need to be encouraged and expanded to consider the full complexity of both ecosystem function, including physical–biological coupling, and diversity of human impacts. One such model is 'Atlantis' (Fulton *et al.*, 2004), which is being used to guide management efforts in Australia. This and other 'agent-based' modelling approaches are opening new avenues to the exploration and synthesis of our knowledge of marine ecosystems. Research on how best to develop these models and how they can be effectively employed to inform management is urgently needed.

Concluding Remarks

Our intent was not to provide a comprehensive synopsis of all the research needs of an EAF. The list would be quite long and it would be difficult to judge which ones are priorities. Rather, our approach considers those issues that have been and are of importance to ocean and fisheries managers in our region. Based upon these management needs and associated research issues, several points are made of more general importance and applicability to the broader marine research community. In illustrating our approach, we encourage others to undertake similar exercises. The EAF is a hot topic in many countries, and a broader set of ecological issues than have been considered under single-species management urgently require attention.

References

Bowen, W.D., McMillan, J.I. and Mohn, R. (2003) Sustained exponential growth of the grey seal on Sable Island. *ICES Journal of Marine Science* 60, 1265–1374.

Brander, K.M. (2005) Cod recruitment is strongly affected by climate when stock biomass is low. *ICES Journal of Marine Science* 62, 339–343.

Choi, J.S., Frank, K.T., Petrie, B.D. and Leggett, W.C. (2005) Integrated assessment of a large marine ecosystem: a case study of the devolution of the Eastern Scotian Shelf, Canada. *Oceanography and Marine Biology: An Annual Review* 43, 47–67.

Cury, P.M. and Christensen, V. (2005) Quantitative ecosystem indicators for fisheries management. *ICES Journal of Marine Science* 62, 307–310.

Cury, P., Shannon, L. and Shin, Y.-J. (2003) The functioning of marine ecosystems: a fisheries perspective. In: Sinclair, M. and Valdimarsson, G. (eds) *Responsible Fisheries in the Marine Ecosystem*. CAB International, Wallingford, UK/Cambridge, Massachusetts, 426 pp.

DFO (2003) State of the Eastern Scotian Shelf Ecosystem. DFO Canadian Science Advisory Secretariat Ecosystem Status Report 2003/04.

DFO (2005) Eastern Scotian Shelf Integrated Ocean Management Plan (2006–2011). Draft for Discussion. Oceans and Coastal Management Report 2005-02, 73 pp.

Duffy, J.E. and Stachowicz, J.J. (2006) Why biodiversity is important to oceanography: potential roles of genetic, species and trophic diversity in pelagic ecosystem processes. *Marine Ecology Programme Series* 311, 179–189.

Fletcher, W.J. (2005) The application of qualitative risk assessment methodology to prioritize issues for fisheries management. *ICES Journal of Marine Science* 62, 1576–1587.

Frank, K.T. and Brickman, D. (2001) Contemporary management issues confronting fisheries science. *Journal of Sea Research* 45, 173–187.

Frank, K.T., Petrie, B., Choi, J.S. and Leggett, W.C. (2005) Trophic cascades in a formerly cod-dominated ecosystem. *Science* 308, 1621–1623.

Fulton, E.A., Smith, A.D.M. and Punt, A.E. (2004) Ecological Indicators of the ecosystem effects of fishing: Final Report. Report No. R99/1546. Australian Fisheries Management Authority, Canberra, Australia.

Harris, L.E., Comeau, P.A. and Clark, D.S. (2002) Evaluation of Cusk (*Brosme brosme*) in Canadian Waters. CSAS Res Doc 2002/104.

Hubbell, S.P. (2001) *The Unified Neutral Theory of Biodiversity and Biogeography*. Princeton University Press, Princeton, New Jersey, 448 pp.

Kenchington, E.L. (2003) The effects of fishing on species and genetic diversity. In: Sinclair, M. and Valdimarsson, G. (eds) *Responsible Fisheries in the Marine Ecosystem*. CAB International, Wallingford, UK/Cambridge, Massachusetts, 426 pp.

Kostylev, V.E. (2002) Challenges in habitat classification and mapping. In: *Proceedings of Benthic Habitat Classification Workshop Meeting of the Maritimes Regional Advisory Process, 2002*. J. Arbour (Chair), V. E. Kostylev (ed), pp. 72–77.

Kostylev, V.E., Todd, B.J., Longva, O. and Valentine, P.C. (2005) Characterization of benthic habitat on northeastern Georges Bank, Canada. In: Barnes, P.W. and Thomas, J.P. (eds)

Benthic Habitats and the Effects of Fishing. American Fisheries Society, Symposium 41, Bethesda, Maryland, pp. 141-152.

Mahon, R., Brown, S.K., Zwanenburg, K.C.T., Atkinson, D.B., Buja, K.R., Calflin, L., Howell, G.D., Monaco, M.E., O'Boyle, R.N. and Sinclair, M. (1998) Assemblages and biogeography of demersal fishes of the east coast of North America. *Canadian Journal of Fisheries and Aquatic Sciences* 55, 1704-1738.

MacCall, A.D. (1990) *Dynamic Geography of Marine Fish Populations*. University of Washington Press, Seattle, Washington, DC, 153 pp.

Mohn, R.K. (1996) Assessing closed areas with a fishery model having several spawning components. In: Burke, D.L., O'Boyle, R.N., Partington, P. and Sinclair, M. (eds) Report of the second workshop on Scotia-Fundy groundfish management. *Canadian Technical Reports in Fisheries and Aquatic Sciences* 2100, 157-169.

Mohn, R. and Bowen, W.D. (1996) Grey seal predation on the eastern Scotian Shelf: modelling the impact on Atlantic cod. *Canadian Journal of Fisheries and Aquatic Sciences* 53, 2722-2738.

Myers, R.A., Bowen, K.G. and Barrowman, N.J. (1999) Maximum reproductive rate of fish at low population sizes. *Canadian Journal of Fisheries and Aquatic Sciences* 56, 2404-2419.

O'Boyle, R. (2001) Meeting on Turtle By-Catch in Canadian Atlantic Fisheries. CSAS Proceedings Series. 2001/17.

O'Boyle, R. and Jamieson, G. (2006) Observations on the implementation of ecosystem-based management: experiences on Canada's east and west coasts. *Fisheries Research* 79, 1-12.

O'Boyle, R., Sinclair, M., Keizer, P., Lee, K., Ricard, D. and Yeats, P. (2004) An integrated management framework for the eastern Scotian Shelf. DFO Canadian Science Advisory Secretariat Research Document Series, 2004/076.

O'Boyle, R., Sinclair, M., Keizer, P., Lee, K., Ricard, D. and Yeats, P. (2005a) Indicators for ecosystem-based management on the Scotian Shelf: bridging the gap between theory and practice. *ICES Journal of Marine Science* 62, 598-605.

O'Boyle, R., Kostylev, V., Breeze, H., Hall, T., Herbert, G., Worcester, T., Ricard, D. and Sinclair, M. (2005b) Developing an Ecosystem-Based Management Framework for Benthic Communities: A Case Study of the Scotian Shelf. ICES CM 2005/BB: 18, 31 pp.

Pauly, D. (1995) Anecdotes and the shifting baseline syndrome of fisheries. *Trends in Ecology and Evolution* 10, 430.

Platt, T., Fuentes-Yaco, C. and Frank, K.T. (2003) Spring algal bloom and larval fish survival. *Nature* 423, 398-399.

Rosenberg, A.A., Bolster, W.J., Alexander, K.E., Leavenworth, W.B., Cooper, A.B. and McKenzie, M.G. (2005) The history of ocean resources: cod biomass using historical records. *Frontiers in Ecology and the Environment* 3, 84-90.

Rutherford, R.J., Herbert, G.J. and Coffen-Smout, S.S. (2005) Integrated ocean management and the collaborative planning process: the Eastern Scotian Shelf Integrated Management (ESSIM) initiative. *Marine Policy* 29, 75-83.

Simon, J.E., Frank, K.T. and Kulka, D.W. (2002) Distribution and abundance of Barndoor skate (*Dipturus laevis*) in the Canadian Atlantic based upon research vessel surveys and industry/science surveys. DFO Canadian Science Advisory Secretariat Research Document Series, 2002/070.

Southwood, T.R.E. (1977) Habitat, the templet for ecological strategies? *Journal of Animal Ecology* 46, 337-365.

Southwood, T.R.E. (1988) Tactics, strategies and templets. *Oikos* 52, 3-18.

Smedbol, R.K. and Wroblewski, J.S. (2000) Metapopulation theory and northern cod population structure: interdependency of subpopulations in recovery of a groundfish population. Canadian Stock Assessment Secretariat. Research Document 2000/087. 30 pp.

Walters, C.J., Christensen, V. and Pauly, D. (1997) Structuring dynamic models of exploited ecosystems from trophic mass-balance assessments. *Reviews Fish Biology and Fisheries* 7, 1-34.

Zwanenburg, K.C.T., Bundy, A., Strain, P., Bowen, W.D., Breeze, H., Campana, S.E., Hannah, C., Head, E. and Gordon, D. (2006) Implications of Ecosystem Dynamics for the Integrated Management of the Eastern Scotian Shelf. *Canadian Technical Report of Fisheries and Aquatic Sciences* 2652, 91 pp.

8 Implementing an Ecosystem Approach to Fisheries Management: Lessons Learned from Applying a Practical EAFM Framework in Australia and the Pacific

WARRICK JEFFREY (RICK) FLETCHER

Department of Fisheries, Western Australia

Abstract

The ecosystem approach to fisheries management (EAFM) is one of a growing number of concepts generated since the late 1990s to describe taking a more comprehensive approach to the management of natural resources. Most countries now recognize the need to deal explicitly with all the ecological consequences of fishing activities as well as their social and economic implications.

Implementing these concepts has previously proven difficult, but since the early 2000s substantial progress has been made in Australia and more recently in the Pacific region. This has been possible through the development of a practical framework for the assessment of wild capture fisheries against the principles of EAFM. This framework includes a four-step, risk-based process that generates an EAFM report on a fishery covering its impact on target species, by-catch species and the broader ecosystem, plus the potential social and economic outcomes produced and the fishery's current governance system.

From experiences applying this framework, particularly within the Pacific, a number of critical lessons have been identified. These include the recognition that EAFM must be undertaken as a risk-based management process, not as the excuse for undertaking more detailed research. EAFM can be started with whatever level of information is available, with the process helping to determine what additional work is really needed rather than simply what is possible. Finally, implementing EAFM should not be used as an excuse to delay dealing with problems that are already well documented. The lack of good governance arrangements, not the lack of ecological data, has been the most commonly identified high-risk issue.

Introduction

The ecosystem approach to fisheries management (EAFM) is just one of a growing number of concepts that have been generated since the late 1990s

 The Ecosystem Approach to Fisheries (eds G. Bianchi and H.R. Skjoldal)

to describe taking a more comprehensive approach to the management of natural resources. The key elements of such an approach within a fisheries context are that it requires the managing agency and the industry deal with all the ecological consequences of fishing and also understand the social and economic implications flowing from their activities (e.g. Scandol *et al.*, 2005; Fletcher, 2006).

The key difference among the various strategies is the scope of the issues they are attempting to address. This can range from a small-scale fishing operation up to all the activities occurring within an ocean. Thus, sustainable development (or ecologically sustainable development (ESD), as it is known in Australia) is best viewed as an overall goal for government for which a variety of strategies, such as EAFM, can be used by different sectors/agencies to assist in meeting this goal (Fletcher, 2006). Within this context, the defining element for an EAFM system is that the scope of issues covered needs to be restricted to those that can be managed, or at least directly influenced, by the relevant fisheries management agency (hence the 'F' component).

Implementing these concepts has previously proven difficult (Garcia and Staples, 2000; Charles, 2001; FAO, 2003). Since early 2000, Australia has been one of the regions where there has been substantial progress. A major reason for these advances has been the requirement for any exporting state-based fishery and all Commonwealth-managed fisheries to submit a comprehensive application to the Commonwealth government's environment agency addressing a set of guidelines for sustainable fisheries in order to receive certification for the ongoing export of their catch (Commonwealth of Australia, 2001). This was a powerful incentive to implement systems capable of providing the information needed across all the ecological elements of ESD.

Within the Pacific region, the Forum Fisheries Agency (FFA) has recently been undertaking an initiative to introduce EAFM to enable a more sophisticated approach to the management of the tuna fisheries in the western and central Pacific region. This process has been designed to assist in implementing the objectives and articles outlined in the Western and Central Pacific Fisheries Commission (WCPFC) Convention (Anon, 2005), which should minimize the likelihood of external criticism of the fisheries potentially affecting their markets. Finally, the Secretariat of the Pacific Community (SPC) has also begun to adopt EAFM methods to aid in their programmes for improving the management of the coastal fisheries of this region. In both cases, the outcomes from the management of these fisheries are integral to the livelihoods of many within these communities. Consequently, a key requirement within the Pacific region has been to deal effectively with the socio-economic elements of EAFM.

While it is clear that external pressures can increase the impetus to begin implementing an EAFM style of management, the real benefits to a fishery must ultimately come from improved management outcomes. If the management systems imposed do not improve conditions at a local level, they are highly unlikely to persist in the longer term. Consequently, the challenge is to create a system that not only produces outcomes that external parties would consider appropriate, but also one that enhances the management outcomes for all stakeholders in the fishery – including the fishers, managers and local communities. Thus, the ongoing motivation for implementing EAFM must ultimately come from within the country/community/industry or it is unlikely to succeed.

In Australia, a national framework for the assessment of fisheries against the principles of sustainable development was developed to help meet the increased assessment needs of EAFM within Australia in an efficient manner. This framework (based upon Chesson *et al.*, 1999) uses a four-step, risk-based process to generate reports on all relevant EAFM issues for a fishery, including impacts on target species and the broader ecosystem, the potential social and economic outcomes and the current governance systems (see Fletcher *et al.*, 2002, 2005 for full details).

The framework was originally conceived in 2001 by a national ESD working group and has been through a number of trials and refinements, having now been applied to more than 30 commercial fisheries in Western Australia (Kangas *et al.*, 2006) and approximately 20 other fisheries across other parts of Australia. It has also been applied to the tuna fisheries of a number of member countries in the WCPFC and most recently it has been trialed on various coastal fisheries and communities of the Pacific (Fletcher, 2007). These experiences have been invaluable to our understanding of how the system operates in varying circumstances. The remainder of this chapter will outline a summary of the EAFM process and highlight some of the more critical elements that need to be understood to enable implementation in an efficient and practical manner.

Summary of the EAFM Process

The EAFM process developed for fisheries is comparable to the processes used in all risk management systems (e.g. Standards Australia, 2004; see Fig. 8.1 for an outline). This reflects the fact that managing a fishery is just a specific application of risk management principles. Consequently, the EAFM process needs to identify all the good and bad things associated with a fishery and develop management plans to control actions to enable performance to be maintained or improved to acceptable levels.

The four main steps involved in the EAFM process are:

Step 1. Determine the scope of the assessment by developing a clear description of what you are trying to manage/assess and the range of societal values that need to be addressed.

Step 2. Based on the scope, identify all the issues across the range of EAFM elements (retained and non-retained species, the ecosystem, community outcomes and administrative systems) and determine the relevant outcomes to be achieved for each issue given the requirements of any convention, country needs, local requirements and global attitudes. The outcomes can, therefore, be based on ecological concerns, economic realities or social attitudes, which often have different implications for the level of actions that need to be taken.

Step 3. Using some form of risk assessment and the precautionary approach, determine if an issue needs direct actions to achieve the outcome(s) wanted.

Step 4. Where direct actions are deemed necessary, a formal management system should be developed that includes having clear operational objectives (based on the outcomes identified in step 2) and a way of assessing performance against these operational objectives. These systems must also include the moni-

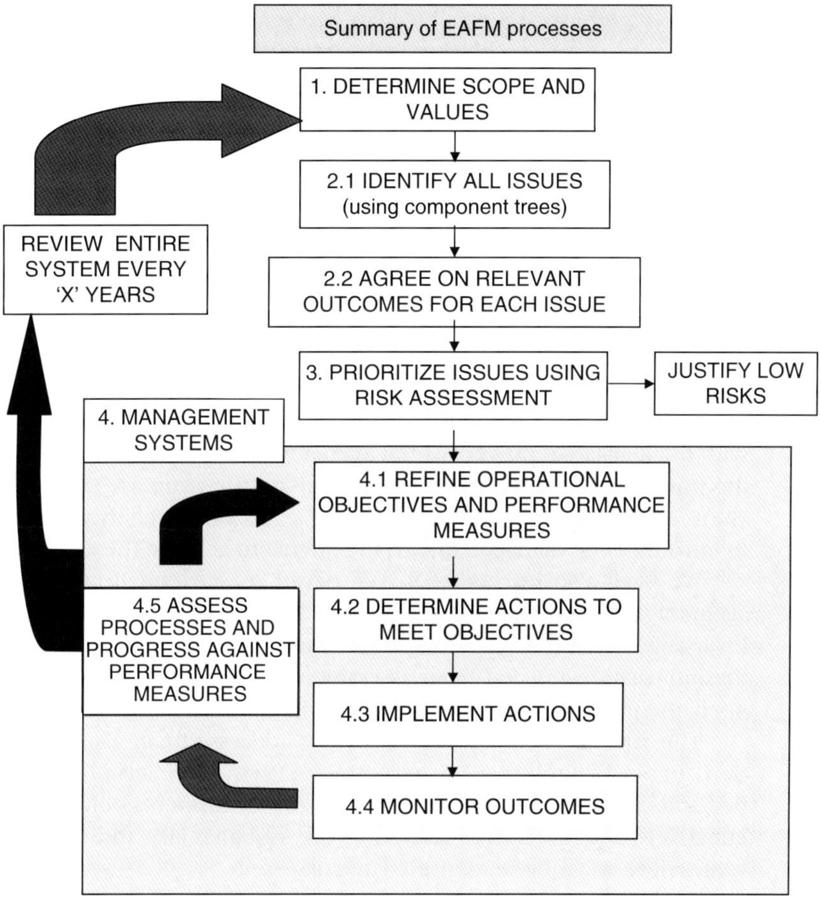

Fig. 8.1. An outline of the various steps involved in the ESD/EAFM framework. The outline and the processes are largely based on the AS/NZS:4360 risk management system (SA, 2004).

toring and review of performance and a plan for what will happen if performance is not acceptable.

These four steps are the basis of all EAFM assessments, irrespective of whether the scope of the activity being examined is a fishery operated by an isolated atoll community or a large-scale industrial fishery that involves many stakeholders. While the basic EAFM approach is the same for all fisheries, the precise methods used for undertaking each of these steps needs to vary according to the situation that is being addressed.[1] This includes recognizing the level of sophistication in management arrangements and processes that are available, the complexity of the problems that are being addressed, the level of information available and the level of formal education of those involved. What can be appropriate in the assessment of a highly industrial fishery will almost certainly be inappropriate

[1] Detailed reports and manuals have now been generated on these processes for use in different circumstances (see http://www.eafm.com.au).

when trying to assess a small fishery in a remote community. A summary of the main lessons to assist with these adaptations is provided below.

Results: Lessons Learned

Step 1: Scope

Clearly determining the scope of what will be addressed is an important step because it affects how the rest of the process will operate. The scope of an ecosystem-based assessment could cover the activities of just one individual, a single fishery, all fishing activities in a region or all activities in a region. A key factor in determining the appropriate scope is that the process can only generate practical outcomes for management when the scope aligns closely with the legislative powers of the agencies directly involved. If you do not have the power to regulate or manage an activity, then you are unlikely to establish enforceable objectives and performance levels and certainly you will not be able to introduce the management arrangements to achieve these.

We have also learned that it is better to start small by getting the management arrangements for the lowest level management units documented. Invariably, when trying to undertake regional multi-fishery assessments, unless the information needed from assessments at the lower levels has been developed, the process stalls or generates trivial outcomes (Fletcher, 2006).

The societal values that need to be addressed can also vary among locations. In assessing the tuna fisheries in the Pacific, five sets of values were identified and used (Table 8.1). By contrast, in the assessments completed in Australia, effectively only objectives related to the sustainability and social acceptability values have so far been identified and assessed.

Step 2: Issue identification

The identification of the relevant issues for a fishery can be greatly assisted by the use of a component tree approach. Each of the five elements of EAFM

Table 8.1. A brief description of the five different societal values that were identified as being potentially relevant for assessing risks within the fisheries operating in the Pacific region.

Potential EAFM values	Description
Sustainability	Keeping biomass levels above B*msy*
Viability	Avoiding extinction for a species (i.e. B*current* can be < B*msy* but > B*extinct*)
Economic	Optimize/maximize economic benefits
Social	Optimize social acceptability
Food security	Ensure subsistence levels of capture

Note: not all values were relevant to every issue.

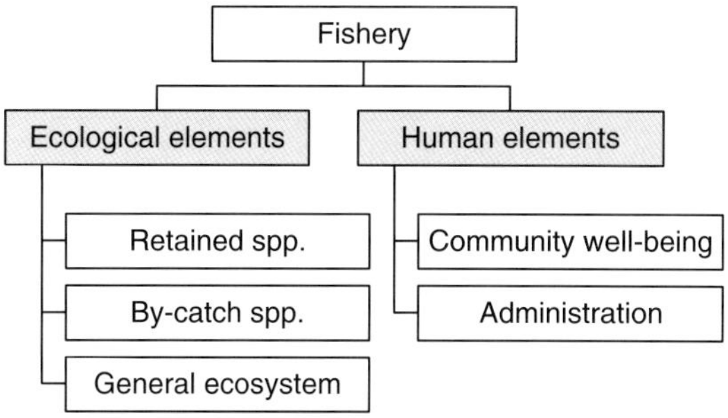

Fig. 8.2. The base framework for EAFM showing the five key elements.

(Fig. 8.2) has a generic component tree that includes the issues that are relevant for many fisheries (Fig. 8.3A). These are, however, only the starting point: each fishery tailors these trees to suit its individual circumstances. This can include splitting issues to provide greater detail, adding issues not present, or removing those that are not relevant (Fig 8.3B). The need to add, remove or alter the trees will depend upon the fishing methods used, the areas of operation, the species involved and the types of community and administrative constraints operating in the region. This approach maximizes the consistency of the outputs and minimizes the chances of missing issues.

(A)

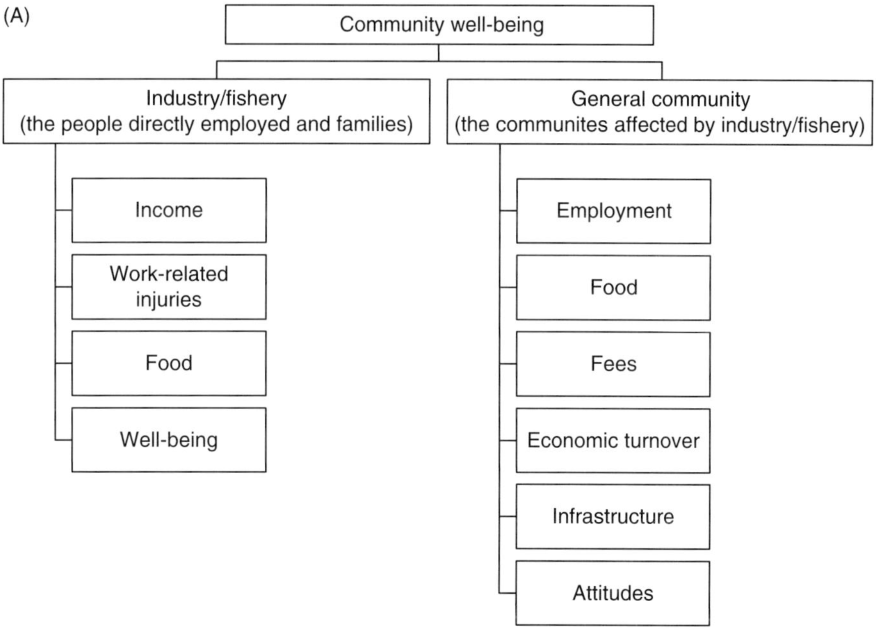

Fig. 8.3. (A) One of the five generic component trees: Community well-being; (B) A community well-being tree that has been tailored to the specific issues for the tuna fishery of a country in the WCPFC.

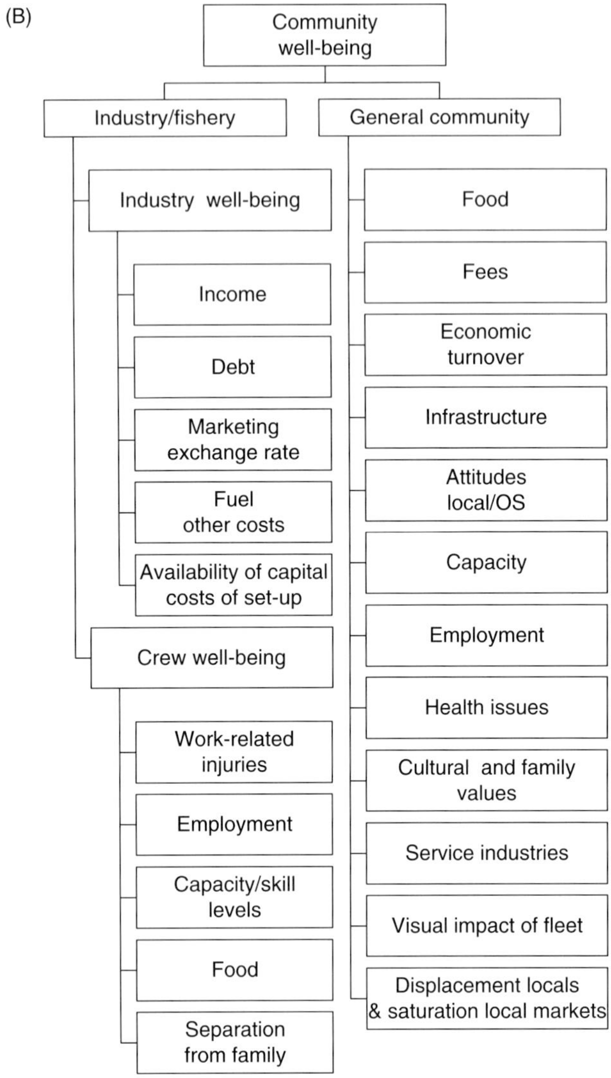

Fig. 8.3. Continued

It is important that the identification of issues is carried out in a workshop format in which stakeholders are part of the process. The more community-based a fishery (i.e. the less industrial), the more appropriate it may be to begin by identifying the community well-being issues, rather than starting on the ecological issues. This helps not only to engage the stakeholders, because you are clearly focusing on the issues most important to them, but also to decide what values the communities want to achieve from the utilization of their resources. Finally, it is important not to impose your values on others; different societies have different concepts of what is acceptable and important.

Step 3: Prioritization

A large number of issues can be identified through this process, even for a small fishery, the importance of which will vary greatly. Consequently, it is vital to have some way of prioritizing among the issues so that only those that require direct actions receive what are usually scarce resources.

To determine the priority of issues, and therefore the appropriate level of management response, the EAFM process uses risk assessment. There are a variety of risk analysis methods that can be used; the complexity, data requirements, cost and timelines required for completion are also variable. A number of qualitative risk analysis tools were developed for use in the EAFM process based upon the AS/NZ Standard (Standards Australia, 2004). These work by assigning a level of consequence (impact) and the likelihood (probability) of this consequence actually occurring to generate an estimate of the risk (low, medium or high) for each issue (see Fletcher, 2005, 2007 for details). Only medium- and high-risk issues require direct management or resources, with high-risk issues probably requiring additional management intervention.

For some situations, such as those where very limited data are available (which includes many socio-economic issues) or where the stakeholders have limited formal education, the use of the consequence × likelihood system has sometimes been found to be inappropriate. In such circumstances, a risk analysis can still be done by directly rating issues into three risk levels.

Irrespective of what type of risk analysis method is used, documenting why a particular value was chosen is the most critical element, and must always be included. It is the 'why' that helps determine 'what' the next step should be.

Different values/objectives

Risk assessment works by helping to determine the chances of not meeting your objectives, but objectives are affected by the values/outcomes being sought. For each issue, therefore, you need to be clear what objectives are relevant because the risk level may change depending upon what objectives are assessed. Consequently, in determining the priority of issues associated with the tuna fisheries in the Pacific, because there were five societal values, often more than one objective was relevant for a single issue.

A good example was assessing the priority for management actions on albacore tuna. Despite a long history of fishing and the high level of catches in recent years, stock assessments indicate that their spawning biomass has not been reduced substantially (Langley and Hampton, 2006). Furthermore, under current rates of exploitation, the total spawning stock is likely to fluctuate around levels well above the level of B_{msy}, indicating that there is only a low risk to their ecological sustainability (Table 8.2).

The abundance of albacore can, however, become locally reduced both from intense fishing in a region and from their migration routes being affected by regional oceanographic conditions. Both of these can affect local catch rates and hence the economics and flow-on social benefits in countries where albacore is a key target species. Consequently, in these countries, the assessment of

Table 8.2. Excerpt from risk assessments conducted on fisheries within the WCPFC.

Issue	Objective	Risk level
Albacore	Stock sustainability	Low
	–Whole of stock	
	Economic	Medium
	–Industrial	
	Social	High
	–Artisanal	
Yellow fin	Stock sustainability	
	– Whole of stock	High
	–Country-level impact	Low
	Economic	
	–Industrial	Low
	–Charter	Medium
	Social	
	–Artisanal	Low

the economic and social objectives associated with albacore generated moderate and high risks (Table 8.2).

The management implications from these assessments are that if only an ecological risk assessment (ERA) had been done (i.e. just using the sustainability objective), then little direct management of fishing on albacore would be needed. However, because of the economic/social risks, the development of explicit management systems and actions to limit local rates of exploitation was needed to help ensure adequate economic and social outcomes. It is highly likely that similar situations occur elsewhere.

Step 4: Management systems

The final step in the EAFM process is to develop a management system for each of the issues where there was sufficient risk to require direct controls and/or further investigation. These management systems can be developed at a variety of levels of complexity, with the most comprehensive system outlined in detail in Fletcher (2007).

The critical elements in any management system are to have operational objectives (what specifically do you want to achieve for this issue and this fishery), indicators (how will you actually measure performance) and performance measures (what levels of the indicator define acceptable performance). Furthermore, each of the management actions should be directly related to achieving the operational objectives. Finally, there must be regular reviews of progress with appropriate alterations to management where performance is not acceptable. Where there are formal feedback loops, these are often now called decision rules or harvest strategies.

In situations where time or resources are limited, shorter EAFM reports can be generated, at least as an interim step, by capturing the critical elements of

Table 8.3. Excerpts from three identified issues for a tuna fishery from the WCPFC using a summary style management system report (Na = not applicable).

Issue	Objective	Risk level	Indicator	Interim performance measure	Management actions
Albacore	Economic (country)	Moderate	Catch rates	>1.25 albacore per 100 hooks (interim)	Determine more robust economic catch rate level with industry
			Catch	<10,000 t/year	Set explicit allocation levels amongst the different fishing sectors Get confirmation from the regional fisheries assessment group this total catch level will not begin to affect local abundance
Billfish	Sustainability	Low	Na	Na	Na
	Economics	Low	Na	Na	Na
Crew – separation from families	Social	High	Number of complaints from wives	No increase	Educate fishers about the issues and implications of extended trips including impacts on family values Encourage fishers to minimize long trips Discuss possible methods of mitigation with the tuna fishing companies

the risk assessment and a summary of what immediate management actions will need to occur (see Table 8.3 for an example). This approach provides a very rapid way to determine and document the main actions, allowing the EAFM process to make substantial progress within a matter of weeks/months. This can be a useful way to gain momentum and engender the 'buy-in' by stakeholder groups, which is needed to have the process continue.

Discussion

Many of the difficulties so far experienced with implementing EAFM appear to result from a mixture of myth, fear and unrealistic expectations about what is needed for this process. This produces a situation where it can seem much too difficult to begin. As a consequence of being placed in a situation of effectively being forced to undertake such a process, in Australia we quickly learned to take a pragmatic, staged approach and to be realistic about what the process needed to generate.

One myth to dispel is that EAFM requires full certainty about all the possible ecological, economic and social interactions and issues associated with a fishery. Instead, to be consistent with EAFM principles, the level of uncertainty associated with an issue must only match the level of precaution that has been taken in determining the management arrangements and settings. In this context, it is advisable to beware of having scientists in charge of EAFM, because it is fundamentally a management process, not a research activity. Given that there will always be uncertainties to operate in a practical manner, EAFM must be undertaken as a risk-based approach, not as an automatic permit to collect more detailed ecological data.

With the large number of issues that can potentially be generated using the EAFM framework and the uncertainties that are often present in the information available, using a risk-based approach enables you to start with whatever data you currently have. This does not mean that new information will never be needed. In many cases the EAFM process will identify certain risks as 'high' owing to uncertainties, when the collection of additional data may reduce these risks to satisfactory levels. However, in some cases the alternative strategy may be to take management actions that reduce the level of impact to a point where the level of uncertainty in understanding matches the risk. Thus, the EAFM process helps to determine what level of management action (or inaction) is appropriate given the level of risk and the current level of knowledge available.

The experiences of undertaking risk analyses across many different types of fisheries in different countries have shown that specific 'ecosystem'-related issues and particularly the lack of detailed ecosystem knowledge have generally not been identified as the most critical problems facing fisheries. The most common issues affecting the overall performance of a fishery are governance issues, especially where shared stocks are being harvested and it is unclear where direct management responsibilities reside. The expectation that a good EAFM outcome may be possible in such circumstances if we just collect more ecological data is dangerous. There is little value establishing additional research programmes to collect more detailed information where the governance and management systems are unable (or unwilling) to use the additional (or even currently available) information. Under such circumstances, resources should first go into improving the administrative processes. Correspondingly, the decision to begin implementing EAFM should not be used as an excuse to delay implementing management for issues already known to be problematic. Undertaking EAFM will not somehow make these difficult issues disappear; rather, it is designed to find and highlight such issues.

Fully implemented, the EAFM process should greatly assist decision making because it provides an overall framework for understanding the full implications of any management decision. In the initial stages of implementation, the main value will come from the identification and assessment of all relevant issues and the establishment of processes to enable their management to be undertaken effectively and efficiently. Another benefit from using this process is that it helps stakeholders recognize their role and impacts, as well as identifying overlap between fisheries, jurisdictions and other activities.

The general methodology that has been developed for use in EAFM has been found to be useful in a variety of other circumstances. A different version of this framework has been developed and applied to a number of aquaculture industries (Fletcher *et al.*, 2004; Marshall, 2006); it has even been adapted for use in helping to manage agricultural activities in the inland regions of Australia (Chesson and Whitworth, 2005). Thus, the basic processes are highly flexible and robust. However, like any system, it does not provide the answers; it merely assists in this process. The identification of issues and the determination of management solutions for a particular fishery must still come from the people responsible for the management of the fishery. If you do not know what you want to achieve, or are unwilling to do the things needed to achieve this, no process will help.

Even using this system, implementing EAFM is not always an easy process. Depending upon the fishery, the initial stages can involve a large amount of resources to bring together the necessary material and to undergo the consultation necessary to define and articulate the outcomes that are to be achieved. The most important lesson we have learned is that the process does not get any easier by waiting; the quicker you start, the fewer unnecessary activities will be completed and the sooner progress will be made towards generating better management outcomes.

Acknowledgements

I wish to thank the organizers of the CIEAF conference for allowing me to present this paper. The basis of this work was undertaken through activities completed within the Fisheries Research and Development Corporation (FRDC)-ESD subprogramme. The recent assistance of FFA staff, Dan Sua, Darren Cameron and Steve Shanks, and many people from the fisheries agencies within the WCPFC have been instrumental in the refinement of this framework for use in the Pacific and, therefore, for many other locations.

This forms publication number 17 for the FRDC–ESD subprogramme.

References

Anon (2005) Convention for the Western and Central Pacific Fisheries Commission. WCPFC, Honiara, Solomon Islands, 27 pp.

Charles, A.T. (2001) Sustainable fishery systems. *Fish and Aquatic Resources* Series 5. Blackwell Science, Malden, Massachusetts, 370 pp.

Chesson, J. and Whitworth, B. (2005) Signposts for Australian Agriculture – Preliminary framework and collation of industry profiles. Bureau of Rural Sciences, Canberra, Australia, 62 pp. Available at: http://www.nlwra.gov.au/downloads/final_reports/BRR31_Signposts_Final_ Report.pdf

Chesson, J., Clayton, H. and Whitworth, B. (1999) Evaluation of fisheries-management systems with respect to sustainable development. *ICES Journal of Marine Science* 56, 980–984.

Commonwealth of Australia (2001) Guidelines for the ecologically sustainable management of fisheries. Department of Environment and Heritage, Canberra, Australia, 16 pp.

FAO (Food and Agriculture Organization of the United Nations) (2003) Fisheries management 2. The ecosystem approach to fisheries. FAO Technical Guidelines for Responsible Fisheries 4 (Suppl. 2), 112 pp.

Fletcher, W.J. (2005) Application of qualitative risk assessment methodology to prioritise issues for fisheries management. *ICES Journal of Marine Research* 62, 1576–1587.

Fletcher, W.J. (2006) Frameworks for managing marine resources in Australia through ecosystem approaches: do they fit together and can they be useful? *Bulletin of Marine Science* 78, 691–704.

Fletcher, W.J. (2007) A Guide to Implementing an Ecosystem Approach to Fisheries Management (EAFM) within the Western and Central Pacific Region. Forum Fisheries Agency, Honiara, Solomon Islands, 89 pp.

Fletcher, W.J., Chesson, J., Fisher, M., Sainsbury, K.J., Hundloe, T., Smith, A.D.M. and Whitworth, B. (2002) *National ESD Reporting Framework for Australian Fisheries: The* 'How To' *Guide for Wild Capture Fisheries*. FRDC Project 2000/145, Canberra, Australia, 120 pp. Available at: http://www.fisheries-esd.com/a/pdf/WildCaptureFisheries_V1_01.pdf

Fletcher, W.J., Chesson, J., Fisher, M., Sainsbury, K.J. and Hundloe, T.J. (2004) *National ESD Reporting Framework: The* 'How To' *Guide for Aquaculture*. Version 1.1 FRDC, Canberra, Australia, 88 pp. Available at: http://www.fisheries-esd.com/a/pdf/ESDHowtoGuideAquaculture.pdf

Fletcher, W.J., Chesson, J., Sainsbury, K.J., Fisher, M. and Hundloe, T. (2005) A flexible and practical framework for reporting on ecologically sustainable development for wild capture fisheries. *Fisheries Research* 71, 175–183.

Garcia, S.M. and Staples, D.J. (2000) Sustainability reference systems and indicators for responsible marine capture fisheries: a review of concepts and elements for a set of guidelines. *Marine and Freshwater Research* 51, 385–426.

Kangas, M., McCrea, J., Fletcher, W., Sporer, E. and Slowik, V. (2006) Exmouth Gulf Prawn Fishery. *ESD Report Series*. No 1. Department of Fisheries, Western Australia, 128 pp.

Langley, A.D. and Hampton, J. (2006) An update of the stock assessment for South Pacific albacore tuna including an investigation of the sensitivity to key biological parameters included in the model WCPFC-SC2-2006 WP, 4–12 pp.

Marshall, J.A. (2006) Environmental management system framework. Tasmanian Farmed Oyster Industry. Compliance guide and risk assessment for ecologically sustainable development. Report for FRDC Project 2004/096.

Scandol, J.P., Holloway, M.G., Gibbs, P.J. and Astles, K.L. (2005) Ecosystem-based fisheries management: an Australian perspective. *Aquatic Living Resources* 18, 261–273.

Standards Australia (SA) (2004) Risk Management Guidelines: Companion to AS/NZS 4360: 2004. Standards Australia. Homebush, New South Wales, 116 pp.

9 Benefits and Costs of Implementing the Ecosystem Approach to Fisheries

ANTHONY CHARLES[a] AND CASSANDRA DE YOUNG[b]

[a]*Saint Mary's University, Halifax, Nova Scotia, Canada;* [b]*FAO, Rome, Italy*

Abstract

The ecosystem approach to fisheries (EAF) has been advocated widely, on the basis of its potential to meet a variety of goals, notably ecosystem health, sustainable resource use and human well-being. However, choices arise in implementation of EAF management, and each choice will produce benefits and costs. Benefits may be in terms of greater protection for a threatened species, greater long-term stability in food supply for a local community, reduced wastage or many other possibilities. Costs could include the direct costs of implementation (e.g. increased management costs) as well as the indirect or induced costs resulting from *how* the EAF is implemented (e.g. reduced employment and revenues in the short term).This chapter emphasizes the importance of a comprehensive assessment of these benefits and costs arising in EAF implementation, in order to improve decision making. The chapter undertakes a preliminary exploration of approaches and issues involved in such an assessment, examining four major aspects: (i) requirements, components and tools of EAF implementation having likely benefit and cost impacts; (ii) distributional implications, i.e. to whom the benefits and costs accrue, among stakeholders, inter-temporally and across spatial or administrative scales; (iii) compilation of potential EAF-related benefits and costs, grouped into ecological, economic, social and management categories; and (iv) the feasibility of various methodologies for assessing the benefits and costs of EAF implementation. The chapter also emphasizes the need for further research and analysis to develop the frameworks required for efficiently assessing the benefits and costs of EAF implementation in practical situations.

Introduction

The ecosystem approach to fisheries (EAF) is an approach to management that 'strives to balance diverse societal objectives, by taking account of the knowledge and uncertainties of biotic, abiotic and human components of ecosystems and their interactions and applying an integrated approach to fisheries' (FAO, 2003). The EAF is advocated based on its potential to help meet a broad range of goals, relating to aquatic ecosystem health, sustainable natural resource use and human well-being, among others (FAO, 2003, 2005; Garcia *et al.*, 2003).

Within these general goals are more specific objectives - both ecological (e.g. biodiversity conservation and benthic quality) and human-oriented (e.g. employment and income generation as a result of rehabilitated ecosystems, reduction in the risk of fishery collapses and aesthetic benefits). Many objectives of the EAF can be related to the various ecosystem values - both use values (those arising through direct use of the ecosystem) and non-use values (those based on the nature of the ecosystem rather than on its exploitation). Some of these are as follows:

Use values:

- Net economic benefits of fishing, including income and employment.
- Food provision and food security benefits.
- Non-fishing use values that arise from fisheries ecosystems, e.g. tourism, aquaculture.
- Values of fisheries ecosystems as mechanisms for social interaction and livelihoods.

Non-use values:

- Cultural benefits of fisheries ecosystems (e.g. for artistic expression or ceremonies).
- Aesthetic and existence benefits (e.g. the value of watching a sunset by the sea, or of knowing that whales are swimming in the sea).
- Option value (i.e. the value of maintaining fisheries ecosystems in terms of possible future benefits that might be realized as a result).

The potential of the EAF to achieve various goals and objectives, and in particular to enhance a range of ecosystem values, makes the desirability of the approach persuasive. However, implementation of EAF management involves significant choices, with respect to the issues to be tackled, the tools to be used and the scope of the overall management framework. Thus, *how* the EAF is implemented in practice can vary greatly from case to case, and this in turn will determine the resulting benefits and costs. As with any management intervention, EAF will have its costs as well as its benefits (Charles, 2001).

For example, a particular scenario of EAF implementation may increase some of the above-mentioned use and non-use values, but may decrease others. It could lead to an increase in a specific ecosystem value in the long term (thus representing a 'benefit'), but that same value could decline in the short term (reflecting a cost). Indeed, just as there are many potential benefits of implementing the EAF (in terms of achieving desired goals and objectives), so too are there many potential costs, depending on *how* EAF management is implemented - from the direct costs of implementation (e.g. increased management costs) to possible indirect or induced costs (e.g. the possibility of reduced employment and revenues in the short term).

Since the manner by which an ecosystem approach is applied, to address a given fishery system or a specific issue, will largely determine the resulting benefits and costs, there may be options with higher or lower ratios of benefits to costs, and differing distributions of those benefits and costs. Therefore, in any implementation of EAF management, it will be useful to fully assess, *a priori*, the possible benefits and costs that may arise. This will help to provide an understanding of the potential impacts of EAF management, so as to: (i) determine the most efficient means to address the issues of concern within an ecosystem context; and (ii) assess the distributional impacts, i.e. the extent to which differences in benefits and costs arise among the individuals and groups involved, so appropriate mechanisms can be determined for redistributing those benefits and costs, or compensating for their differing impacts.

To this end, this chapter discusses some perspectives on the benefits and costs of EAF implementation, exploring: (i) EAF interventions as a source of benefits and costs; (ii) the distributional impacts of the EAF; (iii) possible benefits and costs of EAF implementation, organized within ecological, economic, social and management categories; and (iv) feasibility of the various methodologies for measurement and analysis of EAF benefits and costs.[1]

EAF as a Source of Benefits and Costs

The benefits and costs that could arise in applying the EAF will depend, as noted earlier, on the manner of its implementation, particularly how the relevant considerations and challenges are dealt with, and which management tools are utilized. Indeed, every choice made and every step taken in implementing the EAF will have corresponding benefits and costs. This section lists a sampling of EAF measures and methods that have been proposed, drawing directly on FAO's EAF guidelines (FAO, 2003), and focusing on those that are particularly likely to have significant benefit and cost implications. While each item in the listings may or may not be relevant to a given fishery or ecosystem under discussion, the set as a whole provides an idea of the diversity of actions within EAF implementation needing assessment with respect to benefits and costs.

Table 9.1 lists a range of possible requirements and/or actions to be faced in implementing EAF management, drawn verbatim from FAO's (2003) EAF Guidelines (sections 1.4 and 4.2). Each entry in Table 9.1 is an aspect that would seem to have likely benefit and/or cost implications, but which would need to be examined on a case-by-case basis. For example, under 'technological considerations' (section 1.4.3), the Guidelines state: 'The impact of some fishing gear and methods on the bottom habitat (biotic and abiotic) can often have a negative effect on the ecosystem . . . the introduction of restrictions may be necessary and, where possible, new technologies that mitigate any negative impact

[1] Note that while cost–benefit analysis is one of several potentially useful methods in assessing EAF benefits and costs, this chapter deals more broadly with such benefits and costs, and does not focus on that method per se. Indeed, the principal grouping of benefits and costs used here involves ecological, economic, social and management categories, as opposed to the common approach in cost–benefit analysis of differentiating between 'economic' benefits and costs (of a societal or global nature) and 'financial' ones (from a private perspective).

Table 9.1. Aspects of EAF implementation having benefit and/or cost implications, drawn from the EAF Guidelines. (From FAO, 2003; specific document section as shown.)

The fisheries management process (1.4.1)

... recognizing the broader economic and social interests of stakeholders under EAF, the setting of economic and social objectives will need a broader consideration of ecological values and constraints than is currently the case. This will require a broader stakeholder base, increased participation and improved linkages of fisheries management with coastal/ ocean planning and integrated coastal zone management activities ...

Biological and environmental concepts and constraints (1.4.2)

... to be able to implement EAF at an operational level, delineation of the 'boundaries' is required and can be achieved by a sensible consensus based on proposed EAF objectives.

Technological considerations (1.4.3)

The impact of some fishing gear and methods on the bottom habitat (biotic and abiotic) can often have a negative effect on the ecosystem ... the introduction of restrictions may be necessary and, where possible, new technologies that mitigate any negative impact will need to be developed.

Many ecosystems, especially those in coastal waters, are impacted not only by fisheries, but also by other users, including upstream land-based activities.

Social and economic dimensions (1.4.4)

... as the overarching goal of EAF is to implement sustainable development, the shift to EAF will entail the recognition of the wider economic, social and cultural benefits that can be derived from fisheries resources and the ecosystems in which they occur.

The consideration of a broader range of ecosystem goods and services necessarily implies the need of addressing a wider range of trade-offs between uses, non-uses and user groups.

Institutional concepts and functions (1.4.5)

An effective ecosystem approach will depend on better institutional coordination (e.g. between ministries).

A greater emphasis on planning at a range of geographical levels that involves all relevant stakeholders will be required ...

The challenge to implement improved fisheries management ... may be particularly formidable in small-scale fisheries, where the difficulty and costs of the transition to effective management may outweigh the available capacity and short-term economic benefits derived from it.

Legal (4.2.1)

EAF is ... likely to require more complex sets of rules or regulations that recognize the impacts of fisheries on other sectors and the impact of those sectors on fisheries.

Institutional (4.2.2)

A major problem in EAF development may stem from disparities between the ecosystem and jurisdictional boundaries and these disparities will need to be addressed ...

The number of conflicts will inevitably increase under EAF as the number of stakeholders and objectives increase.

Under EAF, it must be recognized that the access rights system will frequently need to encompass other uses in addition to the use of the target resources.

Continued

Table 9.1. Continued

Educating and informing stakeholders (4.2.3)

Successful implementation of EAF will require that stakeholders (including management agencies) understand and accept the need for this more inclusive approach to fisheries management, and management agencies should actively promote such understanding and acceptance.

. . . scientists and management authorities need to appreciate and use the knowledge of fishers themselves about the ecosystem, along with that of their representatives and communities.

Effective administrative structure (4.2.4)

Administrative structures under EAF . . . will have to be better integrated with more effective roles in auditing or oversight.

will need to be developed.' This suggests a specific type of EAF intervention, one which, if implemented, will certainly have benefit and cost considerations.

Similarly, the EAF Guidelines (FAO, 2003: section 6) describe threats and challenges likely to be faced in implementing the EAF. Some of these are overarching concerns (e.g. equity issues and poverty), while others deal with the relatively complex nature of the 'people processes' in EAF management (e.g. problems in reconciling competing objectives of the multiple stakeholders, insufficient or ineffective participation of stakeholders and the time and cost required for effective consultation with a wide range of stakeholders). Challenges arise due to the inherently more complex nature of the EAF (e.g. insufficient knowledge, biological and ecological uncertainty and problems arising in aligning the boundaries of the ecosystem and the jurisdiction of management authorities). There are also resource issues in terms of a lack of adequate capacity to handle and analyse available information, inadequate monitoring and insufficient education and awareness. The key point here is that dealing with each of these threats and challenges, or indeed a failure to do so, will be accompanied by various benefits and costs, the specific level of which will depend on the context at hand.

The specific fisheries management tools that might be chosen within EAF management also have benefit/cost implications. FAO (2003, section 3.2) lists management tools of widespread use in fisheries and thus of relevance to EAF implementation. Technical measures discussed there include: (i) gear modifications that improve selectivity; (ii) size selectivity of target species; (iii) non-target species selectivity; (iv) spatial and temporal limits on fishing; (v) control of the impact from fishing gear on habitats; and (vi) energy efficiency. Input (effort) and output (catch) controls include limits on overall fishing mortality, capacity limitation, effort limitation and catch limits. Ecosystem manipulation can occur through habitat modifications, prevention of habitat degradation, providing additional habitat, population manipulation and restocking and stock enhancement. Finally, the Guidelines note that there are various rights-based management approaches that can be used. Use of any one of these 'tools' can be expected to have corresponding benefits and costs, and suitable assessment of these would be needed prior to use of the management method.

The EAF Guidelines do, in some cases, describe specific management options available to deal with EAF-related issues. For example, continuing the example earlier in this section of dealing with the negative impacts of fishing gear on the ocean bottom and its 'biotic and abiotic habitats', the Guidelines (section 3.2) note several options: use of towed gear with reduced bottom contact is a technical option in such areas. Prohibition of certain gear in some habitats is another, e.g. trawling in coral reef and seagrass areas. A third option is to replace a high-impact fishing method with one with less impact on the bottom, e.g. trapping, longlining or gillnetting. The choice among these different options clearly illustrates a need to assess the benefits and costs of each proposed option, prior to choices being made between them.

EAF and Distributional Impacts

In implementing EAF management, as important as it is to assess the aggregate benefits and costs involved, equally crucial is the matter of who receives the benefits and who incurs the costs of implementation (Charles, 2001; Mathew, 2003). This is a question of assessing distributional impacts - indeed essentially every possible benefit or cost has distributional issues. Three major types of distributional impacts can be noted: (i) across stakeholder groups at a given point in time; (ii) across time; and (iii) across scales.

The most commonly discussed form of distributional impact is that reflected in differences among the various stakeholders (and others in the fisheries ecosystem) in terms of the benefits and costs each receives *at a given point in time*. Specifically, some individuals or groups may receive immediate benefits from EAF implementation, while others may be disadvantaged. An example of this might be the initiation of a no-take marine-protected area (MPA) in a manner that improves catch levels for those located outside the MPA, but either excludes or increases the costs of those who historically fished within that area.

A second form of distributional impact is *inter-temporal* in nature, involving variations in when the various benefits and costs occur. Any particular benefit or cost that arises in the course of EAF implementation may do so over a range of timescales in the evolution of the fishery. For example, some potential benefits may be realized over a longer time frame (e.g. depending on the rate of recuperation of ecosystem health over time), while some costs of implementation may arise in the short term. Inter-temporal distributions of benefits and costs may affect or constrain the time frame within which EAF management can be implemented, when placed in the context of certain critical realities (e.g. annual food supply considerations and electoral time frames).

A third consideration is the *scale* at which benefits and costs occur. Potential benefits and costs may occur over a wide range of spatial, geographical or administrative scales (e.g. local, national and international). There may, for example, be a benefit that is international in scale (e.g. increased existence value of conserved biodiversity) and a corresponding cost that is local in scale (such as negative impacts on fishers in a specific community affected by a fishery closure). Even within a given fishery ecosystem, the migration of fish and/or larvae may lead to situations in which those incurring the costs of conserving resources or habitats may not be those receiving the benefits (or may be sharing the benefits with others who are not incurring costs).

These various distributional impacts underlie not only the assessment of the human-oriented benefits and costs involved in the EAF, but also the key challenge of implementing EAF management, in terms of having the approach accepted by stakeholders.

Envisioning EAF Benefits and Costs

The preceding sections have described how each EAF intervention is a source of benefits and costs, and in particular will have varying distributional impacts among the affected parties, and across space and time. But what specific benefits and costs may arise in EAF implementation? While recognizing that these benefits and costs will vary according to the context of the fishery and ecosystem, this section seeks to compile a sample of potential benefits and costs that could arise in practice. This includes both those that have actually occurred previously and those that seem logically possible to occur in real-world fishery situations, particularly building on the ideas discussed earlier in the chapter.

Table 9.2 displays a range of possible benefits and costs, grouped into a logical ordering of four sets, within ecological, economic, social and management categories. The possibilities shown are but a sampling – not necessarily the most prominent or the most likely benefits and costs. Some may occur in a wide range of fisheries, while others may arise only rarely, and some may tend to be large-magnitude in impact, while others may typically be more minor in nature. Indeed, since the occurrence of a particular benefit or cost in practice will depend on the specifics of the fishery and of EAF implementation, therefore for any given situation, each benefit or cost will have a certain probability of being realized. In the case of costs, the product of this probability and the magnitude of the cost represent the 'risk' associated with that cost.

The approach of Table 9.2 – a 'thought process' to compile possible benefits and costs of EAF implementation within ecological, economic, social and management categories – can also be applied to consider specific management options. For example, consider the option noted above of implementing MPAs as a tool of EAF management. What are the benefits and costs involved?

Potential ecological benefits could include the replenishment of fish stocks, enhancement of biodiversity and protection of spawning or juvenile fish. Social and economic benefits may arise through increases in the value of particular consumptive and non-consumptive uses within the MPA, and various non-use, existence and option/hedging values. Management benefits may come from more efficient and/or less expensive monitoring, and greater local compliance in the case of community-based MPAs. On the other hand, there will be various costs in establishing and operating the MPA. For example, if fishers are displaced from their usual fishing grounds as a result of the MPA, economic costs may arise in terms of the extra costs of travelling further to reach their new fishing grounds, as well as the reduced time available for fishing, due to the increased

Table 9.2. Some possible benefits and costs of EAF implementation.

Ecological benefits and costs	
Potential benefits	Potential costs
• Healthier ecosystems (directly or with EAF linkages to effective integrated management) • Increased production of goods and services from aquatic ecosystems • Improved fish stock abundance (due to healthier ecosystems) • Reduced impact on threatened/ endangered species • Reduced by-catch of turtles, marine mammals, etc. • Less habitat damage (due to more attention to fishing impacts on ecosystems) • Lower risk of stock or ecosystem collapse • Reduced contribution of fisheries to climate change (if EAF leads to lower fuel usage) • Improved understanding of aquatic systems	• Decreased fish stocks (if funding for fishery management is reduced through EAF, thus making it less effective) • Loss of genetic biodiversity if shift in fishing effort to unprotected areas occurs (if EAF is reliant on marine-protected areas) • Greater highgrading/dumping, and thus more wastage (if incentives created through catch quotas) • Reduced fish catches (if greater abundance of predators, e.g. seabirds or seals, due to better protection)
Economic benefits and costs	
Potential benefits	Potential costs
• Increase in benefits to fishers per fish caught (i.e. bigger fish from a healthier ecosystem) • Increased catches (especially in long term) • Increased contribution of fishery to the overall economy (especially long term) • Reduced fishing costs (if EAF results in reduction in unwanted by-catch) • Increased net economic returns (if EAF reduces fishing effort toward MEY*) • Higher-value fishery (if increased availability of food to top predators increases stock sizes) • Greater livelihood opportunities for fishers (e.g. in tourism, if charismatic species abundances increase through EAF) • Increased non-use (e.g. cultural) and existence values (e.g. from an appreciation of healthier aquatic systems and an increased abundance of aquatic life, etc.)	• Reduced catches (especially in short term, to rebuild stocks and ecosystems) • Loss of income to negatively affected fishers • Increased income disparity among fishers (if EAF impacts occur unevenly) • Reduction of government revenues from licenses, etc. (if EAF leads to reduced effort and/or catch levels) • Reduction in societal benefits accruing to fishers (if less government support for them) • Reduced contribution to the economy (in the short term, due to reduced fishing activity) • Reduced employment, in the short term and possibly the long term

* MEY, maximum economic yield

Continued

Table 9.2. Continued

Social benefits and costs	
Potential benefits	Potential costs
• Positive impacts on food supply in the long term (if greater catches become possible) • Synergistic positive effect of coordinated EAF across fisheries and/or nations • Greater resilience (if EAF emphasizes multiple sources of livelihoods) • Greater resilience (if EAF implementation increases livelihood opportunities) • Reduced conflict (if EAF processes deal effectively with inter-fishery and multi-sectoral issues)	• Negative impacts on food supply in the short term (and risk of this also in the long term) • Greater inequity (if EAF favours those able to invest in appropriate technology) • Greater inequity (if there is misplaced allocation of responsibility for EAF costs) • Increased poverty among those adversely affected by EAF (short term, or both) • Reduced benefits to fishers (if trade-offs within EAF are detrimental to fishers) • Greater conflict (if EAF leads to enforced interaction among a larger set of societal and/or economic players)
Management benefits and costs	
Potential benefits	Potential costs
• Better integration in management across fisheries, and with other aquatic uses • Clear expression of management objectives, leading to more efficient achievement of societal benefits • Better balancing of multiple objectives (due to a broadening of management attention) • Better balancing of multiple uses, leading to increased net societal benefits • More robust management due to broadening from conventional single-species tools to more integrated management approaches • Improved compliance due to more 'buy-in' to management, through better participation	• Increased cost of management • Increased cost of research • Increased cost of data collection/management • Increased cost of coordination across fisheries and other aquatic uses • Increased cost of additional and more participatory meetings • Increased cost of monitoring, observers, etc. • Increased risk of non-compliance (if regulations too complex or unacceptable) • Increased risk of collapse of management system (if too demanding of resources) • Poor management results and loss of support (if EAF imposed or implemented improperly)

travel times. Similarly, social costs could include the impacts of crowding on fishing grounds immediately outside the MPA, and opposition of local people to the displaced fishers entering the 'new' area. Management costs will also be incurred in operating the MPA, e.g. monitoring costs.

In considering an MPA (or any other management intervention), there is a need to assess these various benefits and costs – whether ecological, economic, social or management in nature. Furthermore, all of these will likely be accompanied by distributional impacts, as discussed above, whether across stakeholder groups at a given point in time, across time or across scales. For example, in implementing an MPA, some stakeholders may be directly and immediately impacted if their access to traditional fishing locations is restricted, while others may see no negative impacts, but instead gain the benefits of increased fish stocks and catches as the MPA improves ecosystem health. Such distributional effects of EAF measures need to be assessed so that suitable redistribution or compensation measures can be taken.

Assessing Benefits and Costs in Practice

As has been noted, of the range of potential benefits and costs that could arise in EAF implementation, those that actually arise in practice will depend on the specific fishery situation. Supposing that such a specific case is under examination, and a particular set of benefits and costs has been identified, there remains the fundamental challenge of assessing and analysing those benefits and costs.

While a varied set of methodologies is available to assist in this task, the feasibility of their use will again be context-specific. Consider, for example, methods for assessing social, economic and management-oriented benefits and costs, relating to human aspects of EAF implementation. Possible methods for assessment come from a wide range of disciplinary and interdisciplinary perspectives, and include: (i) direct fisheries measures; (ii) direct governmental accounting; (iii) socio-economic surveys; (iv) social impact assessment; (v) indicator frameworks; (vi) contingent valuation and travel cost methods; (vii) attitudinal and stated preference surveys; (viii) bioeconomic models; (ix) structured or semi-structured interviews; (x) asset mapping; and (xi) national systems of accounts. These methods are further discussed in De Young *et al.* (2008), along with references to relevant literature. Note that some of these methods have a relatively narrow focus, while others are broader in nature (e.g. bioeconomic models, national systems of accounts and indicator frameworks).

In EAF implementation, faced with the need to assess benefits and costs, the choice among the above methods depends on the feasibility of their use. This in turn depends on the specific fishery context, the nature of the method itself and the type of benefits and costs being assessed. Indeed on the latter point, the feasibility of assessing benefits and costs can vary significantly across the different categories of benefits and costs listed in Table 9.2.

Consider, for example, the feasibility of assessing economic, social and management benefits and costs (i.e. human-oriented ones), particularly with regard to issues of data availability.

First, for *Management Benefits and Costs*, most of the benefits listed (e.g. better integration in management across fisheries, clearer expression of management objectives and better balancing of multiple objectives and of multiple uses) seem to be difficult to assess objectively, although some (e.g. improved compliance through better participation) can be assessed based on available data (such as infractions reports). Several of the costs listed (e.g. for management, research, data handling, monitoring and observers) are readily assessed from governmental sources of information, possibly supplemented by suitable surveys, while other costs (such as the risks of non-compliance or of collapse of the management system) are difficult to measure objectively, but may be amenable to modelling, or to surveys and interviewing.

Economic Benefits and Costs seem, to a certain extent, more straightforward to measure. First, some are amenable to standard fishery data gathering - e.g. for benefits or costs at the fisher level (such as changes in income per fish caught, catches and fishing costs), and for those at the sector level (e.g. changes in employment, net economic returns and contribution to the economy). Some measures can be obtained through governmental accounting systems (such as changes in the revenues from licenses, etc.). Still others may require more specific data collection, e.g. through surveys to assess changes in livelihood opportunities for fishers, or in income disparity among fishers. On the other hand, a set of economic benefits and costs that is considerably more difficult to assess involves the non-market and/or non-use values that are increasingly recognized as crucial components of any such assessment.

The *Social Benefits and Costs* in Table 9.2 would seem to be generally very challenging to assess. This is certainly the case, for example, with changes in management efficiency and overall resilience of the human system. Some measures - such as effects on the food supply, poverty levels, levels of inequity and conflict - can draw on objective methods for measurement, typically through appropriate surveys. However, in general, problems with data availability and with implementing practical mechanisms for data collection are likely to pose difficulties in assessing many social benefits and costs.

As noted earlier, all the benefits and costs discussed above - whether social, economic, management or ecological in nature - will probably have distributional impacts, with the benefits accruing more to some fishery participants and the costs more to others. In considering the assessment of EAF benefits and costs in practice, it is crucial to take into account how they affect the various fishery sectors differentially. Such considerations are included implicitly in the above discussion of social benefits and costs - in measuring levels of inequity - but more broadly, the need to assess distributional aspects of implementing the EAF will arise for all the benefits and costs identified. Each of the methods listed can be utilized to provide information on such matters, although it is probable that information needs will be increased considerably when we seek to assess not only the aggregate level of a benefit or cost, but also its break-down across jurisdictions, fishery sectors or even individuals.

Conclusions

While the EAF carries with it some fairly universal attributes - such as a more consistent focus on aquatic ecosystems, a broadening of the scope of fisheries management, and improved coordination between fisheries and other sectors - nevertheless *how* the EAF is implemented in practice varies greatly from case to case. Implementation of EAF management involves significant choices, with respect to the issues to be tackled, the tools to be used and the scope of the overall management framework.

This chapter has sought to focus attention on the need to understand the benefits and costs involved in each choice to be made as part of EAF implementation. First, the chapter explored the various elements and options in EAF implementation, as described in the EAF Technical Guidelines (FAO, 2003), that have underlying benefit and cost implications. Second, the chapter discussed the crucial role of distributional impacts – across participants, across time and across spatial scales – that can result from adoption of the EAF. Third, the variety of EAF-related benefits and costs was examined, as seen through ecological, economic, social and management lenses. Fourth, the feasibility of the various methodologies for measuring and analysing EAF benefits and costs was discussed, particularly in terms of data availability.

In focusing on EAF benefits and costs, this chapter has elaborated on one of the four major elements inherent in the human dimension of EAF implementation, as described by De Young *et al.* (2008). They note that social, economic and institutional factors can reflect:

1. *Driving forces* behind the need for EAF management.
2. *Benefits and costs* arising in the application of EAF.
3. *Instruments* facilitating the implementation of EAF.
4. *Supports or constraints* in the process of EAF implementation.

The driving forces, the supports and the constraints on EAF have all received some attention previously, and instruments facilitating the EAF (such as incentives and institutions) are the subject of much current work (Valdimarsson and Metzner, 2005; these elements, and references to other studies dealing with them, may be found in De Young *et al.*, 2008). At the same time, it seems clear that a focus on the benefits and costs of EAF implementation is equally relevant. Indeed, it is an understanding of these benefits and costs, together with their distributional implications, that underlies how the supports, constraints and instruments related to EAF implementation can be approached in practice.

Furthermore, there are implications of all this for institutional development in fisheries; as fishery management shifts to an EAF framework, consideration of the corresponding benefits and costs – and particularly the distributional aspects – will need to be accompanied by moves to ensure that those affected by management, whether as winners or losers, are involved in the relevant institutions. This may occur at a local scale, in terms of involving all players within a community-based management system, at a national (or sub-national) scale, in terms of the participation of various fishery sectors, or at an international scale, involving affected nations in a regional management context. Such participation is important from a perspective of ensuring fairness and transparency, as well as from the pragmatic perspective involved in building a consensus for EAF management.

Given these realities, it must be emphasized that the present chapter represents only a preliminary exploration of the benefits and costs relating to EAF implementation. There remains a real need to combine research and analysis with effective management mechanisms for decision making – the former to develop frameworks needed for efficiently assessing the relevant benefits and costs, and their distributional impacts, and the latter to take into account such benefits and costs in appropriately implementing the Ecosystem Approach to Fisheries.

Acknowledgements

This chapter was prepared originally as part of a background document for the FAO Expert Consultation on 'Social, Economic and Institutional Aspects in Implementing the Ecosystem Approach to Fisheries' (June, 2006). We are grateful to Kevern Cochrane and to participants in the Consultation for their helpful comments. Charles acknowledges financial assistance from the Natural Sciences and Engineering Research Council of Canada.

References

Charles, A. (2001) *Sustainable Fishery Systems*. Blackwell Science, Oxford, UK, 384 pp.

De Young, C., Charles, A. and Hjort, A. (2008) Human Dimensions of the Ecosystem Approach to Fisheries: an Overview of Context, Content, Tools and Methods. *FAO Fisheries Technical Paper*. No. 489. FAO, Rome, 152 pp.

FAO (2003) The ecosystem approach to fisheries. *FAO Technical Guidelines for Responsible Fisheries* No. 4 (Suppl. 2). Food and Agriculture Organization, Rome, Italy, 112 pp.

FAO (2005) Putting into practice the ecosystem approach to fisheries. FAO, Rome, Italy, 82 pp.

Garcia, S., Zerbi, A., Aliaume, C., Do Chi, T. and Lasserre, G. (2003) The ecosystem approach to fisheries. Issues, terminology, principles, institutional foundations, implementation and outlook. *FAO Fisheries Technical Paper*. No. 443. Food and Agriculture Organization, Rome, Italy, 71 pp.

Mathew, S. (2003) Small-scale fisheries perspectives on an ecosystem-based approach to fisheries management. In: Sinclair, M. and Valdimarsson, G. (eds) *Responsible Fisheries in the Marine Ecosystem*. CAB International, Wallingford, UK, pp. 47–53.

Valdimarsson, G. and Metzner, R. (2005) Aligning incentives for a successful ecosystem approach to fisheries management. *Marine Ecology Progress Series* 300, 241–296.

10 Creating Incentives for the Ecosystem Approach to Fisheries Management: a Portfolio of Approaches[1]

Cassandra de Young[a] and Anthony Charles[b]

[a]*FAO, Rome, Italy;* [b]*Saint Mary's University, Halifax, Nova Scotia, Canada*

Abstract

There is widespread agreement on the need to enlarge our fishery management 'toolbox', thereby increasing the range of creative measures available to suit the panoply of fisheries that managers face. This chapter attempts to broaden the toolbox discussion by focusing on a certain subset of management mechanisms - the use of incentives. A brief introduction by means of a problem statement and the possible role of incentives in addressing these issues is followed by a description and discussion of incentives as categorized into legal, institutional, economic and social incentives. In addition, concise examples are provided as starting points for further investigation. The ultimate goal of this chapter is to stimulate discussion regarding the appropriateness and desirability of including the use of incentives as part of the toolbox within EAF strategies towards sustainable development.

Introduction

There is widespread agreement on the need to enlarge our fishery management 'toolbox'; thereby increasing the range of creative measures available to suit the panoply of fisheries that managers face. The ecosystem approach to fisheries (EAF) management provides a vehicle to accomplish this, but a key challenge lies in determining the right tools to utilize in specific situations. The idea in choosing suitable management tools is not to rely on *a priori* judgments of the rightness or wrongness of any given tool, which would be inappropriate, but rather to develop and build on an understanding of the biological, sociological, economic and political context in question.

[1] The information contained in this chapter is primarily based on De Young *et al.* (2008). An earlier version of this chapter was presented at the 3rd Regional Workshop of the BCLME EAF Project, Cape Town, South Africa, 30 October–3 November 2006.

Discussion of the management 'toolbox' is influenced by current thinking in fisheries management, which is moving towards a set of understandings that include (but are not limited to) the following: (i) tools used in isolation have less of a chance of being effective than a mix of complementary tools used in tandem; (ii) although managing people is complicated, managing fish and ecosystems is even more difficult if not impossible; (iii) new tools are necessary to help us manage in light of our recognition of the uncertainties we face; (iv) if people are included in the management process and understand why it is in their interest to do something, chances of successful implementation of the resulting management are increased; and (v) something needs to change since, on the whole, fisheries management has been neither effective nor efficient.

This chapter attempts to broaden the toolbox discussion by presenting a certain subset of management mechanisms - the use of incentives. A brief introduction by means of a problem statement and the possible role of incentives in addressing these issues is followed by a description and discussion of incentives as categorized into legal, institutional, economic and social incentives. In addition, concise examples are provided as starting points for further investigation. The ultimate goal of this chapter is to stimulate discussion regarding the appropriateness and desirability of including the use of incentives as part of the toolbox within fisheries strategies towards sustainable development.

What Are Incentives Mechanisms and Why Do We Need Them in Fisheries Management?

An incentive in the broadest sense is any factor that affects an individual's choice of action - such factors could range from the price of an input (e.g. fuel) or final product (e.g. fish) to fines for breaking established rules, to social or peer pressure, and religious beliefs. In any given situation, incentives of various kinds will already be in place, but these may not induce the sort of individual decisions that society may desire. This leads to a need to create or introduce more appropriate incentives - a need that arises in particular in marketplaces when the market price of a product does not fully reflect the impacts (either positive or negative) of its production or consumption on society.

This need for appropriate incentives is related closely to the existence of externalities. Examples of negative externalities include nutrient runoff from farms into water bodies (an action taken by farms that impacts negatively on non-farming users of the water bodies) and carbon emissions from electricity production. Examples of positive externalities include those arising from education and health-care provision, which provide benefits beyond those specific sectors of activity. The lack of internalization of any of these costs and benefits by those choosing to produce or consume goods leads to socially suboptimal levels of such activities, i.e. too much of those producing negative externalities, and too little production of goods that provide positive externalities.

In capture fisheries, externalities have been classified into the following five categories (Seijo *et al.*, 1998)[2]:

- *Stock externalities.* The impacts of one fisher's activities on the availability of the target species for other fishers in the fishery (i.e. the activity of each fisher reduces the fish stock available to other fishers).
- *Crowding externalities.* The impacts of vessel aggregation in the fishing grounds on marginal catch costs in the fisheries (i.e. the presence of any one fishing vessel increases the level of 'crowding', thus increasing the costs of fishing for all vessels).
- *Technological externalities.* Similar to stock externalities, but relating to fishing gear impacts on population structures of by-catch species that are targeted species for other fisheries (so one fleet, targeting certain species, produces negative externalities for other fleets).
- *Ecologically based externalities.* Broadened concept of stock externalities that considers ecological interactions between various species targeted by different fisheries. This could include positive externalities (e.g. if one fishery harvests a species that is competing against the species targeted by another fishery), as well as negative externalities (e.g. if the species are part of the same food web, i.e. predator and prey).
- *Techno-ecological externalities.* The impacts of fishing practices/gear on the broader ecosystem (e.g. habitats and biodiversity).

Where such externalities are not managed, overcapacity, overfishing and welfare losses are the predicted results; thereby impacting the ability of fisheries to: (i) contribute to economic development, food security and poverty prevention/alleviation; and (ii) maintain the wide range of services provided by fisheries ecosystems (e.g. income and employment, social, religious and cultural identities, habitats and biodiversity regulation).

As suggested earlier, other externalities impacting ecosystem productivity stem from non-fisheries activities, such as agriculture and aquaculture nutrient runoff, marine transport pollution and tourism-related impacts. Depending on the level of integration within coastal and marine management systems, the fisheries sector may have varying degrees of influence on the management of non-fisheries externalities affecting fisheries ecosystems; however, it remains important that these links be identified and acknowledged as these external factors will certainly impact the effectiveness of any fisheries management system.

The first three categories of externalities mentioned earlier - stock, crowding and technological - are those which would fall under the narrow definition of conventional fisheries management; while a consideration of the final two categories, ecological and techno-ecological, as well as extra-fisheries externalities, would be a broadening of the management concept to the ecosystem approach to fisheries. With this broadening, the scale and scope of benefits and costs related to applying the EAF will also expand. For example, minimizing turtle mortalities due to fishing activities through gear, spatial and temporal

[2] The discussion focuses only on the physical capture of fish and other aquatic organisms.

adjustments may impose costs locally (i.e. to the fisherman), but create benefits globally (i.e. to those holding values for biodiversity). In addition, while the costs associated with EAF implementation are borne often in the short term, benefits accrual may take place quite far in the planning horizon.

Hence, correcting for externalities is one of the major challenges of implementing the EAF. This brings the focus back to incentives, as it implies a need for additional measures of various sorts to induce fishery participants (and others) to change behaviour in keeping with the EAF. Such measures supporting positive behavioural change could be social, economic, legal or institutional in nature; all of which involve the use of 'incentives' towards behavioural change, i.e. considerations that an individual will factor into their decision making and which will lead to a result more in keeping with desired societal directions (in this case, effective implementation of EAF). From an economics perspective, one might view incentives as influencing the profit maximization of a fishery participant (i.e. increasing profits as a result of EAF-compatible actions, and conversely reducing profits for actions contrary to EAF objectives). From a sociologist's perspective, incentives might be social constraints on behaviour (e.g. resulting from peer pressure and cultural institutions) that lead to more desirable outcomes.

This chapter will present and discuss social, economic, legal and institutional incentives in support of the implementation of EAF, as understood as follows:

- *Legal incentives*. Effective legislation creating positive 'carrots' as well as 'sticks' in the form of significant penalty structures with effective enforcement capability.
- *Institutional incentives*. Fisheries management systems and participatory governance arrangements that induce support from stakeholders.
- *Economic/market-based incentives*. Win-win measures that lead to outcomes that are better for both the fisher and the fishery ecosystem, such as the use of some excluder devices in fishing gear, to increase profits by reducing fishing costs, broadening market access, while also reducing by-catch.
- *Social incentives*. Community-based institutions and social environments that create peer pressure on individuals to comply with agreed-upon community rules.

It is clear that incentives can take many forms - some being of quite general applicability, and others being very specific to particular circumstances. Similar to technical management tools, such as spatial restrictions and catch limits, no single incentive will be a panacea for management - a mix of incentives measures that are appropriate to the fisheries and their socio-cultural settings will minimize unintended consequences and increase the likelihood of effective EAF management. The following sections will provide an array of incentive measures within fisheries (generally applicable to EAF management as well as conventional fisheries management, if deemed different), as well as a final section discussing extra-fisheries mechanisms that are in use to improve ecosystem performance and encourage sustainable development.

Legal Incentives

Regulatory frameworks form the legal backbone of fisheries policies and management systems. These regulatory frameworks, which specify the requirements, rights and responsibilities placed on fishery users so as to meet desired policy goals (such as EAF objectives), are usually enunciated within the fisheries legislation. These might include the requirement to hold a fishing license, to undergo environmental and other impact assessments, to develop fleet-specific or local-level management plans or to use specified impact-minimizing gear. In addition, regulatory frameworks can provide the legal basis for EAF by, among other things:

- Setting property rights systems.
- Providing a framework for coordination and integration.
- Defining roles and responsibilities.
- Specifying international norms and requirements.
- Providing a framework for management processes.
- Providing legal mechanisms for conflict resolution.
- Describing the penalty structures for violations of rules and laws.
- Providing for monitoring and control systems.

Such legal backing provides credibility and clarity to management systems, and hence provides incentives for compliance. In addition to direct incentive-promoting content within legislation, certain characteristics of regulatory frameworks would contribute to promoting positive change, including: (i) being flexible and responsive to various changes, for example, to changes in the knowledge base, and biological, ecological and socio-economic changes; (ii) being stable enough to provide continuity; and (iii) being congruous - providing consistency between fisheries and other sectors and between local, national, regional and international regimes.

Institutional Incentives (Including Fishery Rights)

In moving from conventional fisheries management towards EAF management, some changes to current institutional frameworks are likely necessary in order to, *inter alia*, motivate stakeholder buy-in and participation in fisheries management. These changes will likely include providing ways of taking account of and dealing with the increased scope and demands of this management approach, including:

- A need for increased coordination, cooperation and communication within and among relevant institutions and resource users in the planning process as well as in implementation.
- A need for more information regarding the ecosystem and the factors affecting its health and productivity.
- A need for incorporation of uncertainties into the decision making process due to the increase of factors (predator–prey relationships, nearby activities,

such as agriculture, and their impact on the ecosystem, etc.) causing uncertainties.

- A need for ways of truly involving the broadened definition of stakeholders in decision making and management, such as capacity building and multi-directional information dissemination.

Although not incentive mechanisms per se, proper institutional arrangements may generate incentives to assist in the application of the EAF, such as buy-in, cooperation and reducing the race to fish. One commonly advocated institutional approach for creating incentives supportive of policy goals, such as the EAF, is that of rights-based approaches (i.e. assigning, or recognizing, rights over the use and management of a fishery). Two key elements of rights-based approaches are as follows (Charles, 2002).

Use rights – an institutional mechanism by which fishers, fisher organizations and/or fishing communities hold rights and some security of tenure over access to a fishing area, the use of an allowable set of inputs or the harvest of a quantity of fish. If use rights are well established, fishers will have greater security, as there will be increased clarity with respect to who can access the fishery resources and how much fishing each is allowed to do. This can encourage fishers to support conservation measures – since protecting 'the future' becomes more compatible with their own long-term interests. Examples of use rights include territorial use rights (TURFS), customary marine tenure (CMT) and individual quotas.

Management rights – the right to be involved in managing the fishery – reflects the need, as noted in the FAO Code of Conduct for Responsible Fisheries (FAO, 1995), to 'facilitate consultation and the effective participation of industry, fishworkers, environmental and other interested organizations in decision making with respect to the development of laws and policies related to fisheries management'. This has led notably to the emergence of co-management arrangements involving joint development of management measures by fishers, government, local communities and other stakeholders.

Through use rights and management rights, it is hoped that incentives will be improved, increasing the possibility that participants will: (i) adopt a longer-term perspective on the fishery, since their use rights are secure over a longer time frame; (ii) comply with management regulations, since they have been involved in developing those regulations within the management process; and (iii) engage in greater cooperation, since one's well-being may become more closely intertwined with that of others. Of course, introducing a rights system will have accompanying benefits and costs (and varying distributional impacts of each) and so there is a need to assess these aspects (as well as monitoring any negative impacts of the measures).

While use and management rights have been well discussed in the general fisheries literature, there are some specific considerations that need addressing with respect to EAF implementation. In particular, as EAF implies a broader scope of fisheries management (to include multiple species, the aquatic ecosystem, the range of societal objectives and any interactions with other economic sectors, among other aspects), use and management rights within such a context will need to deal with other 'users' of the ecosystem besides the specific stakeholders in the fishery being addressed. Other capture fisheries, recreational fisheries, aquaculture, offshore oil and mining activities, ecotourism and/or coastal tourism, shipping, urban development, coastal industries and other aquatic-based human endeavours all vie for resources and impact the ecosystem along with fisheries. Just as rights may be allocated to use specific fishery resources and to be involved in managing those resources, so too may there be rights arrangements for others - perhaps in the context of integrated coastal and ocean management, or integrated watershed management. While this goes beyond EAF per se, clearly it is a reality that must be taken into account, and which bears very much on the broader goal of ecosystem health and necessarily involves more than just those within the fishery.

In summary, the judicious recognition or adoption of use and management rights can help align incentives to the desired EAF policy, but this is not a simple task, and indeed taking the wrong approach can produce results contrary to the aims of EAF management. Thus, it is important to understand the relationship between rights and incentives, which will vary from case to case.

Example: Multi-stakeholder management rights. (From Pinkerton *et al.*, 2005 and http://www.westcoastaquatic.ca)

An example of broad-based ecosystem-level *management rights* is that of the regional aquatic management board established on the west coast of Vancouver Island, on the Pacific coast of Canada, which was formalized in 2001, as a multi-stakeholder institution for community-based co-management of aquatic-based resources. It is a forum for shared decision making, where coastal communities and others affected by aquatic resource management can work with governments on integrated management, on an ecosystem basis. The Board is made up by equal numbers from governments (federal, provincial, regional and native) and non-governmental representatives (various economic sectors, communities, etc.). Its operation is based on several principles: Shared Responsibility (all participants are jointly responsible and accountable), Inclusivity (all should have the opportunity to participate in management decisions) and Flexibility (structures and processes should be flexible and expected to evolve). Key objectives are: (i) to consolidate information relating to different aquatic resource uses and utilization; (ii) to integrate expertise and knowledge from all sources; and (iii) to ensure opportunities for coastal communities and others affected by the resource management to participate in integrated management, protection and restoration of aquatic resources.

Economic/Market-based Incentives

Economic incentive mechanisms that are created outside of existing markets are based on the idea of establishing a situation in which economic actors/agents are convinced that it is in their private interest to make the socially desirable choices. In this section, the discussion is separated into 'carrot' and 'stick' incentives categories - we refer below to 'economic incentives' as the 'carrots' (positive incentives) that promote desired behaviour, and 'economic disincentives' as the 'sticks' (negative incentives) that penalize undesirable behaviour.

Such categorization is artificially derived to reflect how the mechanisms affect the benefits and costs structures of the economic agents. Therefore, benefit- and cost-sharing mechanisms have been inserted into these pre-defined categories although neither clearly carrots nor sticks.

Economic incentives (the carrot)

The use of positive incentives may be split into three categories: conservation price differentials, best-practice/conservation payments and rights-based incentives. From an economic perspective, all of these seek to shift cost and revenue curves with the aim of attaining a level of fishing activity that is optimal from a societal perspective. In addition, positive economic incentive instruments would, in theory, allow actors to determine for themselves the least cost means of obtaining a given management objective.

Price differential payments occur when consumers demonstrate through the prices they pay the values they hold for ecosystem goods and services; such payments serve as market signals to industry and governments. For example, these payments may take the form of higher prices paid for 'ecolabeled' products, which establish a mechanism for identifying sustainably produced products and may relate to price premiums or export certificates. The impact on the international market has begun to make itself felt as large retailers pick up on the movement and, perhaps, the price differentiation.

Other attempts to affect consumer choices include fair-trade labels, good fish guides,[3] and fish fairs promoting artisanal and local products. Such instruments are geared towards the provision of information to consumers regarding the circumstances leading to the availability of the offered products (e.g. fishing practices, stock status/sustainability and fishery management regimes).

Example: Certification of red rock lobster, Baja California, Mexico. (From Marine Stewardship Council; http://www.msc.org)

In April 2004, the Marine Stewardship Council certified the red rock lobster fishery on the Pacific coast of Baja California, Mexico, as a sustainable fishery. The trap fishery 'is currently exploited by about 500 fishermen belonging to nine fishing co-operatives and spread over ten villages. Fishing legislation for the fishery was first drawn up in the 1940s as a result of which fishing rights were allocated to co-operatives. . . . Management involves a combination of limited entry, strict delineation of co-operatives fishing areas and community-based self-regulatory measures'. The fishery is heavily export-oriented, with 90% of the catch going to markets in Asia, France and the United States. There is thus a clear economic incentive for certification, which provides the potential for better global access to markets, and a higher market price (if a 'price differential' develops relative to non-certified lobster).

[3] These guides present lists of fish products ranked by some measure of biological sustainability and are usually focused on specific markets to assist consumers in their consumption choices of fish products commonly found in local markets or supermarkets.

Best-practice/conservation payments are transfers to the fishing industry, fishing communities and/or fishers directly, from governments, non-governmental organizations (NGOs) or other institutions, to compensate for some or all costs of implementing sustainable fishing practices. Such practices may include the use of best-available technologies (e.g. turtle exclusion devices (TEDs) and vehicle monitoring systems (VMSs)) or restrictions on fishing patterns (e.g. no-take zones or seasons and buyback programmes). These transfers may be considered as payments from those who benefit from conservation or best practices to those who bear the direct costs of their implementation. If the transfers are made by governments, they may be equivalent to environmentally positive subsidies, made on behalf of society as a whole, while if originating with NGOs, foundations, etc., such conservation payments would typically reflect the focus of those bodies.

One form of best-practice/conservation payments is competition to engage and reward the fishing industry in the design of fishery-specific technology. These can complement regulatory mechanisms (that involve the requirement of technological change), allowing industry participation through the design of the most appropriate and low-cost options – they have met with certain success as creativity from within the industry is rewarded and the process tends to increase acceptance of use (see box below).

Example: Win–win by-catch possibilities in Australia's northern prawn fisheries. (From Brewer *et al.*, 2006.)

With the required introduction of turtle exclusion devices (TED) and by-catch reduction devices (BRD) within the northern prawn fisheries in 2000, concerns regarding the economic impact of this conservation-based management tool were voiced within the fisheries. Therefore, industry and scientists worked together to assess these impacts and to improve locally used designs. While commercial prawn catches were decreased by approximately 4–6%, damage to prawns by heavy animals was decreased by over 40%, representing 1–3% of their catches; thereby increasing the catch value. Further reductions in commercial losses are expected with increased familiarization and fine-tuning of the devices. In addition, other benefits such as increased ease in handling and sorting and reduced danger to the crew were associated with the exclusion of larger animals. Large by-catch reductions were identified for sea turtles, sharks, rays and large sponges; while by-catches of sea snakes and small by-catch left much room for improvement.

Conservation payments can occur when the non-use/existence benefits of certain resources are higher than the extractive-use benefits. In these cases, the opportunity costs of not using the resources need to be compensated, either through direct or indirect transfers. This is especially important in small-scale fisheries that depend on the extractive uses for their livelihoods. Direct payments have been used by specific conservation projects in which fishing communities are paid to maintain a given habitat or not to use a resource. Unfortunately, such payments are usually linked to the longevity of the given project and, therefore, once the project has finished, so have the conservation payments. Ferraro and Kiss (2002) present a review of current debates regarding direct payments to conserve biodiversity.

Other conservation payments have focused on indirect transfers, particularly on training or other livelihood diversification methods based on the thesis that reducing fishing communities' vulnerability will naturally increase their ability to sustainably use and manage fisheries resources (SFLP, 2006).

Another market-based conservation mechanism is that of ecotourism development, involving shifts from extractive uses of resources to non-extractive uses. Essentially, the idea is that payments from tourism compensate for lost fishing revenues and may provide for alternative or diversified livelihood sources. However, as is often the case for substitutes, negative impacts on ecosystems may occur (e.g. pollution, crowding and noise from boats and divers); thereby warranting caution in their use. In addition, the demand for ecotourism may not be sufficient and stable enough to guarantee conservation of habitats and commercial species and this demand may only pertain to highly valued species, such as sharks, whales and turtles.

Example: Valuation of whale sharks. (From Graham, 2004.)

Graham (2004) studied this matter, comparing the Taiwan market price for a (dead) whale shark – between US$7,116 and US$21,400 – with the (live) value derived from tourism estimates. She notes that 'Using the 2002 Belize whale shark tourism survey results, each shark is worth at least US$34,906 annually. A similar annual value of US$33,500 for each grey reef shark . . . was recorded in the Maldives. If whale sharks live to at least 60 years old, then an individual might be worth US$2,094,340 over its lifetime providing it repeatedly visits the tourism site.' She concludes that 'the economic argument for protecting whale sharks is undeniable'. This in turn implies the potential for conservation payments to encourage such practices.

Rights-based incentives are the third form of 'carrot' – such incentives, as discussed earlier, typically address the implementation of user rights within a fishery; thus removing, to a greater or lesser extent, the condition of open access and providing an incentive for long-term sustainability. With an effective mix of user rights, the remaining fishing actors may be able to maximize the net present value of the resources, if any future streams of benefits and costs are either integrated into the price/value of the use right (e.g. a permit or a quota), or into the choices made by the 'owners' of the resources. For a review of experiences in the use of property rights and the implementation of quota rights in fisheries management, see FAO (2000a, 2001). FAO (1982) provides a discussion on the conditions affecting the successful creation and maintenance of TURFs. Examples of TURFs may be found across the globe (e.g. Argentina, Chile, Japan, Peru, Philippines, South Africa, the USA and Vanuatu) and have usually, but not always, developed in situations with historic roots of community-based management of natural resources.

> **Example:** Territorial Use Rights in Fisheries (TURFs). (From Castilla and Defeo, 2001.)
>
> When fishers hold use rights, there is more secure access to the fishery, and potentially greater incentives for compliance with management – particularly when there are accompanying management rights. Castilla and Defeo (2001), in a review of management practices in Latin American shellfish fisheries, examined the role of TURFs, concluding that the examples studied 'illustrate the strong potential that the apportionment of TURFs has, when accompanied by a co-management approach. In Chile, the allocation of TURFs among communities that extract benthic shellfish is an efficient tool to cope with overexploitation concerns. . . . Allocation of TURFs to fisher organizations ameliorated the weaknesses of enforcement regulations and the high transaction costs in a country with more than 4,200 km of coastline . . . [and] improved the status of shellfisheries. . . . The formal allocation of TURFs to fisher organizations such as the collectively managed spiny lobster fishery of Punta Allen (Mexico) constitutes another sound example'.

Economic disincentives (the stick)

Economic disincentives within an EAF context mirror the polluter-pays principle (PPP) and the user-pays principle (UPP) used in the allocation of costs of pollution prevention and control measures and sustainable development paradigms.[4] These principles attempt to correct for existing market failures by internalizing into the production function the costs of using natural resources and of negative impacts on the ecosystem. Such principles have become standard policy in treating water, air and hazardous chemicals/waste issues; while their application to the fisheries sector has been slower to materialize. Coffey and Newcombe (2001) have provided a nice analysis of the current and potential use of PPP in European fisheries and the generalized results are presented below. In their work, one can identify several objectives for the use of economic instruments (e.g. taxes, charges and levies[5]) in line with the PPP/UPP: (i) cost recovery for fisheries management; (ii) paying for resource use; and (iii) paying for environmental damage prevention or alleviation.

Cost recovery for fisheries management, generally through taxes/levies, while not really a 'stick' incentive measure, will change the private profit functions of fishing activities and should instil a sense of ownership of the results of management as a direct link is made between the benefits of management

[4] The PPP means that the polluter bears the expenses related to any pollution prevention and control measures; meaning that these costs are reflected in the cost of goods and services, which cause pollution in production and/or consumption. The UPP is a variation on the polluter-pays principle that 'calls upon the user of a natural resource to bear the cost of running down natural capital' (UNSD, undated).

[5] The word taxes will be used interchangeably in the text for taxes, charges, fees and levies.

and their costs (Cox, 2000).[6] However, while explicit research on applying cost-recovery mechanisms elsewhere and implicit use within fisheries co-management regimes are occurring,[7] it must be noted that use of cost recovery mechanisms has, for the most part, been applied only within the OECD countries, and indeed in cases where revenues are collected from fisheries activities, more often than not these revenues go directly to the central government budget. In such cases, the link between benefits and costs of management services cannot be made and fisheries authorities continue to base their management activities on governmental appropriations.

Paying for resource use, often through license/access fees, taxes and tradable or auctioned quotas, is an acknowledgement within the fisheries sector of the value of natural resources, much as in the use of land, water or other natural resources. Historically, access to fisheries resources was free and all profits from the use of these resources were either dissipated, in the case of open access fisheries, or kept by the fishing industry. Governments and, hence, societies, had not insisted on payments for the use of these natural resources. However, with the onset of the Law of the Sea in 1982[8] and the idea of national ownership/stewardship of marine resources, the idea of private individuals paying society for the use of natural resources has gained ethical acceptance and jurisdictional backing.

The level of such payments would depend on the particular fishery and the economic concept of rent, which is the 'bonus' profit[9] from using a natural resource. In an open access fishery, there would be no rent to be had; so no rent extraction is possible. Moving from an open access fishery to a socially optimal fishery, would increase the rent value of the fishery to the private users and, by consequence, those remaining in the fishery would pay for this privilege.

Third country access agreements (i.e. foreign fleets paying for the right to fish in another country's EEZ (exclusive economic zones)) have been used in cases where national fishing fleets do not have the capability of exploiting certain stocks and could benefit from rent extraction through fees and taxes. These agreements have been wrought with criticisms (e.g. IEEP, 2003), but as information sharing,[10] experiences and monitoring

[6] The OECD has proposed further that including industry in the decisions about and in the provision and payment of management services is highly likely to create incentives to improve the fishery's performance and to increase the cost-effectiveness of management services (OECD, 2003).

[7] See, for example, Keizire (2001) analysis of the fisheries management financing in Uganda and the Asia-Pacific Fishery Commission (APFIC, 2005) work regarding the implementation of fisheries co-management.

[8] The United Nations Convention on the Law of the Sea of 10 December 1982. See http://www.un.org/Depts/los/convention_agreements/convention_overview_convention.htm

[9] 'In relation to fisheries, a 'rent' is generally thought of as the difference between total revenues obtained from the fishery and the total costs (estimated at their opportunity costs) of employing the various factors of production that together make up the enterprises participating in the fishery.' FAO (2000b).

[10] To this aim, WWF has created a 'Handbook for negotiating fishing access agreements' (Martin *et al.*, 2001).

capabilities[11] are increased, such agreements may benefit national economies while ensuring sustainable harvest levels.

Paying for ecosystem damage prevention or alleviation, either through bearing the costs of appropriate technology or through paying fines for damages inflicted, is probably the most politically palatable use of economic disincentives as it relates to a given 'bad' action. The fining of actions that have negatively affected ecosystems is quite common; however, these cases tend to involve actors outside of the fishing sector who have damaged habitat, such as through oil spills or dock building.[12] Fishing activities that have harmed the ecosystem (e.g. dynamite fishing, destructive anchoring, discarded by-catch and incidental fishing of marine mammals) tend to be controlled through regulations, such as no-take zones, gear restrictions and by-catch limits.

One economic disincentive in use with respect to harmful fishing activities is the use of trade barriers, such as the blocking of export permits under the Convention on International Trade in Endangered Species of Wild Flora and Fauna (CITES).[13] CITES was developed to minimize the effect of international trade on commercial species either threatened with extinction (Appendix I species) or exploited unsustainably (Appendix II species). Trade in Appendix I species is all but prohibited; while trade is permitted for Appendix II-listed species if the related fishing practices are proved sustainable (i.e. the species 'was legally obtained and if the export will not be detrimental to the survival of the species'). If the potential exporter is unable to prove the sustainability of the fishery, exportation rights are not granted; hence, representing a change in the burden of proof. In theory, sustainable management of fisheries should keep the commercial species from being placed on the CITES listing; however, such trade measures are an acknowledgement of the impacts of market forces on our ability to manage resources.

Example: CITES and queen conch in Jamaica. (From Cascorbi, 2004.)

The Convention on International Trade in Endangered Species of Wild Flora and Fauna (CITES) represents an international legal mechanism that has the effect of creating incentives for positive EAF-related behaviour, particularly through the provision of export permits. As Cascorbi (2004) reports:

'Until 1999, Jamaica was the world's largest producer of queen conch. Most of this was fished on the Pedro Banks, a large undersea area that is the habitat for one of the Caribbean's largest and most important queen conch stocks. In the early 1990s, Jamaica's landings of Pedro Banks conch topped 3,000 t/year. Jamaica also conducted its first conch stock assessments in the early 1990s. Recognizing a decline in the resource, the Jamaican government introduced annual catch and export quotas, implemented in 1994 in Jamaica's first conch fishery

Continued

[11] See FAO (2002) for guidelines on monitoring, control and surveillance (MCS) aspects within access agreements.

[12] See the discussion below on extra-fisheries mechanisms.

[13] See http://www.cites.org/

Example: Continued

management plan. MSY for queen conch was calculated at 700–1,300 t/year. Unfortunately, illegal fishing is now rampant on the Pedro Banks, much of it by foreign vessels that simply ignore Jamaican law. In the years 1999–2002, illegal harvest was estimated to account for 40% of the conch fishing on the Pedro Banks. Jamaica conducted its third conch stock assessment in 2003. Although this stock assessment suggested a total allowable catch of 900 t, Jamaica set its conch export quota at 500 t to allow for some inevitable losses to illegal fishing . . . Based upon the findings of its September 2003 Significant Trade Review, CITES considers Jamaica to have an adequate conch management regime and relatively healthy queen conch populations. Jamaica is one of only two Caribbean conch-exporting nations to earn the CITES designation of 'least concern' for its queen conch resources.'

Other trade barriers at the national level, such as the US dolphin-safe tuna policy, create incentives to implement sustainable fishing practices as defined by the importing country. However, such measures may prove ineffective in creating more sustainable practices if substitute markets are available for the given products.

Although the PPP concept may now be found in more and more national fisheries' legislations, and is applied in relation to the impacts of non-fisheries activities on fisheries habitats, it is difficult to claim a widespread use of such economic instruments. Coffey and Newcombe (2001) presented a few considerations as to why the use of economic disincentives with respect to environmental damage/control may prove difficult. Implicit in the shift towards PPP is the removal of perverse incentives (minimizing 'harmful' subsidies) as, without doing so, fishing effort will continue to rise; making ineffective and inefficient any attempts to internalize the environmental costs of fishing.

Social Incentives

Just as economic and/or market mechanisms can induce individuals to make choices compatible with societal objectives, so too can social factors produce similarly desirable behaviour. Thus, developing and implementing successful EAF strategies requires understanding and working with the social mechanisms surrounding access to resources, institutional organization and decision making, local management and power structures and attitudes and perceptions towards authorities and institutions. In other words, attention is needed to *social incentives*.

Indeed, much of what has been mentioned in earlier sections might be termed social incentives: providing alternative livelihood opportunities or transparent and participatory management approaches, for example, both impact and are impacted by the social framework surrounding the fishery. Properly implemented management systems, from inception to monitoring and control, will affect people's incentives to comply with regulations as well as the strict economic gains and losses related to these regulations. Peer pressure within fishing communities can be harnessed as a social incentive producing more socially desirable choices (e.g. improved compliance). Moral and religious codes, whether fishing is considered a right or a privilege, and knowledge about the ecosystem, will certainly involve social incentives that influence individual and group behaviour (even though these are rarely captured in conventional decision making models).

Recognizing and/or developing social incentives requires suitable understanding of why people act in certain ways, as well as an understanding of the socio-economic context of a given fishery (e.g. employment and livelihood opportunities, fishing traditions, local ecological knowledge and changing demographics). Such an understanding will assist in the identification of potential impacts of management interventions (e.g. where would displaced effort go) and will assist in promoting wanted change – without such knowledge, much information regarding the motivations, interests and priorities of the resources users will be lost and management misguided.

Extra-fisheries Incentive Mechanisms

The above discussion has focused mainly on mechanisms falling within the conventional management sphere of influence. However, as the EAF requires a broader approach to resource management, negative and positive externalities stemming from outside of the fisheries sector are increasingly being incorporated into management, whether promoted by the industry itself or by government instigation. Therefore, a brief description of mechanisms to internalize the benefits and costs of extra-fisheries activities and values is presented below.

Non-fisheries 'polluter pays'

Financing EAF implementation through a 'polluter pays' approach involves collecting revenues from those using the natural resource and/or causing ecosystem damage, and using those funds to finance positive moves to EAF management. In addition to the polluter-pays and user-pays incentive mechanisms described previously, governments and fisheries associations have begun reclaiming the restoration costs in dealing with ecosystem damage inflicted by actors outside of the fishery sector (e.g. upstream activities, changes to habitats, pollution and destructive practices). Individuals convicted of damaging the ecosystem are required to either pay fines, which may or may not be directly related to damage costs, or more directly to repair the damage or pay for work related to the conservation and protection of the affected habitat. Examples of trust funds established within fishing associations to manage funds collected for such restoration work are providing institutional precedence for the transfer of funds into the fisheries sector.

Example: Non-fisheries polluters paying for fishery ecosystem damages.

Canada – under the Federal Fisheries Act, the Department of Fisheries and Oceans (DFO), Canada, uses fines from habitat violations to restore damaged fish habitat. The convicted offender pays money directly to repair fish habitat or enhance fish stocks, often through local non-profit environmental groups. For examples of such convictions, see DFO (2004).

United Kingdom – the Anglers' Conservation Association (ACA) represents its members in court cases against private and public entities polluting British lakes and rivers. Money collected is kept within member fishing clubs and used in rehabilitation trust funds. See http://www.a-c-a.org/whatwedo.html, for examples.

USA – the Columbia River Estuarine Coastal Fund was established in 2004 through the collaboration of the Foundation, the Service and the US Attorneys for Oregon and the Western District of Washington from fines imposed on shipping companies that illegally discharged oily waste into the Pacific Ocean near the mouth of the Columbia River. Conservation and restoration projects will be funded with US$1.2 million in community service payments from polluters. See http://www.nfwf.org

Extra-fisheries 'beneficiary pays'

Note that related to 'polluter pays' is the idea of 'beneficiary pays', in this case implying that those receiving the benefits of EAF implementation should pay the costs required to achieve those goals. Extra-fisheries benefits that can accrue from application of the EAF are being acknowledged through a global increase in environmental awareness (i.e. the recognition of the goods and services provided by ecosystems and the need to minimize damaging impacts on these ecosystems), a desire to improve human conditions (i.e. decreasing hunger and poverty, improving livelihoods), and the hopes of holistic, decentralized natural resource management (i.e. through good governance, participatory processes, community-based management and integrated resource management).

The possibilities for garnering funding from international sources to support EAF are numerous (e.g. donor countries, international trust funds, development banks and funding facilities). Combined, these possibilities may lead to initiatives for the global community to financially support EAF efforts, particularly in jurisdictions that otherwise might be unable to afford such efforts.

However, understanding the various and appropriate sources of funding requires a large and, perhaps, daunting investment on the part of fisheries managers. For example, some funding sources may target sectoral-specific activities, while others may target specific issues, such as biodiversity or marine-protected areas. Accounting systems and even vocabulary may vary significantly across sources and funding sources may or may not be tied to certain conditions, economic or otherwise.

In addition, as EAF management is likely to comprise both development and conservation components, no one source of funding is likely to cover all EAF needs. Hence, a portfolio approach to funding will be necessary; increasing the time and energy devoted to developing and using these funds.

Furthermore, there is a crucial issue of institutional sustainability to consider when utilizing external funds – i.e. ensuring that long-term arrangements are in place so that EAF implementation is not jeopardized when the specific funding period ends.

In recognition of these complexities, guides to finding relevant financing sources have been developed. Importantly, some of these guides provide detailed business planning for marine-protected areas and other skills to assist fisheries managers in planning their financial needs assessments and donor funding requests. The major categories of international funding described in these guides[14] are:

- Bilateral and multilateral donors.
- Biodiversity Enterprise Funds.
- Debt for nature/environment swaps.
- Environmental funds and conservation facilities.
- The Global Environmental Facility.
- Foundations.

In any case, an evaluation of the potential benefits from EAF application, whether at the local, national, regional or international level, would assist in organizing efforts at the appropriate levels.

Example: Financing coastal resource management in the Philippines. (From Salamanca and Luna, 2002; Salamanca, 2003.)

Salamanca and Luna (2002) presented an historical perspective of coastal resource management (CRM) in the Philippines and discussed 'the factors that are thought to have played crucial roles, the formal institutions that underpin its development, and the issues that need to be addressed for CRM to fully succeed.' Within this report and a related background article (Salamanca, 2003), the authors estimated the financial needs and sources of funding for 290 CRM projects and activities from 1974 to 2000. Over this period, approximately US$230 million were spent on activities undertaken to manage the coastal zone and its resources through various implementers (i.e. integrated, multi-sectoral, government-led, NGO-initiated and fisherfolk-led) and various focuses (i.e. livelihood, education, research, advocacy, conservation and population). The authors estimated that approximately US$9,000 per km^2 of coral reef was spent over the 16-year period to protect the nation's 26,000 km^2 of reefs. While studying the financial investments, the authors also investigated the sources of funds and found that 63% of the funding came from 44 international sources (i.e. bilateral and multilateral sources, debt for nature swaps, the international NGO community and international philanthropic organizations); 36% from the Filipino government, and 1% from local donors; thus, highlighting the importance of international sources of funding.

[14] WWF Guides – http://www.worldwildlife.org/conservationfinance/pubs.cfm; Conservation Finance Alliance Guide – http://guide.conservationfinance.org/; Debt for nature/environment swaps guide – http://biodiversityeconomics.org/finance/topics-42–00.htm; GEF funds guides – http://www.gefweb.org/Partners/partners-Nongovernmental_Organ/ngo_guide/ngo_guide.html

Concluding Remarks

In this chapter, we have addressed four forms of incentives: legal, institutional, economic and social for use in fisheries management frameworks following the ecosystem approach. Many of the incentive measures presented fit in quite naturally with existing conventional management strategies (e.g. participatory approaches, good governance and well-established rights systems); however, other measures adjust for the broader understanding of the values that societies have for ecosystem goods and services (e.g. PPP and UPP, extra-fisheries externalities and garnering support for globally distributed benefits with localized costs).

While incentive mechanisms have been presented in this chapter because of our belief in their usefulness for fisheries management, a few caveats are, nevertheless, warranted. The risk of relying on one tool or one subset of tools is, as always, high - unfortunately, there is no such panacea and the broader scope of EAF will require a broader mix of the toolkit. In the same vein, while economic incentives received considerable attention in this chapter, in reality - given the nature of externalities - a reliance on the market to fix all ills may well disappoint. This is why it is also important to work with legal, institutional and social incentives, and furthermore, to recognize that the potential role of government remains strong in implementing EAF management. The use of resources, whether natural, human or financial, will require societal choices and the need to understand the trade-offs among the various choices involved.

This chapter has been based on the concept that understanding human behaviour is at the core of fisheries management. This would include considering both demand and supply sides of fisheries (i.e. markets and trade) - the incentives they create and the impacts they have on our ability to manage - as well as the extra-fishing variables that affect human behaviour.

The mechanisms discussed in this chapter are not new concepts; although, perhaps, there has been relatively little application to the fisheries sector. However, examples exist (within fisheries and among other sectors) that may provide us with some guidance on their use and, like most borrowed tools, creativity may be required to adapt them to the context at hand.

References

APFIC (2005) Asia Pacific Fisheries Commission. *Report of the Regional workshop on mainstreaming fisheries co-management*. Siem Reap, Cambodia, 9 to 12 August 2005. FAO Regional Office for Asia and the Pacific, Bangkok, RAP Publication 2005/23, 48 pp. Available at: http://www.apfic.org/

Brewer, D., Heales, D.S., Milton, D., Dell, Q., Fry, G., Venables, W. and Jones, P.N. (2006) The impact of turtle excluder devices and bycatch reduction devices on diverse tropical marine communities in Australia's northern prawn trawl fishery. *Fisheries Research* 81, 176–188.

Cascorbi, A. (2004) Seafood Report: Queen Conch. *Seafood Watch Seafood Reports*, 30 pp. Available at: http://www.montereybayaquarium.org/cr/cr_seafoodwatch/content/media/MBA_SeafoodWatch_QueenConchReport.pdf

Castilla, J.C. and Defeo, O. (2001) Latin American benthic shellfisheries: emphasis on co-management and experimental practices. *Reviews in Fish Biology and Fisheries* 11, 1–30.

Charles, A. (2002) Use rights and responsible fisheries: limiting access and harvesting through rights-based management. In: A

Fishery Manager's Guidebook - Management Measures and Their Application. *FAO Fisheries Technical Paper* 424. Available at: http://www.fao.org/documents/

Coffey, C. and Newcombe, J. (2001) *The Polluter Pays Principle and Fisheries: the Role of Taxes and Charges*. Discussion paper of the Institute for European Environmental Policy, London. Available at: http://www.ieep.org.uk

Cox, A. (2000) Cost recovery in fisheries management: the Australian experience. In: *Proceedings of the Tenth Biennial Conference of the International Institute of Fisheries Economics and Trade*. July 10–14, Corvallis, Oregon. Available at: http://oregonstate.edu/Dept/IIFET/

De Young, C., Charles, A. and Hjort, A. (2008) Human dimensions of the ecosystem approach to fisheries: an overview of context, concepts, tools and methods. *FAO Fisheries Technical Paper*. No. 489. FAO, Rome, Italy, 152 pp.

DFO (2004) Department of Fisheries and Oceans Canada. *Administration and Enforcement of the Fish Habitat Protection and Pollution Prevention Provisions of the Fisheries Act*. Annual Report to Parliament for the period of April 1, 2003 to March 31, 2004. Available at: http://www.dfo-mpo.gc.ca/

FAO (1982) Territorial use rights in marine fisheries: definitions and conditions. *FAO Fisheries Technical Paper* 227. Available at: http://www.fao.org/documents/

FAO (1995) *Code of Conduct for Responsible Fisheries*. FAO, Rome, Italy, 49 pp. Available at: http://www.fao.org/documents/

FAO (2000a) Use of Property Rights in Fisheries Management. *FAO Fisheries Technical Paper* 404/2. Available at: http://www.fao.org/documents/

FAO (2000b) FAO Fisheries Department. *The State of World Fisheries and Aquaculture (SOFIA) 2000*. Rome, Italy, 142 pp. ISBN 9251044929. Available at: http://www.fao.org/documents/

FAO (2001). Case studies on the allocation of transferable quota rights in fisheries. *FAO Fisheries Technical Paper 411*, FAO, Rome, Italy. Available at: http://www.fao.org/documents/

FAO (2002) Implementation of the International Plan of Action to Deter, Prevent and Eliminate Illegal, Unreported and Unregulated Fishing. *FAO Technical Guideline for Responsible Fisheries Number* 9. FAO, Rome, Italy, 122 pp. Available at: http://www.fao.org/documents/

Ferraro, P.J. and Kiss, A. (2002) Direct payments as an alternative conservation investment. *Science* 298, 29 November.

Graham, R.T. (2004) Global whale shark tourism: a 'golden goose' of sustainable and lucrative income. *Shark News* 16, October.

IEEP (2003) Institute for European Environmental Policy. *Fisheries Agreements with Third Countries - Is the EU Moving towards Sustainable Development?* Report commissioned by the WWF's European Fisheries Campaign. Available at: http://www.ieep.org.uk/

Keizire, B.B. (2001) *Opportunities and Options for Financing Fisheries Management in Uganda*. Final Project under the United Nations University Fisheries Training Programme. Reykjavík, Iceland. Available at: http://www.unuftp.is/

Martin, W., Lodge, M., Caddy, J. and Mfodwo, K. (2001) *A handbook for negotiating fishing access agreements*. World Wildlife Fund. Washington, DC. Available at: http://www.panda.org/

OECD (2003) Organisation for Economic Co-operation and Development. *The costs of managing fisheries*. Paris, France. ISBN: 9264099751. Available at: http://www.oecd.org

Pinkerton, E., Bedo, A. and Hanson, A. (2005) Final Evaluation Report of the West Coast Vancouver Island Aquatic Management Board (AMB). Available at: http://www.westcoastaquatic.ca

Salamanca, A.M. (2003) The cost of action: CRM investment in the Philippines. *NAGA, WorldFish Center Quarterly* 26 April–June.

Salamanca, A.M. and Paz Luna, M. (2002) Institutional Arrangements Affecting Coastal Resources Management Initiatives in the Philippines: Trends, Demands and Issues. Presented at 'The Commons in an Age of Globalisation,' the Ninth Conference of the International Association for the Study of Common Property, Victoria Falls, Zimbabwe, June 17–21, 2002.

Seijo, J.C., Defeo, O. and Salas, S. (1998) Fisheries bioeconomics. Theory, modelling and management. *FAO Fisheries Technical Paper*. No. 368. FAO, Rome, Italy, 108 pp. Available at: http://www.fao.org/DOCREP/003/W6914E/W6914E00.HTM

SFLP (2006) DFIF/FAO Sustainable Fisheries Livelihoods Programme. Reducing fisherfolk's vulnerability leads to responsible fisheries. *SFLP Policy Briefs on Development Issues*. Available at: http://www.sflp.org/

11 Fisheries Assessment and Decision Making: Towards an Integrated Advisory Process[1]

SERGE MICHEL GARCIA

FAO, Rome, Italy

What is needed are mechanisms for performing the science that will guide society in making its decisions, and for building bridges between science and decision making.

(S.A. Levin, 1993)

Abstract

The fisheries governance crisis, the consequent adoption of a sustainable and responsible development framework and the efforts to implement a precautionary and ecosystem approach to fisheries (EAF) have raised the awareness about the systemic nature of fisheries and the uncertainty surrounding the science–policy–decision interface. The consequences include the request for an increased scope of the scientific enquiry, improved cross-disciplinary collaboration and more effective stakeholder participation, both in the scientific and decision making processes. Science and governance have co-evolved for a century and will continue to do so. Trends in social demand and scientific community response are briefly examined, focusing on scientific approaches, the disciplinary puzzle and the ongoing encounter between the normal and post-normal (positivist and constructivist) science paradigms brought about by the formal recognition of uncertainty and the need for greater involvement of social sciences. The contribution of key elements of modern advisory processes (i.e. integration, simulation and participation) is briefly reviewed as well as ongoing practices in fisheries management advisory processes and at other interfaces between science and environmental policy. Following logically, an Integrated Advisory Process (IAP), accounting for the systemic, cross-disciplinary and participative nature of modern fisheries governance, is sketched and its epistemological and operational implications are briefly reviewed.

Introduction

Fishery science and governance have co-evolved for more than a century. The reductionist and disciplinary nature of conventional fishery science when deal-

[1] Paper presented at the Nordic Council Conference on Implementing the Ecosystem Approach to Fisheries, Bergen, Norway, 26–28 September 2006.

 The Ecosystem Approach to Fisheries (eds G. Bianchi and H.R. Skjoldal)

ing with the stepwise increase of the demand complexity have led to the emergence of segmented visions of the sector from viewpoints such as:

1. Resources: the domain of the fishery biologist.
2. Technology: the domain of the gear technologist and engineer.
3. Markets: the domain of the economist.
4. Environment: the domain of the ecologist.
5. Stakeholders and society: the domain of the sociologist.
6. Institutions: the domain of fishery administrators, lawyers and political scientists.

The first three have been closely combined in the scientific support to conventional fishery development and management, at least at the national level. The following three are essential to the new ecosystem approach to fisheries (EAF). The dissociation between these elements, reflected also in the disconnection between strategic development planning and operational management, has led to misfits, delays and contradictions and, ultimately, to a series of interrelated bio-ecological, social, economic and political crises. Their necessary short-term resolution through compensations and technological quick fixes has dragged science into a frenetic race, burning its resources in short-term advice at the expense of the strategic analysis the fisheries crises really needed (Catanzano and Rey, 1997). The successive local, then national, crises ultimately converged into an international fishery governance crisis affecting all its sub-sectors, from coastal small-scale fisheries to high or deep sea large-scale ones. The crisis is slowly expanding from capture fisheries to aquaculture. The time seems to have come, therefore, for a change in the way science is implemented and integrated in the governance process.

With the emergence of frameworks for the use of sustainability indicators, the precautionary approach and the EAF, the need to increase the scope of the scientific enquiry and to improve the role of stakeholders in the decision making process has gradually increased, to the point of becoming a leitmotiv, particularly in the area of co-management, community-based management, etc. (Chopra *et al.*, 1989; Pollnac, 1994; Pomeroy, 1995; Cochrane, 1999, 2002; Jentoft, 2000, 2005, Ostrom, 2000; Wiber *et al.*, 2004). As a result, a number of questions have emerged regarding the nature, methods, processes and performances of fishery research and its relation with the decision making process, the stakeholders and the public at large.

In fisheries, the large body of literature available, the impressive developments in a few developed countries and the work of the well-endowed regional fishery commissions give a general impression that fishery management decision making processes are strongly science-based institutions.[2] This is not yet the case, however, in a large number of countries or regional organizations, where scientific capacity is still insufficient and/or the science–policy interface is deficient.

[2] This impression is reinforced by the UNCLOS requirement for the 'best scientific evidence available' that has definitively linked the exercise of fishing rights to science-based processes.

In general, the expectations about the role of science in the resolution of societal issues has grown following the United Nations Conference on Environment and Development (UNCED) and the emergence of large-scale environmental issues such as climate change and greenhouse effects, freshwater systems acidification, desertification and drinking-water degradation and shortage. Boosted by the growing societal concern, growing advocacy and the boom in information and communication technology (ICT), the processes of knowledge generation, communication and social learning are becoming central issues with significant consequences on science.

In the fisheries arena, overfishing has spread to the point of becoming also a global societal issue and the interface between fishery science and decision making has been progressively changing, in a similar direction, at least in a few leading countries. During the preparation of the Conference on Implementing the Ecosystem Approach to Fisheries (Bergen, Norway, October 2006), it was realized that the event would provide an opportunity to discuss the consequences of EAF for fishery science and for the formal advisory and decision making processes in which it had been involved for more than a century.

This chapter was prepared at short notice to provide an opportunity for a possible initial discussion at the Conference. In preparing it, I have had to venture out of my discipline into social sciences, an area with which I am not familiar, well aware of the risk of creating confusion and unnecessary controversies when extrapolating one's experience beyond its legitimate boundaries. However, encouraged by a recent joint work with Professor A.T. Charles on systemic implications for research and governance (Garcia and Charles, 2007) in which integrated assessment (IA) was addressed, I tried to elaborate further on this last aspect. I did my best to understand and accurately represent views and trends in the relevant areas of social learning and fisheries modelling, conscious of the complexity and numerous nuances I might miss. This chapter intends only, therefore, to present my current reading and reflections without any claim of originality and to draw the fishery science community's attention to an issue and some of its implications that I consider fundamental for the future of fishery science and governance.

The chapter examines briefly the trends in social demand and the corresponding scientific community response, focusing on scientific approaches, the disciplinary puzzle and the ongoing encounter between the normal and post-normal (positivist and constructivist) science paradigms brought about by the formal recognition of complexity and uncertainty and of the need for greater involvement of stakeholders and social sciences. The contribution of key elements of modern advisory processes (i.e. integration, simulation and participation) is then briefly reviewed together with the ongoing practices in the scientific advisory processes in fisheries management as well as at other interfaces between science and environmental policy. Following logically, an Integrated Advisory Process (IAP), accounting for the systemic, cross-disciplinary and participative nature of modern fisheries governance, is sketched and its epistemological and operational implications are briefly reviewed.

Co-evolution of Fishery Science and Governance

During the last century, fishery science and governance have co-evolved progressively, albeit perhaps not smoothly, adapting the capacity, modus operandi and approaches to the requirement of the other, in response to societal requirements (Garcia, 1994, 1996a; Catanzano and Rey, 1997; Rice, 2005). EAF is a response to new and more complex societal requirements and it implies a further step in that co-evolution. The following sections look briefly at that evolution and its possible next steps.

Evolution of societal demand

During the last century, societal demand has progressively shaped the co-evolution of fishery science and management institutions at least since the Second World War (Garcia and Reveret, 1991; Garcia, 1996a, 2005; Catanzano and Rey, 1997; Rice, 2005). Until the end of the 1960s, the demand was for information in support of fisheries development and expansion, corresponding to an active phase of scientific exploration, discovery and assessment of resources potentials and technological developments. Since the early 1970s, the evolution of fisheries has been affected by the outcome of numerous international conferences and the adoption of a number of instruments[3] reflecting the growing societal concern about the unsustainability of ecosystem use and related livelihoods, the related persistent poverty and the recurrent problems with food security and safety. These instruments have progressively broadened the parameters against which fisheries performance is judged, rebalancing the societal requirements for socio-economic development and environment conservation or, as generally stated, between human and ecological well-being.

In the process, societal objectives for fisheries shifted from the sustainable development of fisheries to the contribution of the sector to national sustainable development. This has increased the complexity of planning and management strategies, calling for a rapid adaptation of decision making and its supporting research (Garcia and Charles, 2007).

Scientific community response

Evolution of approaches

Since the early 1900s, fishery research has developed its understanding, organizing it in progressively more complex fishery models in an attempt to match the emerging socio-ecological complexity of fishery systems, including more

[3] For example, the 1972 Stockholm Conference on Human Development, the 1982 Law of the Sea Convention, the 1992 UN Conference on Environment and Development, the 2000 Millennium Summit and the 2002 World Summit on Sustainable Development.

components, variables and effects, and involving more disciplines along the way. Concerned initially with systematic discovery, description and inventory of resources, fishery sciences went through a phase of rationalization of single stocks exploitation with the adjunction of neoclassical economics under a classical reductionist paradigm usually coined as Cartesian or Newtonian.

A more systemic approach, sometimes coined as Prigoginian, has been advocated quite early in the process and for decades (Rothschild, 1971; Prigogine and Stengers, 1979; Walters, 1980; McGlade and Allen, 1984; Hilborn and Ludwig, 1993; Gallopin *et al.*, 2001). It started developing significantly only during the last 10 years (Barreteau *et al.*, 2001; Garcia and Charles, 2006). As demand for advice became more complex and knowledge improved, the simpler biological or bio-economic representations of the 1940s and the 1950s[4] progressively evolved, in the 1970s and 1980s, into multi-species and ecological models,[5] assisted in this by the computer revolution. During the last decade, these models evolved into more systemic models[6] combining human and ecological representations in fairly realistic simulations. The agent-based models (ABMs) used nowadays by some teams are among the most flexible ones available today, with the ability to operate across scales and to combine quantitative and qualitative information, offering for the first time the opportunity to integrate the biophysical, economic and social parameters of fisheries (Le Fur, 1996, 1998; Shin *et al.*, 2004; Little *et al.*, 2006).

Gallopin *et al.* (2001) indicate that despite its importance for the science–policy interface, the use of systemic science for decision making is still very limited. Fisheries are no exception and an important part of the scientific advances have remained in literature domain with little practical application in management (Garcia and Charles, 2006). The gap between scientific understanding and governance led to a failure of the science–decision interface as the governance paradigm and the constraints imposed by rigid institutions restricted the flux of knowledge between its generators and its potential users (Catanzano and Rey, 1997). The uptake seems to have accelerated during the last decade, mainly at national level and in a few countries, reflecting perhaps what Rice (2005) refers to as an *asynchronous co-evolution* of fishery science and governance.

Paralleling but largely independent of this process, social research on fisheries proceeded through, *inter alia*, ethnographic studies, analysis of the functioning and dynamics of fishing communities and sociological studies of various aspects the fishery system, providing important understanding about the human component of the fishery systems. Working more generally at the interface between society and decision making in environmental and ecosystem management, social science developed operational research methods and expert systems before turning to simulations using ABMs (also referred to as multi-agent systems, MAS) to simulate the interactions between social behaviours and resources dynamics and to support collective decision processes.

[4] For example, by Graham, Schaefer, Ricker, Beverton, Holt, Gordon and Clark.

[5] For example, by Anderson and Ursin, Laevastu, Polovina, Pauly and Christensen.

[6] For example, by Cochrane, Butterworth, Smith, Punt, Fulton, Shinn and Cury and Le Fur.

The disciplinary puzzle

The main scientific challenge for all the disciplines involved is in developing coherence between the disciplinary projections of the multidimensional fishery system on its bio-ecological, techno-economic, socio-cultural, institutional and other planes. The insights to be gained from the combination of the various disciplinary angles have been stressed since the early 1960s (Guimaraes Pereira and Funtowicz, 2003a) and well argued since then (Flinterman *et al.*, 2001; Gallopin *et al.*, 2001; Ludwig *et al.*, 2001) including, sometimes the advantages of integrating non-scientific, religious and ethical sources in the process.

However, despite recurrent warnings since the origin of fishery science,[7] the failure to mobilize the contribution of all relevant disciplines has affected the science capacity to provide adequate advice for policy development to management, contributing to severe miscommunication and misperceptions, leading to policy and governance failures (Garcia and Grainger, 1997; Sutinen and Soboil, 2003). The development of social modelling applications, largely parallel with and apparently unnoticed by conventional fishery science, despite obviously converging objectives and approaches, indicates that the great opportunity for disciplinary integration offered by ABMs for a decade has not yet been fully utilized.

Like in the Indian parable of the blind men describing an elephant (used by Funtowicz, 2002; Ramchandran, 2004; Loughlin, 2005) bio-ecologists looking at a 'sick' fishery see the bio-ecological syndrome, prescribing technical measures such as mesh-size regulations, closed seasons, reductions of mortality and marine protected areas (MPAs). Economists see the economic syndrome and solutions (e.g. taxes, incentives and rights). Sociologists see social problems and propose social solutions (e.g. participatory management). Lawyers and institutional experts also see legal and institutional problems and propose new legal instruments and institutions. While with time some degree of integrated diagnostic is developed, and with few notable exceptions noted later in this chapter, the need for greater integration is obvious (Degnbol *et al.*, 2006). It should be obvious that, in order to correct this historical functional defect, innovative decision-support systems need to involve a broader range of sociological disciplines (e.g. geographers, anthropologists, historians and sociologists, as appropriate), as important components of decision making support (Jentoft, 1998; Charles, 2001; Ludwig *et al.*, 2001).

The shift to more systemic research and advisory frameworks has already started in some leading countries and needs to be accelerated, completed and generalized. In this ongoing process, assumptions of equilibrium, reversibility and universality are progressively relaxed. The inherent uncertainty of complex systems, their potential for self-organization, the existence of multiple relevant scales, cross-scale interactions and multiple, partly separable causes are recognized. More efforts are needed also to reduce the chronic disconnection between operational management and strategic development planning and to improve

[7] Michael Graham and Harold Thompson, founders of modern fisheries science, already cautioned scientists, in the late 1940s 'against making management recommendations based solely on fish biology "without regard to the technology and the livelihood of fishermen"' (Kesteven, 1996).

the participation of stakeholders and the communication with the public at large (Garcia and Charles, 2006). The form of science needed, sometimes referred to as *sustainability science* (Gallopin *et al.*, 2001) is described as combining historical, comparative and experimental approaches at multiple scales, accepting multiple lines of evidence and sources of knowledge as well as integrative modes of enquiry, accounting for the multiplicity of objectives and uncertainty, and actively looking for societal consensus (Holling, 1993; Ludwig *et al.*, 1993, 2001).

In fisheries, however, the integration of social sciences is still marginal. In the heated debate over the role and nature of science for fisheries sustainability in the early 1990s (Levin, 1993),[8] there was hardly more than a passing reference to social research and its potential contribution (Hilborn and Ludwig, 1993). These authors recognized that the new stream of fishery science required the study of human motivations and responses as part of the system to be studied and managed and had the most natural connection to social sciences. Social sciences, however, represent a galaxy of schools of thought with differing potential to assist or unnecessarily complicate the problems faced in fisheries. As Ludwig *et al.* (2001) put it: 'we need to think carefully about which aspects of the social sciences can contribute (or at least not make things worse)'.

From normal to post-normal science?

As highlighted above, the biophysical and sociological disciplines have evolved in parallel with little contact, except for neoclassical economics. As shown by the emergence of bio-economics and ecological economics, disciplines may cross-breed into trans-disciplines. However, the perspective of a broadening of fishery science to better integrate sociologists in a process largely dominated by biophysical sciences and economics, brings up the perspective of a strong interaction between the 'normal' science currently dominating fishery research and the 'post-normal' science or 'second-order' science proposed by social scientists for the interface between science and environmental policy (Funtowicz, 1989; Funtowicz and Ravetz, 1991, 1995). Recently, Gibbons *et al.* (1994) and Novotny *et al.* (2001, 2003) have described the recent trend in science as a rapid shift, over the last decade, from what they called Mode-1 to Mode-2 research.[9] Although involving other criteria related to source of funding and intellectual property rights, the criteria used to describe these modes overlap significantly with those of normal and post-normal science.

The issue may not be trivial because these streams of science, with different disciplinary membership, relate respectively to two apparently antagonistic

[8] Levin dedicated a whole issue of Ecological Applications to the issue.

[9] The debate about the reality and extent of that shift may not be closed (Auranen, 2005; Leshner, 2005).

[10] Other analogous dichotomies of science criticized by Gould (2005) as potential sources of irrelevant '*science wars*' between scientific, experimental (empirical) knowledge and other forms of knowledge (e.g. myths, religions, social wisdom) include: ancient versus modern (17th century); science versus religion (19th century); positivists versus constructivists (19th and 20th centuries); and realists versus relativists, modern versus post-modern, normal versus post-normal and mode-1 versus mode-2 in the 20th century and 21st century.

paradigms: positivism and constructivism.[10] The first has provided the philosophy behind the modern quest for the rigorous elaboration of scientific knowledge needed to unravel the true 'Laws of Nature' and is the foundation of most of the modern science and technological developments (Gould, 2000). The second questions the existence of such fundamental, immutable laws and of scientific objectivity and provides a philosophy for effective social learning through collective processes (Bousquet and Le Page, 2004) in which science is only one of the sources of knowledge. The question is of direct relevance to this chapter because fishery scientists have traditionally focused on the production of scientific understanding and advice for decision, while many social scientists working on environmental management issues seem to have focused on the ways in which social groups integrate individual knowledge and experience, values and perspectives to construct the agreed knowledge on the basis of which they act.

This difference is of interest in that the relative failure of fisheries governance may well reside, at least in part, in the lack or dysfunction of the institutional bridge needed to transform scientific conclusions into actionable social knowledge (see Levin's statement quoted at the beginning of this chapter). A better integration between conventional fishery and social sciences is advisable, but one might wonder how deep that integration might be considering the differences in the underlying paradigms. The issue is examined below starting with a brief description of the paradigms concerned as described by various authors with different degrees of contrast (or extremism).

Normal science and positivism

Normal science is the science philosophy under which most of the traditional fishery research described above has been undertaken. It is predicated on the assumption that the scientific community knows what the world is like and 'certified' by disciplinary peer review, evolving through paradigm shifts (*sensu* Kuhn, 1962). It is the form of science that evolved through centuries of efforts to unravel the 'Laws of Nature' and struggled for centuries to distinguish objective scientific knowledge from religious and esoteric beliefs, leading to the technological developments of modern times.

Positivism is one of the names given to the paradigm under which this science developed. Derived from the 18th century Enlightenment philosophy, which advocated rationality as a means to establishing an authoritative system of ethics and logic, the concept was developed by Auguste Comte, the inventor of 'sociology' (that he called social physics), at the beginning of the 19th century. It is strongly related to the Aristotelian, rationalist and empiricist foundations of the western 'hard' sciences, honoured *inter alia* by Sir Francis Bacon (1561–1626), Galileo Galilei (1564–1642), René Descartes (1596–1650), John Locke (1632–1704), Isaac Newton (1643–1727), John Stuart Mill (1806–1873), Claude Bernard (1813–1878) and many other famous scientists and engineers of modern times who led the technological advances of the last two centuries.

Under these paradigms, the normal scientist intends (pretends) to discover and establish the objective truth, i.e. the immutable laws of nature, a reality taken for granted and the findings are assumed to be independent of his/her social position, background or personality. The scientific knowledge is considered as the only authentic knowledge and it can be scientifically verified, i.e. confirmed or falsified by empirical observations of reality and experimentation. Facts must be assembled in a transparent way and are analysed in a reproducible manner. Conclusions are peer-reviewed and published for general access and public scrutiny and held as 'best scientific evidence'[11] until new data or improved analysis replace them with better ones.The concept has been strongly associated with reductionism and assumptions of equilibrium and reversibility.

These premises are considered by post-normal science proponents and constructivists as inadequate for dealing with environmental issues of societal dimensions (Funtowicz and Ravetz, 1995; Gallopin *et al.*, 2001; Ravetz, 2003). Their critiques relate to the alleged rigidity of the paradigm and difficulty to change it, the strong disciplinarity and frequent disconnection with people, the dangers of 'inbreeding' of strictly disciplinary peer-review processes, the illusion of absolute scientific truth[12] and excessive trust in scientific constructions, and the fact that scientific knowledge is taken as the only valid basis for policy decisions.The possible misrepresentation, denial or otherwise misuse of science by decision makers has added to distrust, as in the cases of the HIV-AIDS-infected blood transfusion scandal in France or in the 'mad cow' disease[13] in Europe. Additional confusion may come from the effect of the internet – that increased greatly the access to grey and pseudo-scientific literature – and from the use and abuse of the media to divulge 'findings' before peer review for the sake of publicity or advocacy.

The science–decision process that developed under the positivist paradigm is also considered by its detractors as largely linear, with information elaborated by scientists and communicated to managers for the decisions with which fishers will ultimately have to comply. An example of this process would perhaps be that of scientific determination of the total allowable catch (TAC) in an international fishery, its negotiated subdivision in national quotas and subsequent harvesting by the fishery.

These views of normal fishery science might, however, appear as a caricature to many modern fishery scientists working at the interface with national policy and management (particularly in small-scale fisheries) who, by necessity, have already drifted towards less classical (fundamentalist) forms of science and decision making.

POST-NORMAL SCIENCE AND CONSTRUCTIVISM. By opposition, post-normal science is defined by its proponents as one in which the 'normal' premises listed above are not assumed. It is described as:

> the science needed as a bridge between science and policy when "hard" decisions need to be made on information that is irremediably "soft" . . . when most problems in practice have more than one plausible answer . . . and many have

[11] As required by UNCLOS.

[12] As all 'scientific' findings are seen as affected by context, social origin, perceptions and other biases of the scientist.

[13] Bovine spongiform encephalopathy (BSE).

> no answer at all . . . a new approach to problem-solving strategies in which the role of science, still essential, is appreciated in its full context of the uncertainties of natural systems and the relevance of human values.
>
> (Funtowicz, 2006)

It is considered appropriate when stakes are high, values are in dispute, facts are uncertain or incomplete and decisions are urgent and with significant social and economic consequences. Under these conditions, 'hard' reasoning and processes must be combined with 'soft' ones (Funtowicz, 1989, 2006; Funtowicz and Ravetz, 1994a, 1995). It is considered particularly adapted when the relation between the observer and the observed (i.e. the scientist and the research object) changes at a speed that exceeds the speed 'at which the mechanism of social interaction can validate the shared perception used as input for science' (Giampietro, 2003).[14] In other words, post-normal science is advocated when things change too fast to allow the 'normal' check and balances processes to take place and when negative impacts on people may be felt before the scientific conclusions have been validated. Post-normal science is less concerned with 'internal' (disciplinary) coherence and quality than with coherence and relevance of the scientific outputs to the social context within which they are to be used. In fisheries, reference to post-normal science has been made in relation to the precautionary approach to fisheries (Garcia, 1994, 1996b).

These views are explicitly connected by Funtowicz to postmodernism, a philosophical movement emerging during the mid-1950s in the fields of arts and science. The movement reflected a negative reaction to the (excessive) role of (arrogant) science and an opposition to the legitimacy of scientific knowledge. It holds that no scientific communication is devoid of myth, metaphor, cultural bias and political content. These views are also related to the constructivist philosophy of science elaborated by Berger and Luckman (1966). This philosophy assumes that the social 'reality' (or perceived truth) is socially constructed as a result of a collective learning process and does not necessarily reflect any external, objective, 'transcendent' reality. This position relates also to truth relativism, the doctrine that no absolute, universal truth (i.e. true in all possible contexts) exists, but that truth is always relative to some particular frame of reference, such as language or culture. As a consequence, the proponents of post-normal science advocate that any assessment has to take into account the subjective perceptions and individual framings of actors (Pahl-Wostl, 2005), an *a priori* problematic proposal for normal science.

Post-normal science proponents argue that, contrary to the positivist assumptions, most complex problems have more than one solution and some may have no solution at all. They hold that scientists are necessarily influenced by their life and work experience and their theoretical background, particularly when the solution to a problem is not entirely determined by the facts available. They consider that normal science has failed to provide as satisfactory explanations for social life as it could offer for other natural phenomena and would be unable to deal effectively with the reality of complex and chaotic socio-ecological systems. In addition, the shift from a positivist to a constructivist paradigm displaces the emphasis from the production of specialized scientific

[14] Gallopin *et al.* (2001) refer to a sort of 'Heisenberg' effect where the act of observation and analysis become part of the system under study, and so influence it in various ways.

findings to the process of constructing, through social learning, the socially robust knowledge on the basis of which social action (e.g. improved behaviour, compliance) can be expected.

The negation of the existence of, or the possibility to discover, the absolute truth may be shocking for some conventional fishery scientists, e.g. those trained in physical or biological oceanography, ecology, etc. It should be stressed, however, that the postmodern perspective on truth and society was developed as a reaction to the unjustified 'true' stand taken in the unfalsifiable social theories of Freud, Marx, Keynes and others, including Darwin. The fact is, however, that when fisheries are recognized as the complex socio-ecological systems they really are,[15] the full understanding and predictability of the fishery system (its 'absolute truth') is out of reach, placing 'normal' fishery scientists, *de facto*, in a post-normal situation.

The need to develop a closer association between the two types of science may raise a problem in fisheries because UNCLOS requires that decisions be based on the best scientific evidence available when the postmodern movement negates the objectivity of such evidence. However, post-normal and normal sciences may not be mutually exclusive or antagonistic. Post-science is not always seen as an attack on, or replacement of, normal science and it has been proposed as an extension, assistance and enhancement to it (Funtowicz, 2002). Similarly, Auranen (2005) argues that the evolution of modern research may not be a dramatic shift from Mode-1 to Mode-2, but a progressive shift in balance between two existing forms of research. Gould (2005) also criticized the artificial dichotomization and polarization of what he considers as a continuum between forms of science. The evolution of fishery research described in the first section would support these interpretations.

Towards an alliance?

With the full recognition of the systemic nature of fisheries[16] and of their governance failures, fishery science faces the implication of its necessary broader opening to both ecology and social sciences. It faces a series of interconnected challenges related to complexity and connectivity within the sector such as delayed responses; remote control (in time and space); feedback control; existence of multiple interconnected scales; self-organization; loss of universality; collapse of classical assumptions of linearity, equilibrium, reversibility, predictability and human controllability; and differing stakeholders' perspectives. The prescription for dealing with these challenges includes the use of more complex models combining qualitative and quantitative information; interdisciplinarity;

[15] Implying incomplete knowledge available, loss of universality, fallacy of the equilibrium, reversibility and linearity concepts, multiplicity of objectives and constraints and the potential for time-related change, self-organization and surprises.
[16] A central issue during the last ICES Symposium on Fisheries Management Strategies, Galway, Ireland, 27–30 June 2006 (http://www.ices06sfms.com/).

reconnection of strategic with operational issues and processes; integration of informal knowledge[17]; improved communication with stakeholders and society; consideration of societal values; expectations and perceptions in decision making; and more generally, improvement of the science–policy–stakeholder interface (Garcia and Charles, 2006).

These challenges and solutions head conventional fishery science towards the adoption of a 'softer', more post-normal attitude. The need to broaden the scientific enquiry process, in particular on the human components, brings with it the need to develop a hybrid inquiry system combining 'normal' science and neoclassical economics components with post-normal components, particularly in the social dynamics and environmental economics areas.

The implications of this shift go beyond research into governance. As already noted above, the normal and post-normal paradigms seem to correspond logically to different concepts of governance the caricatures of which could be as follows:

- The conventional approach to governance could be considered 'normal' in that it tends to be reductionist, mechanistic, top-down, prescriptive, command-and-control and authoritarian. The scientific observation and analysis produce advice given to and used by decision makers to select/justify management measures conditioning fishery performance that the actors need to comply with. This approach has proven its weakness, time and time again. In improved variants of this 'caricature', in a very small number of fisheries and countries, fishers are involved in the decision making process and may provide some knowledge – in addition to data – to the scientific enquiry.
- The evolved governance required under EAF could be considered as post-normal in that it should rather be synthetic, context-sensitive, bottom-up, optional, incentive-based and participatory. The scientific observations include informal knowledge, and participative analyses produce management options that are elaborated and discussed with the stakeholders before being negotiated between the latter and the management authority.

In reality, the top-down and bottom-up modes need to be combined in order to deal properly with the different time and space scales at which operational management and strategic planning need to be conducted (Garcia and Charles, 2006).

The need to combine the advantages of both normal and post-normal science and to avoid their respective shortcomings is therefore also reflected in the evolving fishery management paradigm, pointing towards the need for an objective alliance. The following section provides some reflections on the current advisory processes of relevance to the design of an improved IAP.

[17] Adding to the database what social scientists call 'extended facts' i.e. those relevant elements of reality that are brought in by stakeholders' participation and would, otherwise, not have been part of the scientific process.

Reflections on the Advisory Process[18]

Fishery research has been involved for more than a century at the interface between science, policy and decision making. Its role has always been to:

- Understand and explain the fishery system, suggesting strategic and tactical benchmarks, identifying emerging problems, assessing their causes and contributing to their social recognition.
- Identify and analyse related management options, clarifying the possible courses of action, their cost, benefits and other potential outcomes.
- Monitor and assess implementation performance, ensuring the feedback loop necessary for adaptive change.

Shaken by repeated management failures, the fisheries governance paradigm has been modified with a more forceful intrusion of stakeholders, non-governmental organizations (NGOs), the media and the courts in a governance process historically monopolized, at least in appearance, by science and policy makers (Garcia and Grainger, 1997). Conventional, top-down, command-and-control management systems tend now to be considered by most scholars as obsolete and ineffective. The participation and activism of stakeholders has increased, affecting the interaction between science and decision making (Garcia, 2005). The recent political requirement for an EAF has added stress, requiring a speeding up of the evolution of the science–governance system.[19]

The convergence of all these factors results in the fact that, perhaps for the first time in history, people are witnessing, through the media, the consequences of a large-scale failure of science-based management[20] and the attempts being made, in scientific and management institutions, to correct the situation, mitigate the negative effects and improve future performance. As the measures are very likely to affect the livelihoods of many of them, directly or indirectly, the demand has been growing for a stronger involvement of stakeholders in the whole science–policy process. In parallel, a debate had started about the role of science, its relative responsibility in the failure, its real degree of understanding of the fishery system and its capacity to predict the consequences of the advice it provides. These developments have occurred together with, and have possibly been accentuated by, the adoption of participative forms of governance such as co-management and community-based management, triggering a (still limited) adoption of the participatory research processes already used for decades in agricultural research, e.g. for rapid appraisal.

[18] This section draws heavily on a paper on fishery systems and the implications for research and governance by Garcia and Charles (2006).

[19] This is particularly the case with the ecosystem approach to fisheries (Smith *et al.*, in preparation).

[20] The argument made by many scientists that the failure is due in large part to the non-implementation, by managers, of scientific (biological) advice is an irrelevant fallacy as one of the reasons for not following the advice is indeed the biological narrowness of the advice for a manager confronted with a complex reality.

Overall, the progressive recognition of fisheries as socio-ecological complex systems has led to the emergence of three important interacting elements in the evolution of the science–decision process: integration, simulation and participation. These will be further elaborated below. There are many interpretations of the concepts of integration and participation and combining the two leads to a wide range of meanings and interpretations (Darier *et al.*, 1999). There is also a very large body of scientific literature on participation, communication, transmission of knowledge, social learning, etc., some of which relates to fisheries. The following sections do not pretend to present an exhaustive analysis of the issues and should be considered as a preliminary analysis to be more fully addressed in an interdisciplinary environment.

Integration

Biophysical sciences appear to consider fisheries as natural resources systems oscillating around long-term positions subject to human disturbance (e.g. fishing pressure). Social sciences, on the other hand, appear to consider them as social systems evolving in competitive mode towards maximum economic benefits under natural resources constraints. Both representations are only partly true and a truly systemic view on fisheries should finally recognize fisheries for what they really are: a plexus of complex natural and human sub-systems co-evolving in a globalizing environment (Bousquet and Le Page, 2004; Garcia and Charles, 2006). Dealing with them correctly requires a higher degree of integration between these representations than the present practice reflects in most countries[21] and in all regional fishery bodies.

Integration is usually defined as the process used to unite, coordinate or blend components into a functioning or unified whole or larger unit. The concept has been abundantly used in fisheries to signal a broadening of scope and coupling of processes. It has also been used in reference to cross-sectoral considerations, e.g. in the context of integrated coastal areas management. It reflects the perceived need for synthetic assessment between various fields of expertise (Darier *et al.*, 1999). The EAF refers to the need for an IA (e.g. ICES, 2000; Garcia *et al.*, 2003) and the terminology is also used by social scientists (Pahl-Wostl, 2005).

Looking at fisheries in a systemic perspective, Garcia and Charles (2006) took a broad view of the term, stressing that integration was needed between: (i) science and policy, for improved advice; (ii) policy and society, for improved objectives and strategies; (iii) relevant disciplines, joining forces to tackle complexity; (iv) scientific and other forms of knowledge, to make full use of the information available; (v) quantitative and qualitative analyses, to increase analytical power, combining 'hard' and 'soft' evidence; and (vi) scientific facts, values and perceptions in decision making to improve social acceptance and compliance. This list

[21] Countries such as Australia, New Zealand, Canada, South Africa, the USA and some Nordic countries have already started paving the way, with varying degrees of operationalization of the concepts.

illustrates the fact that integration is neither simple nor easy. It is relevant both in the scientific arena (e.g. for modelling) and the governance arena (for communication and decision making processes). Achieving it requires the development of the interfaces between science and policy and between them and the society.

Simulations

Simulation models may not be a panacea, but they have the potential to structure the interface between science, policy and society, at least in areas where the related capacity is available. Numerous tools are available to deal with issues related to policy making and decision making under conditions of complexity (e.g. Funtowicz *et al.*, 1999, for a review). During the last three decades, computer simulations have become the backbone of complex systems analyses, particularly in the area of environmental and ecosystem management to analyse the interactions between natural and social dynamics.[22] They have the potential to promote an operational and strategic alliance between disciplines, as intermediaries facilitating interdisciplinary reflections, triggering the development of common languages, symbols and metaphors in support of converging paradigms. They can be used to reformulate questions, materialize understanding and communicate with policy makers and stakeholders, particularly if the latter have contributed to their development. Computer simulations are particularly useful to recreate historical situations, understand dynamic processes, test theories and assumptions and assist in developing foresight (e.g. through scenarios) and improving precaution.

Simulation models have been regularly used in fisheries since the early 1970s in an attempt to reproduce their biophysical and economic reality with the view to foresee their potential reaction to natural events (e.g. climatic oscillations) or human intervention (e.g. management strategies and pollution). The most recent developments have seen the use of individual-based model and ABMs (see section on evolution of approaches).

Sociologists see simulation models as part of the dialogue between science and society and as a help in structuring it (Pahl-Wostl *et al.*, 2000). They have developed a 'soft systems' approach in which simulation models are not used to predict and control the social system but to mobilize and guide the potential for change, taking account of the subjective perspectives of the relevant actors. The model development is participative and iterative and accounts for the fact that the actors may change views in the interactive process (Ramanath and Gilbert, 2004; Pahl-Wostl, 2005). Simulations have been used to produce *role games* in which stakeholders are involved with the view to: (i) acquire knowledge from stakeholders; (ii) validate model structure and rules; (iii) provide a process of mediation[23] in collective and adaptive decision making;

[22] Cf. Journal of Artificial Societies and Social Simulation. Available at: http://jasss.soc.surrey.ac.uk/JASSS.html

[23] Mediation is defined as the act or process of mediating, the intervention between conflicting parties to promote reconciliation, settlement or compromise (Merriam-Webster online dictionary, 2006).

(iv) investigate the potential of the simulated systems for self-transformation[24]; and (v) assist in future scenario-building, contributing actively to social learning (Barreteau *et al.*, 2001; Bousquet and Le Page, 2004).[25] Simulation models are thus used to help in transforming complicated scientific findings into socially robust knowledge[26] and in the emergence of more collective decision making. Such game roles were used, for example, to deal with irrigation issues in Senegal, local plant genetic resources management in Madagascar, agroforestry development planning in Southern France, agricultural dynamics and land use in North Vietnam, etc., using digital as well as non-digital analogues more adapted to low-tech environments in rural areas (cf. Bousquet and Le Page, 2004). New initiatives of this type are apparently planned in fisheries using the ECOPATH family of models (Christensen, 2006).[27]

Participation

Participation of stakeholders is used for their information and education, consultation, collaboration, decision making, etc. This section focuses on some aspects of the participation of non-scientists in the scientific enquiry process and does not pretend to summarize the huge literature available on participation, for instance, in Chopra *et al.* (1989), Jentoft (2000, 2005), Hisschemöller *et al.* (2001), Guimaraes Peireira and Funtowicz (2003a), Wilson and Delaney (2005) and UNU-IAS (2006). The ways in which decisions and governance takes place and the level of involvement of social actors affects their reaction to them. It is generally agreed that without community participation, socio-ecological problems and their solutions cannot be defined in their human-relevant ways, reducing the relevance and legitimacy of policy initiatives.

Participation is advocated on the basis that it can improve stakeholder ownership of the process; relevance and legitimacy of politically and socio-economically difficult decisions; moral force and political influence of the actors; consensus and mobilization; knowledge of the sector functioning and expectations; problem formulation and identification of solutions; transparency and public scrutiny; conflict resolution power and equity; and the potential for environmental stewardship. It can also potentially reduce enforcement costs as well as social and economic risks. More interesting in the perspective of IA is the view of Checkland (1981) who places participation at the interface between society and science, on the one hand, and between the reality and the models of it, on the other, playing the two-way role of quality-check and acceptability-check.

[24] Self-transformation is an expected characteristic of complex socio-ecological systems (Garcia and Charles, 2006).

[25] See the concept of 'Companion Modelling' at http://cormas.cirad.fr/fr/reseaux/ComMod/Charte.htm

[26] Scientific knowledge must be not only reliable but also 'socially robust' (Nowotny *et al.*, 2001).

[27] In the latter case, however, the purpose seems to be the more conventional one of 'marketing' fully developed ecological research theories to policy-makers.

Obtaining the benefits has a cost. It is usually implicitly assumed, by participation advocates, that the exercise (e.g. the extended peer review by the community) involves actors fully committed to resolving the issue through 'fair negotiation'. However, freeriders, bad faith and narrow interests are a daily reality and management solutions fully satisfying all stakeholders are likely to be the exception rather than the rule. If the willingness and commitment to fair negotiation are not available, the situation may be beyond a post-normal, participative, non-coercive approach (Ravetz, 2003). Hisschemöller *et al.* (2001) give an account of the difficulties to expect in an IA, stressing that the major problem resides in the consequences of participation, i.e.:

- The difficulty for a highly participative process among diverse groups of interests to produce a clear consensus towards a decision (a serious problem for a decision maker!).
- The dependency of the process outcome on the composition of the stakeholders group such that broadening the group or exchanging some individuals might sometimes change the outcome.
- The uneven level of commitment to resolving the issue at stake, depending on the expected distribution of costs and benefits (the presence of freeriders or elements benefiting from a lack of resolution).

The natural 'conservatism' (risk-aversion) of stakeholder groups and tendency to agree on common denominator solutions may impede the adoption of more radical, but possibly more effective ones. The low capacity of some groups, to communicate effectively their perspective is, on the other hand, an important impediment to reaching true consensus.

Stakeholders' participation in the scientific advisory process, however, is still very limited in fishery science, as far as one can judge from the literature. There are notable exceptions (e.g. in Australia and New Zealand) where the introduction of fishing rights has led in some cases to the development of a research capacity by the industry itself. In social sciences, however, participation is an obvious necessity and it has received a lot of attention in the areas of policy making, operational management of natural resources use and other environmental issues (Chambers, 1994; Berkes, 1999; Jentoft, 2000; Berkes *et al.*, 2001; Eckley, 2001; Bousquet and Le Page, 2004). In this research environment participation is not advocated to 'market' research findings and to convey policy decisions to people concerned but to involve them, from the onset, in the research and policy debate (Guimaraes Pereira and Funtowicz, 2003). The participatory processes involve the use of instruments such as citizen panels, in-depth groups, focus groups, actors' platforms, citizen juries, stakeholder analysis, participatory analysis, electronic public conferences and other modes of interaction (Funtowicz, 2002; Engels, 2005), the differences between which are not always that clear. The organization and running of participatory processes seems to be a field of expertise in itself and the most adequate approaches in each case will probably be culture- and path-dependent.

In the caricature of a conventional ('normal', top-down and command-and-control) decision making model of fisheries management, decisions result from rational calculations (choices) on the part of a decision maker more or less fully informed, including by fishery scientists, of the situation and of the potential consequences of the available alternatives. In more participative (democratic, post-normal) systems, decisions result from a series of interactions between stakeholders with different backgrounds and objectives and varying degrees of influence, facilitated by the decision maker, supported by scientific analyses. In practice, the difference between the two approaches might not be as clear-cut and hybrid approaches are likely to be necessary combining top-down and bottom-up governance to ensure both high-level responsibility and convening power and local operational effectiveness. If explicitly provided with feedback control mechanisms, both processes could be seen as part of adaptive management[28] (Walters and Hilborn, 1976; Holling, 1978). In addition, through more active participation, the informal knowledge - the use of which is promoted actively by social sciences (Berkes, 1999) - can be scientifically validated and integrated into the 'best scientific information available' requested by the 1982 Law of the Sea Convention (LOSC) as a basis for decision making.

In the fisheries arena, participation in decision making is usually agreed as necessary with degrees in stakeholders' decisional power depending on local culture and political systems. Participation of stakeholders in the scientific enquiry, however, is still in its infancy. Social scientists argue that knowledge used in addressing societal issues should be recognized as valid by the stakeholders community concerned (Guimaraes Pereira and Funtowicz, 2003a, b). The conventional action in that direction has been in efforts towards increasing stakeholders' understanding and improving global access to information. The emergence of ICTs, the web, computer models, games, etc., as well as NGOs and the media have played a key role in that respect[29] and are opening new avenues.

Participation of stakeholders in the fishery advisory process takes many forms:

- *Contribution of raw data* on the fishery without which fishery science would have probably never developed. Fishers are probably by far the economic sector most hard-pressed for operational data, but improvements are necessary and possible. This will not be discussed further in this chapter.
- *Contribution of informal knowledge* on the fishery system, the ecosystem and the resources obtained by fishers through personal experience, intra-generational exchange of information, transmission by elders; etc. The aim of collecting such knowledge is the co-production of better strategic and operational knowledge that can be validated and integrated in the 'best scientific evidence' available. The difficulties are in: (i) obtaining, usually for free, knowledge which, most often, is part of the fishers' assets; and

[28] Adaptive management is a process concept developed in ecology. While it could, conceptually, be used in a top-down mode, it is usually understood as participative and requiring close interaction with stakeholders for the generation and interchange of knowledge. Its adoption raises the profile and contribution of stakeholders in the management process and contributes to the growing demand for sharing, decentralisation or devolution of management powers.

[29] Even if the latter has dangerously blurred the boundary between science, advocacy and propaganda.

(ii) separating beliefs from facts and facts from empirical interpretation. In rapidly changing situations, e.g. in reaction to economic or climatic evolution, informal knowledge might be essential for timely responses.

- *Contribution of perceptions, values and expectations* may appear as more problematic to the 'hard' scientist. They are evidently relevant in the decision making process in helping to identify the multiple interests (and objectives), perspectives and expectations to be accounted for in the decisions. They can also be important, before that, for science, to identify key structural elements of models, to identify and test assumptions, to select the various management or development alternatives to be assessed, to confront model outcomes with local reality and to foresee the potential reaction of the stakeholders to policy choices and management measures.
- *Participation in modelling and scenario-building*: Through the above, as well as through targeted interaction, stakeholders can contribute to the process of knowledge representation, issue-framing, option identification and scenario-building with the view to create common grounds for decisions, *ex ante*.
- *Quality assurance*: The concept, developed by social scientists,[30] is that of a process to ensure that the knowledge used to take decisions affecting livelihoods is both scientifically sound (through disciplinary peer review), understood and accepted, increasing decisions' legitimacy and people's trust. Stressing the dangers of inbreeding implied by disciplinary peer review, Funtowicz and Ravetz (1990) stressed the need for extended peer review, and quality assessment through institutional frameworks, networks and partnerships, involving scientists and stakeholders, as a way to deliver more socially robust knowledge (Gibbons, 1999). Funtowicz (2002) recognizes, however, that establishing extended peer-review mechanisms will not be easy or devoid or errors.

An IAP to be used at the science/policy interface of complex socio-ecological fishery systems needs to combine the 'ingredients' mentioned above, combining participation and simulation, with strong interdisciplinary science support, to reach the level of integration and legitimacy needed to increase foresight, relevance and compliance. The perspective of a substantial increase in public participation in scientific advisory processes raises three questions:

- How can participation of non-scientists in knowledge-building be increased while maintaining and increasing the rigour of scientific analysis[31]?
- How can scientists communicate complex facts and conclusions in an understandable way to the managers, stakeholders and the public?

[30] Starting from the constructivist premises that no scientific demonstration can be an objective absolute truth, the 'normal' concept of scientific 'truth' is replaced by that of quality assurance (QA) ensured by a proper process, as science's ultimate regulative principle. Defined in terms of uncertainties and decision-stakes, QA encompasses public interest, citizen and vernacular science. QA is undertaken by an extended peer community replacing the 'normal' collegial community.

[31] Sustainability science should be even more rigorous than 'normal' science by being better informed (Gallopin *et al.*, 2001).

- How can the interface between social sciences and the disciplines presently involved in fishery science be better integrated?

Based on the considerations in the first and second sections, it can be argued that the process needed, characterized by a higher degree of integration and participation, should promote the continuous interaction between science, policy and society. The stakeholders should be able to provide data and knowledge to the scientific enquiry process from its early steps, contributing to the development of models that they should be able to understand and, to the largest extent possible, accept as realistic. Without interfering with the analytical scientific process, stakeholders should also be confronted with conclusions and implications, providing the feedback needed to improve the models' realism and relevance. Finally, they should be able to negotiate with policy makers, the most acceptable alternative, possibly satisfying both their requirements and the societal goals and values. Their current interaction with fishery science and decision making, when such interaction exists, is usually direct, e.g. fishers sit directly at the negotiation table, but it could also be mediated.[32] The purpose of the interaction is generally limited, at the end of a largely one-way process, to 'market' the scientific conclusions and the options retained for management.

There are precedents that we will examine below before describing what such a process might be in a systemic (e.g. ecosystemic) perspective. The following sections will examine the present practice in participatory advisory processes in fisheries management and in environmental management before concluding with a proposal for a fully integrated advisory process.

Participatory Practices in Fishery Management

Following decades of extensive use of computer simulations, progressively more participatory processes have been developed in fishery research during the last decade together with the introduction of agent-based simulations, e.g. for *ex ante* assessments of fisheries policy, Management Strategy Evaluation (MSE) and the development and testing of Operational Management Procedures (OMPs)[33] (Butterworth *et al.*, 1997; Little *et al.*, 2006). The first involves the testing of management strategies development and implementation, in the long term, simulating the adaptive recurrent process. Its outcomes are used as background, additional information in the decision making process. OMPs, on the other hand, use simulations to test, also in the long run, the entire adaptive management process, but the outcome is the joint selection, by the

[32] Mediation is a conflict resolution method involving the intervention of a neutral third party to promote agreement (compromise, reconciliation or settlement) among parties involved, in which each party's views are translated for the others (Babin and Bertrand, 1998) and Merriam-Webster online dictionary 2006). Scientists might play a useful role in such mediation.

[33] OMPs are defined by Butterworth *et al.* (1997) as a set of clearly defined decision rules, pre-agreed among the parties, specifying: how the regulatory mechanism is set; what data are needed; and how these will be analysed and used.

management authorities and the stakeholders, of a set of management rules and measures (the OMP proper) which, by tacit agreement, will hold until they are formally changed through another run of the participatory exercise. The first one aims at 'landscaping' the decision making field. The second is more prescriptive and intends to 'fix the rules of the game' through a pre-agreed course of action.

The modus operandi at the interface between fishery science, policy and decision making has taken different forms depending on local culture, political set-up and history. In many countries where a functional administration exists, the interface between stock assessment and management involves the activity of expert groups in charge of elaborating the best scientific evidence available. Their conclusions and advice are considered by advisory committees with the role of helping in the translation of the scientific advice into acceptable decisions and measures. For MSE, a two-stage process has been used. During the first largely qualitative stage, with strong stakeholders' involvement, potential management scenarios are collectively developed and evaluated *ex ante*. During stage 2, the options considered worthy enough are then quantitatively refined, assessed and tested for robustness against uncertainty using complex simulation models (Smith *et al.*, 2007). For OMPs (as well as for MSE) the process involves intensive computer simulations coupled with intense stakeholder contribution and feedback.

In general, the analytical and modelling frameworks of the natural system have been fairly extensively developed as well as the process of direct consultation with the sector and other stakeholders. However, active participation in the whole process, from model conceptualization to scenario simulation and evaluation does not seem to be the rule and formal participation seems to be largely limited to providing reactions towards the end of the analytical and modelling processes. In addition, the social analysis and modelling of the human subcomponent of the fishery system is still very embryonic and have had little application, if any, in real decision making processes (Pido *et al.*, 1997; Pahl-Wostl, 2002). In addition, the negotiation involved in the final decision making process does not seem to involve any mediation by social scientists.[34]

Participatory Practices in Environmental Management

In the area of environmental management, and in addition to intense participation in decision making, some branches of social sciences have apparently successfully developed a practice of *participative modelling* in which stakeholders are intensively involved. Their contribution ranges from model scoping and building – exploring the stakeholders' mental model of the system – to the running and analysis of scenarios simulations. In this process, scientists and

[34] Jentoft (1998) argued, indeed, that fishermen (contrary to fish) could talk by themselves and did not need (contrary to fish) to have the biologists representing their interest at the negotiation table. He advocated instead an auditing role, assessing governance performance as a neutral, non-involved party.

stakeholders conceive, develop and assess jointly the sustainability of the various policy and management scenarios (cf. examples in Bousquet and Le Page, 2004). The rationale is that, when developed through participative scenario construction,[35] community learning is more effective than learning through demonstration of scenarios based on academic theories - and the process leads to greater procedural legitimacy, facilitating consensus and compliance.

Developed by social scientists, IA and participatory integrated assessment (PIA) have been proposed as operational frameworks for post-normal science applications to environmental issues (Toth, 2003). Combining overlapping definitions, IA could be defined as:

> a scientific 'meta-discipline' and structured process of dealing with complex problem domains that integrates knowledge from various scientific disciplines and/or stakeholders and makes it available for societal learning and decision making processes.
>
> (Rotmans, 1998; Rotmans and Van Asselt, 2001).

The key goal of such an IA is to integrate the knowledge from different disciplines about an environmental problem along the whole chain of causes and effects to provide useful information for decision makers. Combining analytical and participatory methods, it involves an interdisciplinary participatory process combining scientific and informal knowledge, and explicitly accounting for uncertainty, societal values, perceptions and preferences. Moving away from the participative use of models as a means for unilateral transfer of scientific knowledge to the public, they propose instead a process in which human participants are interacting with one another and with expert knowledge in a structured and decision-oriented setting and in which expert and local forms of knowledge are integrated while citizens are included into the model-building process at an early stage, using techniques such as focus groups and actors platforms (Pahl-Wostl *et al.*, 2000). The Integrated Assessment Society[36] was established in 2003. The *International Journal of Integrated Assessment*, published by Springer, The Netherlands, was created in 2000.

Integrated Advisory Process

Conceptual frame

Because both the strictly scientific assessment and the science/policy advisory processes require intense stakeholder participation, there is a need for an IAP, combining scientific assessment and advice development through which participation can be optimized. Developed in support of strategic policy and decision making in highly uncertain contexts with high risks to society, the types of participative IA as described above appear to have all the ingredients needed to deal effectively with most complex fishery situations, including their management in cross-sectoral environments and or the management of small-scale

[35] Particularly when the scenarios are based on stakeholders' perceptions.

[36] http://www.tias.uni-osnabrueck.de/integrated_assessment.html

fisheries in an ecosystemic perspective. Conversely, the most advanced procedures used to develop and test fishery management procedures appear to have followed a similar path with, perhaps three differences:

- Less use of the stakeholders' knowledge.
- Little (explicit) input from stakeholders into the early stage of model conceptualization.
- Little or no inputs from sociologists (as opposed to economists).

Building on the practices that have evolved in fisheries and in environmental decision making, an ideal IAP could look like the one represented in Fig. 11.1, combining a 'hard' analytical sub-process with a 'softer' negotiating process.

The analytical sub-process is science-based, 'hard', using the best quantitative and qualitative scientific information available about the system, to model it. This sub-process: (i) uses the ***objectively true*** knowledge obtained through conventional peer-review process[37]; (ii) seeks actively for relevant informal knowledge relating to model structure, linkages and assumptions and integrates it after validation; (iii) is informed about societal and policy objectives as well as limits or constraints of an environmental, social, economic and ethical nature; (iv) produces policy and management options to be considered in the following sub-process; and (vi) provides monitoring and iteratively revises its advice based on performance assessment. This process is more effective when it uses computer

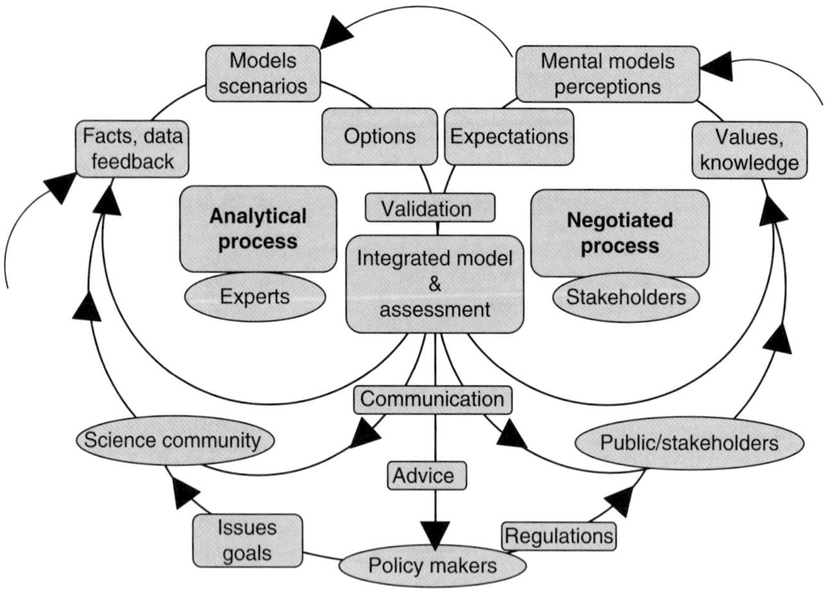

Fig. 11.1. Integrated advisory process. Modified from Garcia and Charles (2006). Inspired by and redrawn from Pahl-Wostl (2002).

[37] Ravetz (2003) stresses that conventional peer-review lacks the safeguard of external assessment and is very vulnerable to the state of morale of each scientific community.

simulation models through which the consistency of system representations can be checked and sensitivity to uncertainty can be evaluated. This sub-process might be more efficient in the *ex ante* and *ex post* evaluation of effectiveness, cost and benefits of policy and management options. It would benefit from a well-bounded interaction with stakeholders in the constitution of the knowledge base and model conceptualization.

The negotiated sub-process is people-based, possibly mediated, 'soft', involving the policy and managing institutions and stakeholders concerned - including scientists. It helps deal with aspects not easily amenable to formal modelling such as ill-defined areas, views and perception, evolution of preferences, reaction to surprises, etc. The process: (i) identifies the views and perceptions of the main stakeholders; (ii) agrees on applicable societal objectives and constraints (fed into the analytical hard process); (iii) contributes knowledge and participates to the validation of the models used in the analytical process; and (iv) considers and ranks the options identified before feeding them into the political decision making process. In this process, knowledge mediation may play an important role. Knowledge representations and mediation become an issue when different sources of knowledge have to be integrated to fit in a common analytical or discursive framework, i.e. in interdisciplinary work or multi-stakeholder debates. In this case, mediation may be needed to forge consensus among different stakeholders with different interests or visions as much as to assist in the interface between stakeholders and the science/policy tandem. This sub-process would be well suited for the *ex ante* evaluation of alternative implementation trajectories with different socio-economic implications.

This dual process - which obviously needs also to be iterative, using feedback, to fit into an adaptive management process - is certainly not totally new to the fisheries arena and may be prefigured by the processes used to elaborate and agree on OMPs and for *ex ante* MSE.[38] These processes could be strengthened (with stronger participation) and need significant generalization. A widespread adoption, with a well-developed social research interface, would provide the much needed platform for a stronger and faster integration of the respective assets of social and fishery science and facilitate a progressive move towards an integrated, area-based, inter-sectoral management framework.

Strategic and Operational Implications

The implementation of an IAP requires the development of an enabling environment within which the different streams of information, presently developed separately in different institutions and processes, meet. However, developing an effective two-way participatory science–policy interface for strongly participative governance is a challenge (Engels, 2005). Difficulties to be expected in the analytical process relate classically to the modelling process itself: e.g. selecting

[38] Although no detailed descriptions of these processes have been found (except in Smith, 2006) particularly regarding their social interface.

a type of model; deciding on its scope, scale and sensitivity and combining qualitative with quantitative information.[39]

Difficulties can also be expected in the negotiated process: e.g. adequate representation, diversity of perceptions and objectives, freeriders, risk of miscommunication and changing perspectives, etc. The operationalization of an IAP will naturally meet with these as well as with new problems emerging from the integration between the two processes (different experiences, horizons, languages and perceptions; need for two-way 'translations' and mediation; degree of authority allocated by the State to stakeholders; decision process, etc.). There is not enough room in this chapter to deal with all of these and many more are probably still to be discovered.

A comprehensive treatment of the issues - beyond the paradigmatic ones already discussed in the first section - that emerge in the social exercise of decision making is beyond the scope of this chapter and my disciplinary competence. In the following sections, I shall, therefore, only briefly review those most obvious to me.

Functional requirements

Analysing science-based decision processes and their outcomes in the environmental management arena, characterized by high environmental risk, uncertainties and political stakes, Jasanoff (2004) stresses that scientists and advisers are involved 'in a hybrid activity that combines elements of scientific evidence and reasoning with large doses of social and political judgement'. She concludes that an effective process requires:

- Agreement by scientific advisers involved in *experts groups* to participate also in the negotiating process leading to decisions, i.e. interacting with the *Advisory Committees* and assisting in the decision making process.[40]
- A dual decision making process: (i) among scientists, to resolve scientific uncertainties or divergences that carry political weight (and societal costs); and (ii) between policy makers and stakeholders, including scientists, to decide on the best course of action. The process is ineffective in a context of scientific disagreements, disparate social and political values or when occurring in an adversarial (judicial) context.
- The defence of 'strict boundaries' around the scientific process to preserve the independence and objectivity necessary for the social and political acceptability of the advice. This point is crucial in a system in which non-scientists and scientists are called to closely cooperate and where the risk for each of them to 'cross the line' is high.[41]

[39] This is very important as part of the information will be numerical, easily integrated in conventional mathematical models and the part coming from the operators will often be qualitative (e.g. integrated in the model form of decision rules).

[40] Recognizing that final decisions are a matter of societal choice.

[41] With stakeholders tempted to interfere with scientific interpretation of facts and scientists tempted to play a role in objective setting or decision making.

- Commitment of all actors to moderate their views towards an acceptable societal position. This requirement recognizes that freeriders or stakeholders with no willingness to reach agreement may stall the process. One could add that, as mentioned earlier, one should be aware of the fact that, despite their willingness, some actors may lack the capacity to express themselves and efforts are therefore needed to detect and correct such biases.

Jasanoff notes in addition that the outcome of the process should be a state of knowledge that satisfies the test of scientific acceptability and supports reasoned decision making while ensuring those exposed to risk have not had their interests sacrificed to scientific uncertainty. The existence of a formal and transparent process of this type, in the long term, may produce scientifically robust knowledge (*sensu* Gibbons, 1999) and help maintain credible and relevant scientific excellence, while reducing the need for 'underground' political pressure.

Working across disciplines

The occurrence of a co-evolution of science and governance requires the simultaneous existence of a supply of science and a demand for governance. This implies that the policy makers and managers request explicitly - and provide the conditions for - a more comprehensive form of advice. This also implies that the present purely operational horizon of management is complemented by a strategic one, with a more complete set of objectives, a multi-scale and multi-stakeholder vision, and a more democratic process. Finally, this implies a change in fishery research development policy, aiming at a closer collaboration if not integration between social and biophysical sciences, e.g. changing the recruitment patterns in fishery research centres, providing incentives for interdisciplinary strategic analysis (to attract academics in the decision making area) and to foster the joint development of comprehensive models (including agent-based simulation models and games).

These changes do not need to happen all at once. Progressive changes are more pragmatic and more likely to be adopted, as shown in the countries where processes of this nature have already started to function. A wide interdisciplinary collaboration around simulation platforms and IAPs may lead to the development of a trans-discipline (*sensu* Flinterman *et al.*, 2001), but the transition to that ideal will necessarily be pragmatic.

Stakeholders

Because of the interconnectedness within and between ecosystems, the number of stakeholders potentially involved could be overwhelming. Stakeholders include researchers, managers and decision makers, policy makers, representative organizations (e.g. NGOs) and, obviously, end-users. A high level of participation of the latter is essential for a democratic process. User-centred simulations allow the end-users to actively participate in reruns of the simulations exploring differing scenarios, proceeding usually by iteration.

Funtowicz and Ravetz (1990) argued for participation in the process of all those with a desire to participate in the resolution of the issue, a proposal raising non-trivial problems of interaction cost and effectiveness. A central problem is that of striking a balance between the broadest possible representation and affordable interaction costs. Once the stakeholders have been defined, it is important to define the roles they are called (and willing) to play (e.g. right holders, stewards, providers of data and traditional knowledge, scientific 'assistants' in model development, actors in a simulation game) in line with the *extended peer-review* concept. As these roles are demanding, however, it is important to ensure that the stakeholders involved are motivated and are capable of participating effectively in order to maintain their commitment to the process.

The role of scientists

The role of scientists in the analytical process is classical and needs no development here. It is also clear that scientists do not have any role in the final decision making, in which only stakeholders and the authorities in charge negotiate which implementation option, among the ones elaborated through the IAP, is the 'best' or most acceptable. However, Jasanoff (2004) calls for participation of scientists as stakeholders in this last sub-process in order to provide the explanations and clarifications that stakeholders may require about the system 'reality' or best scientific understanding, in order to build consensus among groups of stakeholders with different understanding and objectives. This is a role that social scientists may be comfortable with, but which natural scientists are usually reluctant to play, concerned as they are – at least rhetorically – to keep the science process clear from political interference. Jasanoff (2004) insists on the fact that it is indeed fundamental for the success of the integrated process that the '*boarder work*', i.e. the work undertaken by the scientist to keep the scientific arena free from non-scientific influence, be conducted in a credible manner.

How much is too much?

The present generalized crisis of fisheries indicates that the status quo is not an option. But the situation is different in different countries and the need for an IAP varies among fisheries. There is an obvious gradient of increasing complexity from the high seas to the coastal zone where so many fisheries, aquaculture systems, other economic industries and societal requirements interact. A similar gradient may exist between sparsely populated mountains and coastal areas, lake shores or flood plains. Complexity is probably maximal in small-scale fisheries because of the intricacies of the web of activities ensuring and threatening the livelihood of small-scale fishing communities.

But how much effort should be made to face the consequences of complexity? The question was already raised in relation to the systemic approach to fisheries (Garcia and Charles, 2006). It is equally relevant here. The form of governance that would fit with a constructivist view of the world is highly integrative and participative. However, interaction costs can become prohibitive and stall decision making mechanisms (see section on participation).

Recognizing these difficulties and adding the problems hindering interdisciplinarity, how far should the integration process go? One could wonder (with Strand, 2003) to what extent the introduction of new embryonic approaches and instruments, the effectiveness of which is still to be fully tested, is preferable to continued use of the present well-tested approaches and methodologies, patching the system to mitigate its shortcomings.

Time frames

The perspective of adopting an IAP has implementation cost implications that are far from trivial. Based on examples related to climate change, acidification and air traffic issues in The Netherlands, the process of resolving major policy issues, with significant societal stakes, at the national level, required many months, possibly a year, of preparations. The full cycle of resolution needs many months, at least, and possibly a few years (as in the air traffic issue which required 4 years (Hischemöller *et al.*, 2001, p. 67).

One might argue that fisheries sustainability is a mature enough issue to be dealt with within shorter time frames. The issue is well established. Its causes have been abundantly described, analysed and agreed. A number of approaches to resolving the problem have already been tested under various conditions. A global scale agreement is available through the Code of Conduct for Responsible Fisheries. The ecosystem and precautionary approaches have already been adopted. However, the resolution of the issues at local, national and regional levels, where real decisions are made, using an ecosystemic or an integrated coastal management framework, raises a number of non-trivial issues the resolution of which, in a highly participative environment would certainly require time. As a consequence, an IAP would probably be best suited for elaborating multi-year strategic frameworks for fisheries, within which the more OMPs would be implemented.

Coherence with the Law of the Sea Convention

The Convention requires that decisions be based on the best scientific evidence available, a requirement adopted when fishery science was essentially equated with natural science, and considered in post-normal science as an 'elitist' mode of operation (Toth, 2003). Although a number of subsequent instruments, explicitly related to it, have added the requirement to add other forms of knowledge to the foundation for decision making, the fundamental requirement for the scientific nature of the information remains. As a consequence, while necessarily drifting towards post-normal modes of operation, the enquiry process will need to remain demonstrably scientific if a collapse of the decision making process is to be avoided. The point is forcefully made by Jasanoff (2004).

Checks and balances

Closely involving stakeholders in the complex exercise of fisheries assessments for decision making has obvious advantages already mentioned (increased legitimacy, compliance, etc.). The danger of manipulation of the advisory processes by industry or the central administration exists as illustrated, for example, by Likens (1992) in the case of acid rain and by Ludwig *et al.* (2001) in the case of the functioning of the US Environmental Protection Agency (EPA), Saville (1979) and Hutchings *et al.* (1997) for fisheries. Leaving aside the possibility of outright scientific fraud, the possibility exists for scientists to inadvertently 'manipulate' the system. The very close relationship between scientists and managers for decades led managers to 'understand' scientists to the point of adopting their paradigm (e.g. the reference to MSY in UNCLOS and the use of operational command-and-control measures). Shifting the management paradigm towards using fishing rights and incentives, following a stronger involvement of economists is happening. Other examples are given by the increased consideration given to protected areas with the increased involvement of ecologists and to various forms of participatory management following the increased involvement of social sciences. On the one hand, this illustrates the co-evolution of science and governance (Garcia, 1994a, 1996; Catanzano and Rey, 1997; Rice, 2005). On the other hand, the process reflects the danger for science objectivity of getting too deeply 'involved' in the decision interface.

In theory, the joint operation of the relevant disciplines on common models and processes should reduce the risk. However, the unexpectedly high success apparently obtained by participative exercises using complex simulation models on difficult societal issues (Bousquet and Le Page, 2004; J. Weber, Paris, personal communication) could raise a concern. As simulation-based integrated processes are institutionalized, is there not a risk to lose again the critical sense needed on both sides of the participatory process as the stakeholders are 'bought in' to believe in the reality of the models?

As complex models and participation are both indispensable, the solution to this dilemma may be in the introduction of additional checks. Two avenues can be seen:

- The repetition of the participative modelling exercise at regular intervals, e.g. in line with the adaptive management principles, to detect unexpected and undesirable changes.
- The use of additional peer review, e.g. by panels composed of both scientific and industry experts external to the IAP.

Regional and high seas issues

The issues raised above are particularly conspicuous in a regional framework where a tradition of scientific collaboration usually exists, but where the role of social sciences (including economics) in decision making is close to non-existent. The stakeholder issue is highly sensitive and the identification of genuine stakeholders very preliminary (UNU-IAS, 2006). Where resources are not covered by conventions the stakeholders to global commons are a fuzzy group. Where conventions exist, flag States parties are obvious stakeholders and the participation is reduced formally to the national representation, which may involve some representatives of industry or environmental NGOs. The role of other fishery stakeholders potentially significant in a broader perspective (port States, conservation conventions, environmental NGOs, etc.) is evolving, e.g. through the 1995 UN Fish Stock Agreement. Freeriders are common in the form of fishers from non-parties to commissions and fishers practicing rampant illegal fishing. Can they be considered stakeholders? Can solutions be found without involving them? If not, how can they be involved? The issue is even more sensitive in relation to straddling stocks because of coastal State sovereign rights.

The mechanisms usually used to calculate allowable catches and distribute quotas in regional fisheries management organizations (RFMOs) are already overburdened. Could the needed, more integrated, processes be afforded and practical? A solution might be found in a combination of operational (year-to-year) processes with reduced interaction and strategic (multi-year) processes involving a more complete representation and comprehensive debate. The latter may hopefully occur in European waters through the newly established Regional Advisory Councils (RACs).

Other issues

Many other issues cannot be dealt with here such as: (i) participation of non-collaborative stakeholders; (ii) confusion of facts and values; (iii) unrealistic expectations of participants; (iv) moderation (facilitation) techniques during participative exercises; (v) ICTs (e.g. for email conferences); (vi) confidentiality issues; (vii) management of the iterative process; (viii) capacity-building in the developing world; (ix) development of open, web-based, simulation platforms; (x) role of scientists as facilitators of the debate and consensus-building, including among stakeholders; and (xi) archiving the outcomes of the processes. Details on some of these and other issues may be found in Engels (2005).

Overall, the use of an IAP is likely to be more difficult than the present processes as it will highlight the usually ignored complexity and uncertainty in the issues at stake. While no cost-benefit analysis of the issue seems to be available yet, one would hope that the added efforts needed, through such systems, to reach decision will, hopefully, be compensated by better compliance, increased sustainability and reduced externality cost to society.

Discussion and Conclusions

The developments in the science-public-policy interface in the climate change issue are an excellent example of what happens when the societal demand and debate expands faster than the available information base. The debate develops both locally and internationally, sometimes with questionable coherence. The information available appears to be at best, incomplete and, at worst, totally debased by multiple definitions and usages. The resulting uncertainty stems from prejudices, perspectives, incompleteness of information, differences in accuracy and sensitivity of forecasting tools. Excessive levels of uncertainty mask useful signals, calling for uncertainty reduction. The latter may quash unconventional, but ultimately valuable, ideas (Hisschemöller *et al.*, 2001). Balancing the various tensions, interests and perceptions is difficult and the science-policy interface has a key role to play in the process.

Contrary to climate change, fisheries sustainability has been a local, national, regional and global issue for more than a century and institutions have been established to deal with it at all these levels. The UNCLOS requirement for 'best scientific evidence available' as the basis for decision making has been confirmed in all modern international fishery agreements. However, doubts or distrust of science (or of processes involving science) have grown during the last two decades, particularly in relation to large-scale environmental and ethical issues laden with strong socio-cultural and economic consequences.

The recognition of the need for more effective fishery governance has led to the recognition of a need for change, *inter alia*, in the nature and content of the scientific input (i.e. in relation to the ecosystem and the people) as well as the process leading to decision making and, more specifically, for a much more effective and decisive participation of stakeholders. Unavoidably, this evolution leads also to a debate about the place of science in a democratic society and the interaction between them (Prigogine and Stengers, 1979; Walters and Hilborn, 1976; Holling, 1993; Funtowicz and Ravetz, 1995; Ravetz, 2003).

The recognition of the highly systemic nature and complexity of fisheries highlights the fact that the scientific 'evidence' is obtained at the cost of significant reductions and approximations (Garcia and Charles, 2006). The repeated calls for precaution (FAO, 1995; Garcia, 1996b) and adaptive management (Walters and Hilborn, 1976; Holling, 1994) amount to a formal recognition that the *objective truth* about fisheries can only be progressively obtained as the system is tested and that such truth will probably never be complete enough to totally avoid any risk. Such recognition brings conventionally 'hard' fishery science closer to its 'softer' social science counterparts with which it has to develop stronger interactions. The progressive recognition of the 'soft' aspects of 'hard' sciences seems to be indeed pervasive across the whole range of sciences, including in atomic and astronomical physics where, following on the 1927 Uncertainty Principle of Heisenberg, Loughlin (2005), a Nobel Laureate in physics, indicates that 'we are living the end of reductionism, the false ideology that promised to Mankind the control of everything'.[42]

In response, the science-governance tandem has already evolved, in some countries, to progressively more complete and integrated forms of scientific enquiry, as well as to more significant participation of broadening circles of stakeholders in policy development, decision making and management (e.g. in co-management and integrated management).

However, as science deals more closely with heavily value-laden societal issues and risk (including risk provoked by science itself) a broader part of the society wants to have a say in science-governance, shaping its course, fixing its priorities, checking its relevance and accuracy. This 'new' development, which may indeed be a recurrent one, would probably sound to most 'hard' scientists as an anathema. However, when scientists are called, and accept, to deal with questions that are not entirely or satisfactorily answerable by science, a stronger interaction with society, its perceptions and values (ethics) may be unavoidable (Leshner, 2005). Keeping independence and objectivity in that process and perhaps building up more humility will be the challenge of the next few decades (Garcia and Charles, 2006).

Having looked at the changing societal demand, the evolving response of science to such demand, in the biophysical and social sciences, and the evol-

[42] Translated from the French version of the book.

ution of the interface between science, policy and management, in fisheries and in other arenas, the chapter suggests generalizing the institutionalization of an IAP. Such a process is already well advanced in very few countries where fishery participatory management has been institutionalized, e.g. around OMPs and MSEs. Where this has happened, efforts are still needed to: (i) fully integrate social scientists in the process; and (ii) strengthen the role of stakeholders in model development and scenarios analyses. Where this has not yet started, significant institutional reform and capacity-building is needed to allow the development of such a process.

The necessity to develop more complete and more participative science-based advisory systems (e.g. to comply with the LOSC requirements) is evident and so is the need to link social sciences more closely to the process. This requirement for a closer association may open the postmodern ideological trap of the real value of science and the scientific process in societal knowledge-building and decision making.

The philosophical debate between normal/positivist and post-normal/constructivist scientists may not be closed (Gould, 2000; Ludwig *et al.*, 2001; Novotny *et al.*, 2003; Leshner, 2005). It is only a part of the debate between natural and social sciences, 'exact' and 'non-exact' sciences, and on the role of non-scientific knowledge and subjective perceptions in decision making that has been going on for centuries. The growth of societal concern (if not distrust) with science and technology since the late 1950s and the new arising of religious fundamentalisms has led some observers to feel that postmodernism attempted to drag science from its pedestal of objective truth by declaring it to be pure social construct, opening the way to a new eco-fundamentalism riding workhorses such as the precautionary principle.

The developments in the fisheries arena, in the last decade, may have demonstrated that, while uncertainty may be the Trojan horse of postmodernism, it not only exists, but it is often large, with significant consequences, and it can be scientifically taken into account (in line with the FAO guidelines on the precautionary approach to fisheries). In the end, as recently stated by one of the fathers of constructivism:

> the basic fault lines today are not between people with different beliefs but between people who hold these beliefs with an element of uncertainty and people who hold these beliefs with a pretense of certitude. There is a middle ground between fanaticism and relativism.
>
> (Berger, 1997)

The conclusion that these fanaticisms have little relevance in the field is reflected in Gould (2000) who argues that the 'science wars' between hard and soft sciences are an academic debate than can only exist in the minds of those not involved in the real problem solving on the ground. Gould argues against the fundamentalist dichotomized typology of science[43] as being 'doubly and deeply

[43] Gould (2000) refers specifically to the 'war' between '"realists" (including nearly all working scientists), who uphold the objectivity and progressive nature of scientific knowledge, and "relativists" (nearly all housed in faculties of the humanities and social sciences within our universities) who recognize the culturally embedded status of all claims for universal factuality and regard science as just one belief among many alternatives'.

fallacious, wrong as an interpretation of the nature and history of sciences, and wrong as an example of our deeper error in parsing . . . natural continua as struggles between opposing sides'. A similar conclusion was reached regarding the division of science in Mode-1 and Mode-2 and the theory of a massive shift from the first to the second by Auranen (2005) who argues instead for a progressive shift in the proportion of these two 'types' in the scientific landscape. The call for an alliance emerges also in ecology in general, where the need to integrate scientific knowledge with political, economic, social, ethical and religious insights, tempered with the respect for human dignity and for the biosphere is recognized (Ludwig *et al.*, 2001).[44]

While working towards the establishment or strengthening of IAPs, the postmodern ideological trap should therefore be consciously identified and avoided, pragmatically moving towards closer collaboration with social sciences in a societal context, which calls for a more precautionary, hence more 'post-normal' attitude of science and a higher degree of participation by stakeholders. Following the advice of Gould (2000), the alliance between conventional and social sciences should aim at both the continuing social construction and the growing empirical adequacy of scientific knowledge.

Following the conclusion drawn by Hisschemöller *et al.* (2001) in the climate change issue, the improvement of the state of fisheries demands improved communication among the stakeholder communities involved. The IAP would play an important role in this respect, but the implementation difficulties are not trivial.

The timing implications and the interaction costs involved tend to indicate that the process would be most useful and more appropriate for strategic planning of the fisheries economic development and its environmental, ecosystemic, management. The smaller scales and shorter time frames required by operational fisheries management call for simpler science-decision. The need for better interaction with stakeholders at that level remains, but one might hope that, if fisheries operational management processes were undertaken explicitly within strategic planning frames agreed through an IAP, such simplification might be possible, reducing the recurrent interaction costs.

Based on the experience and analyses available (including Jasanoff 2004) a successful IAP will require *inter alia* (see also section on strategic and operational implications): (i) agreement by scientists to participate also in the negotiating process; (ii) a dual decision making process: among scientists and among stakeholders; (iii) involvement of policy makers and stakeholders from the onset of the process; (iv) the defence of 'strict boundaries' between science and 'politics' and between the advisory and decision process (see also Gallopin *et al.*, 2001: 228); and (v) commitment of all actors involved to reach agreement. I would add that the need to strengthen checks and balances requires: (vi) an open information system accessible to all stakeholders where the institutional memory is conserved despite changes in the scientific, administration and stakeholders' institutions; and (vii) a formal calendar for a cyclic (adaptive) process

[44] As a mathematician, Ludwig can hardly be considered as an environmental fundamentalist.

recognizing formally the possibility to revise/renegotiate positions as experience is gained or conditions change.

In closing, I will stress with Gallopin *et al.* (2001) that the necessary shift towards a more 'post-normal' attitude is certainly not a call for relaxing our scientific rigour; on the contrary.

Acknowledgements

The elaboration of this chapter led to numerous discussions with many of my FAO colleagues which helped me focus and track specific difficulties. I am grateful, in particular for the sharp comments, useful suggestions and infinite patience of Gabriella Bianchi and Kevern Cochrane. I owe also some of the thoughts behind this chapter to Tony Charles with whom I prepared a paper on fishery systems and linkages on which this chapter draws heavily (Garcia and Charles, 2007). However, all errors or misinterpretations are mine only. Last but not least, I must also thank the two reviewers for their very useful comments and suggestions.

References

Auranen, O. (2005) Social capital in Mode-1 and Mode-2 knowledge production: a Finnish case study. Paper prepared for the Fifth Triple Helix Conference 'The Capitalization of Knowledge: Cognitive, Economic, Social and Cultural aspects'. Turin, Italy, 18–21 May 2005.

Babin, D. and Bertrand, A. (1998) Managing pluralism: subsidiarity and patrimonial mediation. *Unasylva* 49, 19–25.

Barreteau, O., Bousquet, F. and Attonaty, J.M. (2001) Role-playing games for opening the black box of multi-agent systems: methods and lessons of its application to Senegal Valley irrigated systems. *Journal of Artificial Societies and Social Simulations* 4(2). Available at: http://www.soc.surrey.ac.uk/JASSS/4/2/5.html

Bavinck, M., Cluenpagdee, R., Diallo, M., van der Heijden, P., Kooiman, J., Mahon, R. and Williams, S. (2005) *Interactive Fisheries Governance: Guide to Better Practice*. Eburon Academic Publishers, Delft, The Netherlands.

Berger, P. (1997) Epistemological modesty: an interview with Peter Berger. *The Christian Century* October 29, 972–978.

Berger, P. and Luckman, T. (1966) *The Social Construction of Reality. A Treatise in the Sociology of Knowledge*. Anchor Books, Garden City, New York, pp. 51–55, 59–61.

Berkes, F. (1999) *Traditional Ecological Knowledge and Resources Management*. Taylor & Francis, Philadelphia/London.

Berkes, F., Mahon, R., McConney, P., Pollnac, R. and Pomeroy, R. (2001) *Managing Small-Scale Fisheries. Alternative Directions and Methods*. International Development Research Centre, Ottawa, Ontario, Canada.

Bousquet, F. and Le Page, C. (2004) Multi-agent simulations and ecosystem management: a review. *Ecological Modelling* 176, 313–332.

Butterworth, D.S., Cochrane, K.L. and De Oliveira, J.A. (1997) A. Management procedures: a better way to manage fisheries? The South African Experience. In: Pikitch, E., Huppert, L.D.D. and Sissenwine, M.P. (eds) *Global Trends: Fisheries Management*. American Fisheries Society, Bethesda, Maryland, pp. 83–90.

Catanzano, J. and Rey, H. (1997) La recherche halieutique entre science et action:

reflecxions sur fond de crise. *Nature, Science et Société* 5, 19–30.

Chambers, R. (1994) The origins and practice of participatory rural appraisal. *World Development* 22, 953–969.

Charles, A. (2001) Sustainable Fishery Systems. In: *Fish and Aquatic Resources Series*. Blackwell Science, Oxford, UK, 370 pp.

Checkland, P.B. (1981) *Systems Thinking, Systems Practice*. Wiley, Chickester, UK.

Chopra, K., Kadekodi, G.K. and Murty, M.N. (eds) (1989) Peoples' Participation and Common Property Resources. *Economic and Political Weekly* 24, 189–195.

Christensen, V. (2006) Turning the tide is easy enough. The Sea Around Us. Newsletter, 33, 5 pp.

Cochrane, K.L. (2002) A fishery manager's handbook. Management measures and their application. *FAO Fisheries Technical Paper* 424, 231 pp.

Darier, E., Gough, C., De Marchi, B., Funtowicz, S., Grove-white, R., Kitchener, D., Guimaraes Pereira, A., Hackley, S. and Wynne, B. (1999) Between democracy and expertise? Citizens' participation and environmental integrated assessment in Venice (Italy) and St Helens (UK). *Journal of Environmental Policy & Planning* 1, 103–120.

Degnbol, P., Gislason, H., Hanna, S., Jentoft, S., Raakjær Nielson, J., Sverdrup-Jensen, S. and Wilson, D.C. (2006) Painting the floor with a hammer. *Marine Policy* 30, 534–543.

Eckley, N. (2001) Designing effective assessments: the role of participation, science and governance, and focus. Report of a workshop co-organized by the European Environment Agency and the Global Environmental Assessment Project, Copenhagen, Denmark, 1–3 March 2001. 23 pp.

Engels, A. (2005) The science-policy interface. *The Integrated Assessment Journal* 5, 7–26

FAO (1995) Precautionary approach to fisheries. Part 1: guidelines on the precautionary approach to capture fisheries and species introductions. FAO, Rome, Italy.

FAO (2003) Communication and natural resource management. FAO, Rome, Italy.

Flinterman, J.F., Teclemariam-mesbah, R., Broerse, J.E.W. and Bunders, J.F.G. (2001) Trans-disciplinarity: the new challenge for biomedical research. *Bulletin of Science, Technology and Society* 21, 253–266.

Funtowicz, S. (1989) Post-normal science: a new science for new times. *Scientific European* 20–22.

Funtowicz, S. (2002) Post-normal science. Science and governance under conditions of complexity. *Environment Preservation* 17, 63–74.

Funtowicz, S. (2006) Post-normal science. Environmental policy under conditions of complexity. EC-JRC/ISIS, Ispra, Italy, 8 pp.

Funtowicz, S. and Ravetz, J.R. (1990) *Uncertainty and Quality in Science for Policy*. Kluwer Academic Publishers, Dordrecht, The Netherlands.

Funtowicz, S. and Ravetz, J.R. (1991) A new scientific methodology for global environmental problems. In: Constanza, R. (ed) *Ecological Economics: The Science and Management of Sustainability*. Columbia University press, New York, pp. 137–152.

Funtowicz, S. and Ravetz, J.R. (1994a) Uncertainty, complexity and post-normal science. *Environmental Toxicology and Chemistry* 13, 1881–1885.

Funtowicz S. and Ravetz, J. (1994b) Risk management, uncertainty and post-normal science. In: Reinertsen, H. and Haaland, H. (eds) *Sustainable Fish Farming*. Proceedings of the First International Symposium on Sustainable Fish Farming, Oslo, Norway, 28–31 August 1994. A.A Balkema, Rotterdam/Brookfield, pp. 261–270.

Funtowicz, S.O. and Ravetz, J. (1995) Risk management, uncertainty and post-normal science. In: Reinertsen, H. and Haaland, H. (eds) *Sustainable Fish Farming*. Brookfield, Rotterdam, 261–270 pp.

Funtowicz, S.O., Martinez-Alier, J., Munda, G. and Ravezt, J.R. (1999) Information tools for environmental policy under conditions of complexity. European Environment Agency. Environmental Issues Series, No. 9, 35 pp.

Gallopin, G.C. (2003) In: Guimaraes Pereira, A., Aquera Cabo, M. and Funtowicz, S. (eds). International workshop on Interfaces Between Science and Society. Collecting experience for good practice. Milan, Italy,

27–28 November 2003. Book of abstracts. Knowledge Assessment Methodologies.

Gallopin, G.C., Funtowicz, S., O'Connor, M. and Ravetz, J. (2001) Science for the twenty-first century: from social contract to scientific core. *International Journal of Social Science* 53, 219–229.

Garcia, S.M. (1994) The precautionary principle: its implications in capture fisheries management. *Ocean & Coastal Management* 22, 99–125.

Garcia, S.M. (1996a) Fishery research and management: virtues and constraints of a symbiosis. In: Meyer, R.M., Chang, Z., Windsor, M.L., McCay, B.J., Hushack, L.J. and Muth, R.M. (eds) Theme 2. Fisheries resource utilization and policy. Proceedings of the World Fisheries Congress. Oxford and IBH Publishing Co., New Delhi and Calcutta, India, pp. 174–195.

Garcia, S.M. (1996b) The Precautionary Approach to Fisheries and Its Implications for Fishery Research, Technology and Management: An Updated Review. FAO Fisheries Technical Paper 350/2. Precautionary approach to fisheries: FAO, Rome, Italy, 1–75 pp.

Garcia, S.M. (2005) Fishery Science and Decision Making: Dire Straights to Sustainability. *Bulletin of Marine Science*, 76, 171–196.

Garcia, S.M. and Charles, A.T. (2007) Fishery systems and linkages: from clockworks to soft watches. Paper presented at the ICES International Symposium on Fisheries Management Strategies, Galway, Ireland, 27–30 June 2006. *ICES Journal of Marine Science*, 64(4).

Garcia S.M. and Grainger, R. (1997) Fisheries management and sustainability: a new perspective of an old problem? In: Hancock, D.A., Smith, D.C., Grant, A. and Beumer, J.P. (eds) *Developing and Sustaining World Fisheries Resources: The State of Science and Management*. 2nd World Fisheries Congress. CSIRO, Australia, pp. 631–654.

Garcia, S. and Reveret, J.P. (1991) La recherche face à la pêche artisanale. In: Durand J., Lemoalle, J. and Weber, J. (eds) *Recherche et structure des pêches artisanales: paradigmes et méthodes de recherche. Une introduction*. ORSTOM-IFREMER, Paris, pp. 253–268.

Garcia, S.M., Zerbi, A., Alliaume, C., Do Chi, T. and Lasserre, G. (2003) The ecosystem approach to fisheries: issues, terminology, principles, institutional foundations, implementation and outlook. *FAO Fisheries Technical Paper* 443, 71.

Giampietro, M. (2003) The crash of reductionism against the complexity of reality. In: Guimaraes Pereira, A., Aquera Cabo, M. and Funtowicz, S. (eds) *International Workshop on Interfaces Between Science and Society. Collecting experience for good practice*. Milan, Italy, 27–28 November. Book of abstracts: 61. European Communities.

Gibbons. M. (1999) Science's new social contract with society. *Nature* 402, 81–84.

Gibbons, M., Limoges, C., Nowotny, H., Schwartzman, S., Scott, P. and Trow, M. (1994) *The New Production of Knowledge. The Dynamics of Science and Research in Contemporary Societies*. Sage, London/ New Delhi, India, 170 pp.

Gould, S.J. (2000) Deconstructing the 'Science Wars' by reconstructing an old mold. *Science* 287, 253–261.

Guimaraes Pereira, A. and Funtowicz, S. (2003a) Quality-assurance by extended peer review: tools to inform debates, dialogues and deliberations. *Technikfolgenbschatzung. Theorie und praxis* 2, 74–79.

Guimaraes Pereira, A. and Funtowicz, S. (2003b) Methods for citizen involvement in new governance. Reflections based on three empirical cases. *Technikfolgenbschatzung. Theorie und praxis* 2, 57–62.

Guimaraes Pereira, A., Aquera Cabo, M. and Funtowicz, S. (eds) (2003) International Workshop on Interfaces Between Science and Society. Collecting experience for good practice. Milan, Italy, 27–28 November. Book of abstracts.

Hilborn, R. and Ludwig, D. (1993) The limits of applied ecological research. *Ecological Applications* 3, 550–552.

Hisschemöller, M., Tol, R.S.J. and Vellinger, P. (2001) The relevance of participatory approaches in integrated environmental assessment. *Integrated Assessment* 2, 57–72.

Holling, C.S. (1978) Myths of ecological stability: resilience and the problem of failure. In: Smart, C.F. and Standbury, W.T. (eds) *Studies on Crisis Management*. Institute for Research in Public Policy, Montreal, 97–109.

Also in *Journal of Business Administration*, 9, 97–109.

Holling, C.S. (1993) Investing in research for sustainability. *Ecological Applications* 3, 552–555.

Holling, C.S. (1994) New science and new investments for a sustainable biosphere. In: Jansson, A., Hammer, M., Folke, C. and Costanza, R. (eds) *Investing in Natural Capital. The Ecological Economics Approach to Sustainability*. Island Press, Washington, DC, Covelo, California, pp. 57–73.

Holling, S. (1998) Two cultures of ecology. *Conservation Ecology* (online) 2(2), 4.

Hutchings, J., Walters, C.J. and Haedrich, R.L. (1997). Is scientific enquiry incompatible with government information control? *Canadian Journal of Fisheries and Aquatic Sciences* 54, 1198–1210.

ICES (2000) Report of the study group on ecosystem assessment and monitoring. ICES CM/2000/E:09. International Council for Exploration of the Sea. Copenhagen, Denmark.

Jasanoff, S. (2004) *The Fifth Branch: Science Advisers as Policy Makers*. Harvard University Press, Cambridge, Massachusetts.

Jasanoff, S., Colwell, R., Dresselhaus, M.S., Goldman, R.D., Greenwood, M.R.C., Huang, A.S., Lester, W., Levin, S.A., Linn, M.C., Lubchenco, J., Novacek, M.J., Roosevelt, A.C., Taylor, J.E. and Wexler, N. (1997) Conversation with the community: AAA and the millennium. *Science* 278, 2066–2067.

Jentoft, S. (1998) Social science in fisheries management: a risk assessment. In: Pitcher, T. (ed) *Reinventing Fisheries Management*. Kluwer Academic Publishers, London, pp. 177–184.

Jentoft, S. (2000) The community: a missing link of fisheries management. *Marine Policy* 24, 53–59.

Jentoft, S. (2005) Fisheries co-management as empowerment. *Marine Policy* 29, 1–7.

JRC Team, Guimaraes Peireira, A., Blázquez, J., Corral Quintana, S. and Funtowicz, S. (2005) The GOUVERNe project at the Joint Research Centre. Tools to inform debates, dialogues and deliberations. European Community. Available at: http://alba.jrc.it/gouverne

Kesteven, G.L. (1996) A fisheries science approach to problems of world fisheries or: three phases of an industrial revolution. *Fisheries Research* 25, 5–17.

Kuhn, T.S. (1962) *The Structure of Scientific Revolutions*. University of Chicago Press, Chicago, Illinois.

Le Fur, J. (1996) Simulation de la dynamique d'une exploitation à l'aide de modèles multi-agents: le cas de la pêche artisanale au Sénégal. Programme environnement, vie et sociétés. Lettres des programmes interdisciplinaires de recherche du CNRS, 15, 16–20.

Le Fur, J. (1998) Modelling fishery activity facing change: application to the Senegalese artisanal exploitation system. ORSTOM.

Leshner, A.I. (2005) Where science meets society. Editorial. *Science* 307, 815.

Levin, S.A. (1993) Forum. Science and sustainability. *Ecological Applications* 3, 546–547.

Likens, G.E. (1992) The ecosystem approach: its use and abuse. In: Kinne, O. (ed) *Excellence in Ecology*, vol 3. Ecology Institute, Luhe, Germany, 166 pp.

Little, L.R., Fulton, E.A., Gray, R., Hayes, D., Lyne, V., Scott, R., Sainsbury, K. and McDonald, A.D. (2006) Multiple Use Management Strategy Evaluation for the North West Shelf: Results and Discussion. *North West Shelf Joint Environmental Management Study Technical Report* Vol 18. CSIRO, Hobart, Tasmania.

Loughlin, R.B. (2005) Un univers différent. In: Fayard (ed) *Le temps des choses*. Paris. Translated from: A different universe. Reinventing physics from the bottom up. Basic Books. New York.

Ludwig, D., Hilborn, R. and Walters, C. (1993) Uncertainty, resource exploitation and conservation: lessons from history. *Science* 260, 17.

Ludwig, D., Mangel, M. and Haddad, B. (2001) Ecology, conservation and public policy. *Annual Review of Ecology and Systematics* 32, 481–517.

McGlade, J.M. and Allen, P.M. (1984) The fishing industry as a complex system. In: Mahon, R. (ed) *Towards the Inclusion of Fishery Interactions in Management Advice*. pp. 209–216.

Gibbons, M., Limoges, C., Novotny, H., Schwartzman, S., Scott, P. and Trow, M. (1994) *The New Production of Knowledge: The Dynamics of Science and Research*. Sage, London.

Novotny, H., Scott, P. and Gibbons, M. (2001) *Re-Thinking Science, Knowledge and the Public Age of Uncertainty*. Polity Press, Cambridge.

Novotny, H., Scott, P. and Gibbons, M. (2003) *Introduction. Mode-2 Revisited:*

The New Production of Knowledge, vol 41. Kluwer Academic Publishers, Minerva, pp. 179–194.

Ostrom, E. (2000). Private and common-property rights. In: Bouckaert, B. and De Geest, G. (eds) *Encyclopedia of Law and Economics*. Edward Elgar, Cheltenham, England, pp. 332–379.

Pahl-Wostl, C. (2002) Agent-based simulation in integrated assessment and resources management. In: Rizzoli, A.E. and Jakeman, A.J. (eds) *Integrated Assessment and Decision Support*, Proceedings of the First Biennial Meeting of the International Environmental Modelling and Software Society, Lugano, Switzerland, Vol. 2, SEA, pp. 239–244.

Pahl-Wostl, C. (2005) Actor-based analysis and modelling approaches. *The Integrated Assessment Journal* 5, 97–118.

Pahl-Wostl, C., Schlumpf, C., Bussenschutt, M., Schonborn, A. and Burse, J. (2000) Models at the interface between science and society: impacts and options. *Integrated Assessment* 1, 267–280.

Pido, M.D., Pomeroy, R.S., Garces, L.R. and Carlos, M.B. (1997) A rapid appraisal approach to evaluation of community-level fisheries management systems: framework and field application at selected coastal fishing villages in the Philippines and Indonesia. *Coastal Management* 25(2), 18–20.

Pollnac, R.B. (1994) Research directed at developing local organizations for people's participation in fisheries management. In: Pomeroy, R.S. (ed) *Community Management and Common Property of Coastal Fisheries in Asia and the Pacific: Concepts, Methods and Experiences*. ICLARM, Manila, The philippines, pp. 94–106.

Pomeroy, R.S. (1995) Community-based and co-management institutions for sustainable coastal fisheries management in Southeast Asia. *Ocean & Coastal Management* 27, 143–162.

Prigogine, I. and Stengers, I. (1979) *La nouvelle alliance. Métamorphose de la science*. Gallimard, Paris, France, 302 pp.

Ramanath, A.M. and Gilbert, N. (2004) The design of participatory agent-based simulations. *Journal of artificial societies and social simulations* 7, 12. Available at http://jasss.soc.surrey.ac.uk/7/4/1.html

Ramchandran, C. (2004) Teaching not to f(in)ish? A constructivist perspective on reinventing a responsible marine fisheries extension system. Central Marine Fisheries Research Institute (CMFRI). Kochi, Kerala, India, *Responsible Fisheries Extension Series*, vol 6, 114 pp.

Ravetz. J. (2003) Science and society interfaces: process. In: Guimaraes Pereira, A., Aquera Cabo, M. and Funtowicz, S. (eds) *International Workshop on Interfaces Between Science and Society. Collecting experience for good practice*. Milan, Italy, 27–28 November 2003. Book of abstracts: 3. European Communities.

Rice, J. (2005) Implementation of the ecosystem approach to fisheries management – asynchronous co-evolution at the interface between science and policy. *Marine Ecological Progress Series* 300, 265–270.

Rothschild, B.J. (1971) Systems analysis and fishery management. *Proceedings of the Indo-Pacific Fisheries Council* 14(II), 13–23.

Rotmans, J. (1998) Methods for integrated assessment: the challenge and opportunities ahead. *Environmental Modelling and Assessment* 3, 155–179.

Rotmans, J. and Van Asselt, M.B.A. (2001) Uncertainty in integrated assessment modelling: a labyrinthine path. *Integrated Assessment* 2, 43–55.

Saville, A. (1979) Discussion and conclusions of the symposium on the biological basis of pelagic fish stock management. *Rapports et Procés-verbaux Reunion des Experts du Conseil International pour l'Exploration de la Mer* 177, 513–517.

Shin, Y.-J., Shannon, L.J. and Cury, P.M. (2004) Simulations of fishing effects on the Southern Benguela fish community using an individual-based model: learning from a comparison with ECOSIM. *African Journal of Marine Science* 26, 95–114.

Smith, A.D.M., Fulton, E.J., Hobday, A.J., Smith, D.C. and Shoulder, P. (2007). Scientific tools to support practical implementation of ecosystem-based fisheries management. ICES Symposium on Fisheries Management Strategies. Copenhague, ICES. *ICES Journal of Marine Science* 64, 633–639.

Strand, R. (2003) The uncertainty and complexity of managing uncertainty and complexity. In: Guimaraes Pereira, A., Aquera Cabo, M.

and Funtowicz, S. (eds) *International Workshop on Interfaces Between Science and Society. Collecting experience for good practice.* Milan, Italy, 27–28 November 2003. Book of abstracts: 67. European Communities.

Sutinen, J.G. and Soboil, M. (2003) The performance of fisheries management systems and the ecosystem challenge. In: Sinclair, M. and Valdimarsson, G. (eds) *Responsible Fisheries in the Marine Ecosystem*. FAO/CAB International, Rome, Italy/Wallingford, UK, pp. 291–309.

Toth, F.L. (2003) State of the art and future challenges for integrated environmental assessment. *Integrated assessment* 4, 250–264.

UNU-IAS (2006) Implementing the ecosystem approach in open ocean and deep sea environments. An analysis of stakeholders, their interests and existing approaches. United Nations University Institute of Advanced Sciences. Yokohama, Japan, 39 pp.

Walters, C.J. (1980) Systems principles in fisheries management. In: Lackey, R.T. and Nielsen, L.A. (eds) *Fisheries Management*. Blackwell, Oxford, Boston, 422 pp.

Walters C.J. and Hilborn, R. (1976) Adaptive control of fishing system. *Journal of Fisheries Research Board of Canada* 33, 145–159.

Wiber, W., Berkes, F., Charles, A.T. and Kearney, J. (2004) Participatory research supporting community-based fishery management. *Marine Policy* 28, 459–468.

Wilson, D.C. and Delaney, A.E. (2005) Scientific knowledge and participation in the governance of fisheries in the North Sea. In: Gray, T. (ed) *Participation in Fisheries Governance*. Kluwer Academic Publishers, Dordrecht, The Netherlands.

12 The Ecosystem Approach to Fisheries Management: an Industry Perspective

GRIMUR VALDIMARSSON

FAO, Rome, Italy

Abstract

The ecosystem approach to fisheries management (EAFM) is not a new, radical way of managing fisheries. It is a clarification and expansion of the list of many things that fisheries management already takes into consideration while trying to regulate fishers and avoid overfishing. Unfortunately, the management of far too many fisheries around the world still does not even result in this most fundamental goal of avoiding overfishing and frequently results in the dissipation of the potential economic benefits that fisheries could produce – all before considering additional fine-tuning.

The fishing industry is experiencing the same demands that other sectors of the food business are facing: constant calls for lower prices, better quality, safer foods, longer shelf life and foods that are more convenient to prepare. Impressively, by and large, the food industry has delivered on all of these counts.

Consumers are now perceived as demanding fish products that come from well-managed fishery resources that do not cause major disruptions to the ecosystem. Because customer perception is the reality of the marketplace, food retailers are complying with this new consumer demand and mainstreaming ecolabelled fish products, as part of various market-driven labels. Ecolabelling is becoming a vehicle for gaining an advantage in the marketplace. It is good business to be 'green'.

Over recent decades, the fish processing sector of the industry has gone through a significant change in philosophy on how to respond to ever more demanding product safety and quality regimes. Largely, the successful approach has been to move away from centralized government controls towards making the industry responsible for implementing 'self control' systems that are verified and audited by governments. Such systems require well-specified objectives and ample record-keeping for industry to be able to prove due diligence.

For the fishing industry, the framework provided by the ecosystem approach to fisheries management (EAFM) is essentially yet another Quality System with which industry will have to comply. Indeed, many already call it the Environmental Management System (EMS). When EAFM/EMS are operational, they will require the collection of far more information regarding fishing activities than hitherto provided by fishermen. There will also be calls for transparency in presenting such information to the authorities and society at large in the form of 'social and environmental reporting'. Unfortunately, within 'Olympic' or 'Derby' fishery management regimes, where management measures already increase the costs of fishing by limiting seasons and gears, such additional and costly information gathering will be difficult to achieve. Fortunately, under management regimes where fishing rights are secure and foster long-term stewardship of the fishery ecosystem, fishermen will be able to realize the value of ecolabelled quality products as part of the good business of being green.

> The problems associated with open access systems have been discussed previously and, in order to avoid these problems, the allocation of various forms of explicit, legally enforceable fishing rights is integral to EBFM. . . . Further, the right to fish must carry with it the obligation to fish in a responsible manner, so as to ensure ecosystem conservation.
>
> (FAO, 2003)

Introduction

In terms of employment, some 40 million people are recorded as fishers and aquaculturists around the world, most of whom belong to the small-scale sector (FAO, 2007). Fish and fishery products have been a success on the international markets; they enjoy a good reputation for culinary qualities as well as providing excellent, nutritious food. Fishery products have become the most traded food product internationally, with some 37% (by volume) moved across national borders.

Developing countries provide more than half of this trade, which has been a major success for their economies. However, the surge in fish exports from developing countries has been such that concerns have been raised over its implication on food security in these countries. A recent FAO study (FAO, 2005) showed that generally this export has generated so much revenue that the effect on food security has been positive. The study, however, supported the notion from many other sources that effective fisheries management is necessary for this fish trade to be sustainable.

The fishing industry is experiencing the same demands that other sectors of the food business are facing: constant calls for lower prices, better quality, safer foods, longer shelf life and foods that are more convenient to prepare. Impressively, by and large, the food industry has delivered on all of these. It is fair to say that the incentives for the industry come from both ends, i.e. by demand of governments, but increasingly by an ever more demanding market. It should also be borne in mind that, even if fishery products are generally in high demand and enjoy a good reputation on the markets, the fish industry is in direct competition with other proteinous, low-cost foods, such as chicken.

Globalized markets implicitly call for economy and competition. Therefore, industry is inherently sceptical when new demands are being made regarding new process demands, not the least when these are made by government agencies. In the quest for a more 'level playing field' when it comes to international trade, abandoning subsidies and other support measures, governments are now inclined to charge industry for the services it renders to them. When numerous different government agencies have to inspect a processing facility to certify everything from hygienic standards to the safety of electrical installations, this can become a substantial cost to business. Today, some governments are even recovering a significant part of the cost of fisheries research even though such support is not considered as a subsidy. Thus, the competitive edge for companies or industries is easily distorted.

Industry is also very worried about extremism when it comes to the effects of their operations on the environment. A good example is the suggestion that in fisheries management we should go beyond the precautionary principle and apply the 'Reversal of the Burden of Proof' for environmental legislation without any practical guidance on how that can be done in practice. By applying that principle the onus would be on the industry to prove that they will not affect the environment with their operations (Dayton, 1998).

New demands are coming from the markets: consumers are now perceived as insisting that fish products should come from well-managed fishery resources without causing major disruptions of the ecosystem. Because customer perception is a reality in the marketplace, food retailers are complying with this new consumer demand and mainstreaming ecolabelled fish products. Ecolabelling is becoming a vehicle for gaining an advantage in the marketplace. It apparently is becoming good business to be visibly concerned by the environment.

The question before us, then, is how to implement 'effective' ecosystem-based fisheries management. This has to be seen in the context that the management of far too many fisheries around the world still have not dealt with the most fundamental goal of avoiding overfishing and, as a result, the potential economic benefits that fisheries could produce are dissipated.

The purpose of this chapter is to lay out some industry perspectives regarding the ecosystem-based approach to fisheries management, why the adoption of such an approach could prove beneficial to the fishing industry and what needs to occur for the industry to warrant implementing it.

Who Are 'The Industry' and What Do 'They' Do?

To this author, the 'fishing industry' represents the many different groups of people whose objective is to gain financially (and nutritionally) from fishing activities. Thus, as a corollary, all scales of fishing activities – from large-scale or industrial and commercial to small-scale, artisanal and even subsistence fisheries – represent economic activity.

Despite the concept of 'the fishing industry' as a sector, fishers all over the world are not highly organized and, by and large, do not have significant influence on government decisions regarding their sector. Most of them work independently, and the image is still much of the 'rugged individualist' a kind of antithesis to the organized industrialized worker. The sector attracts individuals that prefer risk and excitement rather than organized predictable activities. Indeed, when fishermen around the world are asked what the government or the central authority can do for them, the answer seems to be a fairly universal, 'Get off our backs!'

However, fishermen in all of these sectors do respond to economic forces. They respond to incentives, commercial pressures and markets - and their responses to these can be summed up as 'How can I gain the most?' or 'How can we catch the most with the least effort and cost?' Swordfish boat captain Linda Greenlaw famously wrote, 'He who catches the most the fastest wins' (Greenlaw, 1999). This statement should not be surprising; it reflects the basic economics of competing to capture a share of a common property resource.

Ineffective Fisheries Management

The main message coming from a vast bulk of analyses around the world is this: The methods by which the world has chosen to govern fisheries are largely ineffective.

Many authors have pointed out that today's management objectives are often unclear and even contradictory (Cochrane, 2000; Cochrane and Doulman, 2005). In addition, when cultural values or socio-economic objectives of fisheries are also taken into account, management indeed becomes complicated. This, of course, makes fisheries management more difficult than for most other production systems which simply concentrate on producing goods that the market wants at competitive prices.

When the need for limiting the amount of fish caught first became widely acknowledged in the 1950s and the 1960s, fisheries agencies focused on the need to ensure that enough fish remained in the water to keep reproducing. However, this biological mandate expanded as new instruments were developed. The Rio Declaration, the Agenda 21 of the Summit on Sustainable Development, the Convention on Biological Diversity, and the Code of Conduct for Responsible Fisheries are all instruments which recognize the nutritional, economic, social, environmental and cultural importance of fisheries and the interests of all those concerned with the fishery sector - in addition to the need for biological considerations. In summary, contemporary thinking focuses not only on the biological sustainability of the fishery sector, but also on its contribution to the economy and society as a whole.

This author believes that the poor management regimes have somewhat echoed the seductive inexhaustibility idea, i.e. some restrictions to fishing may be necessary but that it is not necessary to be too pedantic about it as 'long gives the ocean'. Exact landing figures are really not necessary - keep the accountants away. Most countries have excellent fisheries laws but it is common that they do not anticipate the need for severely restricting access to the resources. Ironically, the main lesson that we have learned - or should have learned - about fisheries over the last 50 years is that, sooner or later, the open or semi-open access fisheries will suffer from overfishing.

Although we have massive literature and persistent media attention highlighting the symptoms of poor fisheries management policies and texts describing where we want fisheries to be, there has been much less attention given to the fundamental flaws in current management policies and to what is at the heart of getting to sustainable fisheries. The fisheries management failures, largely the institutional ones, were neatly summarized by Garcia (2005) as:

- The free and open nature of fisheries (lack of enforceable rights).
- Perspectives of short-term political or financial gain or losses.
- Poor decision making processes (in Regional Fisheries Management Organizations (RFMOs)).
- Poor participatory nature of most systems (top-down systems).
- Lack of transparency and accountability.
- Weak enforcement (both at national and regional levels).
- Scientific uncertainty (affecting the precision of the advice) and errors (affecting the accuracy of interpretations).

It is worth reflecting upon why fisheries management, also in the developed world, is still commonly based on an ineffective approach that has been abandoned in most other sectors of the modern economy; namely centralized, top-down, 'command-and-control' measures that impose increasingly more restrictive and more costly controls on fishers. Despite the well-documented limitations of such approaches, they are the systems of choice by most government fisheries management agencies (FAO, 2003b).

Practically, the only incentive for the fishers to comply with centralist-imposed controls is the risk of being caught. However, when the chances of detection are low and the penalties associated with illegal fishing are light, avoiding regulations is not seen as a serious infringement, and detected non-compliance is simply a business cost. Recent news regarding illegal, unreported and unregulated (IUU) fishing indicates that vessels are now even equipped with anti-aircraft guns to avoid arrest (The Economist, 2006), which clearly indicates how strong the economic incentives driving IUU fishing are.

The consequences of trying to rely solely on top-down controls are well known: fleets exceed recommended limits, rules relating to fishing gears and closed areas are not respected, overfishing occurs, managers impose more restrictive regulations and the cycle repeats itself. Even when catch limits such as total allowable catch (TAC) are set, the fisheries are inevitably conducted as a race or 'Olympic fisheries' resulting in economic waste, overcapacity and poor quality of the products.

Fortunately, the ineffectiveness of current fisheries management practices – to restrict fishing inputs (such as gear, time and area) and to impose total catch limits – is gaining recognition, not the least in light of the technological advances in every sphere relating to finding, locating and capturing fish.

What is even more important in terms of ecosystem-based approaches to fisheries management is that, in such competitive circumstances, the fishers have a very real commercial interest in *not* providing information about their true activities – something that is of fundamental importance when it comes to undertaking more sophisticated reporting – which is what any form of EBFM will require.

What Is at the Heart of Successful EBFM?

> Achieving objectives in EBFM requires suitable management measures. Again, the general principles used in conventional single-species management will still apply but will need to be extended. . . . It should be well understood that broadening of the fisheries management approach does not call for any revolution. Adding ecosystem considerations to present methods can be done gradually.
>
> (FAO, 2003a)

Yet, despite voluminous literature on the ecosystem approach in fisheries, there is no universal agreement on what the EAFM entails.

There is agreement, however, as described by FAO (2003a) about the following: 'Implicit in all initiatives for management of the ecosystem is recognition that man cannot manage the ecosystem as such, but only the human activities using it.' The Reykjavík Conference on Responsible Fisheries in the Marine Ecosystem reinforced the point that ecosystems, as such, cannot be controlled. They are simply too complicated. Stefansson acknowledged that '[a]s the models become more detailed and complex, they are able to address more issues that are of concern to managers, but at the same time it becomes ever more difficult to interpret results' (Stefansson, 2003).

So, if we cannot manage ecosystems, per se, what can we do?

We can only hope to manage the activities of the humans engaged in fisheries - and this key point is at the centre of any successful application of the EAFM. Successful EBFM will link biological and conservation objectives *with* the incentives that make people behave the way they do. The importance of human incentives is being strongly emphasized by economists in explaining the successes and failures of different governance regimes (Easterly, 2006).

The Reykjavík Declaration has a clause on the importance of incentives. Article 2 states: 'There is a need to introduce immediately effective management plans with incentives that encourage responsible fisheries and sustainable use of marine ecosystems.'

To do this, there are two major categories of roles and responsibilities that need to be clarified:

- Governments should be responsible for setting the framework for utilization of aquatic resources.
- The fishing industry should attend to the operational issues of fishing and managing their actions.

Finding the Balance: The Food Processing Industry Example

Indeed, this division and balancing of roles and responsibilities is how management has evolved in the food processing industry as it moved from centralized government control systems to a system where the government sets the framework for processing standards and where clear responsibilities are placed on the industry for its own actions.

In the early 1980s 'food safety controls' consisted of government inspectors taking samples from the end products for inspection and analyses in a laboratory. This led to the notion that industry was not really responsible for the safety and quality of its products and that the responsibility really was that of the government agencies in question.

As the safety and quality issues became more demanding and more complex, a radical change occurred. Instead of relying solely on end product sampling undertaken by those outside the production process, a system was put in place that had the objective of preventing the disease agents from getting into the product in the first place. Thus, the authorities became the ones to verify and approve the plans for preventative measures - the Hazard Analysis Critical Control Point (HACCP) systems that companies were obliged to adopt - but it became the responsibility of the companies for operationalizing and running the HACCP plans.

With the HACCP system in place all the actors in the chain - from the primary producer to the final distributor - have had to shoulder their share of ensuring compliance throughout the entire production and distribution chain. As a result, far more information has been collected in the plants regarding the production processes, the results of which were being fed back into adjusting the processing conditions. There is no doubt that strengthening 'own controls' by the industry has had very good overall results in terms of increased food quality and improving the food safety record, as well as being cost-effective.

Today, industry operates various Quality Management Systems (QMSs) that deal with issues from safety, quality, environment and animal welfare; for example, how to avoid catching turtles or dolphins. All of these schemes have one thing in common; they set well-defined objectives on what is to be achieved against which the outcomes are measured.

Pressure to Participate in EBFM

So, why should the industry buy into a new and more complicated way of managing fisheries? The obvious questions the industry asks are: what is in it for us? Does it make sense? Does it improve anything? What happens if we do not comply? The answers to these questions are usually a combination of 'how big is the stick and/or how big is the carrot'. Traditionally, it has been governments that have driven the agenda on how fisheries are managed, but this is changing fast. Under the influence of what has been termed the 'Corporate Social Responsibility Movement' companies increasingly see marketing opportunities in being ecologically responsible. Thus, recycling packaging materials, limiting pollution and using water and energy efficiently (to name only a few) are just some of the ways in which retailers are now demonstrating their environmental responsibilities to consumers.

In the fishery sector there is a similar movement. The requirement that fishermen provide sustainably sourced or ecolabelled fish products is just one of the retail sector's responses to perceived customer preferences. Companies are 'going green' for commercial and economic reasons. While there are still sceptics who label these activities 'green-washing', the fact remains that leading retailers have committed their companies to selling only sustainably sourced fish products.

This signifies an important turning point for industry as it has declared that it accepts to fulfil certain new significant criteria regarding the source of its products. Not only will it have to live up to this promise, but it will also have to be able to prove to the customers that it actually has. That is indeed a strong commitment. If the fishing industry wants to be able to supply eco-friendly products to the 'green' retailers it must find ways to do so at competitive prices.

The Cost Implications of EBFM

Any industry is inherently cost-conscious, and fishers in fisheries of all sizes are not different. They have every reason to review all new demands by governments and buyers with an eye to costs and the impact of regulations on their ability to earn money.

Moreover, increasingly competitive markets, the strong drive to eradicate subsidies and the 'polluter pays' principle enshrined in Agenda 21, i.e. internalization of environmental costs (Scandol *et al.*, 2005), are shifting more and more costs from governments to the fishing sector.

As discussed above, the food processing sector was faced with a similar dilemma. Government regulations tended to be ever more prescriptive without due concern for the cost of imposing controls and maintaining records. As the 'own checks' systems have developed, governments have been more involved with setting minimum standards and objectives for food safety and quality by showing due diligence and setting 'Food Safety Objectives' and letting industry find ways to comply with these in a cost-effective manner. Experience so far suggests that the food industry has been very creative in finding ways to make the products safer.

Whatever way the 'Ecosystem Approach to Fisheries Management' will develop, it is clear that it will require far more information to be collected about the fishing operations than hitherto, and that such information will have to be presented to the authorities and even society at large in a manner that is transparent and verifiable. To prove compliance with ecosystem-related standards, fishing operations would have to address and report such things as amount of by-catch, incidental catch of seabirds, turtles and dolphins, to name only a few. Ultimately, as with other QMSs, the fish producers will have to be able to prove that they have complied, through auditing and verification by independent inspection bodies.

As with other QMSs, industry likes to see clear objectives with the EBFM so that the need for information gathering is defined and how that information will be used and by whom it will be accessed. This underlines all the important issues of incentives for such an undertaking and the cost implications (Valdimarsson and Metzner, 2005), as well as the oft-repeated notion that humans are part of the ecosystem.

For fisheries to become eco-friendly the regular collection and analysis of information is a fundamental requirement, and as with other complex production processes, constant information feedback will be necessary. So, every vessel would have to submit detailed information to the fisheries administrations after each and every fishing trip, and the data would be used together with survey data to detect changes in the ecosystem and to drive responses from management authorities and the industry. This is much in line with what Sissenwine and Mace suggested:

> The fishing industry increasingly recognizes that it must govern itself in an appropriate manner for there to be responsible fisheries. . . . We hope the fishing industry will do the following: Accept responsibility for providing fisheries information. . . . Embrace collaborative research. . . . Be informed participants in the fisheries management decision process. . . . Comply with fisheries management regulations and not tolerate violations. . . . Avoid waste and destructive fishing practices. . . . Be respectful of other stakeholders. . . . Develop training programmes.
>
> (Sissenwine and Mace, 2003)

How then, can we expect the industry to embrace ecosystem-based approaches to fisheries management?

The Importance of Fishing Rights To Achieve EBFM

For all of these to happen, it is clear that fishers - the industry - must have a clear incentive, a real commercial reason, to provide that data. That incentive requires clear fishing rights that eliminate the 'race' for the last fish.

It is being acknowledged that rights and responsibilities go hand-in-hand (Garcia and Boncoeur, 2004) and that without rights there is little reason for fishers to engage in responsible fishing (France and Exel, 2000). Defined and secure fishing rights are the core of what is good fisheries governance (Sinclair *et al.*, 2002). Fish industry leadership is realizing that it needs to make a call for allocation of clearly defined fishing rights. In the 'Industry Perspectives' at the Reykjavík Conference, it was noteworthy that industry representatives expressed the view that property rights are a key issue in establishing sustainability as a key corporate objective (Sinclair and Valdimarsson, 2003).

Without rights, fishing operators cannot focus on minimizing the costs of their activities, nor do they have much incentive to provide value-added products.

The need for fishing rights was described in the FAO paper at the Reykjavík Conference:

> Introducing rights-based management raises the thorny issues of resource allocation, with the selection of the fishing right holders and deciding on characteristics of the rights (exclusivity, security, permanence and transferability). These necessary decisions, with significant long-term benefits for the State, the right-holders and the consumer, can have short-term economic and socio-political costs, which many politicians find hard to face. The shift to EBFM may not resolve the problems of short-term economic and socio-political costs but doing so heightens the urgency for addressing them.
>
> (FAO, 2003a)

Experience shows that making a change towards fishing rights is not easy – not politically, legally, socially or technically (FAO, 2002). We now know that it requires a well-managed process involving all the stakeholders. It is a time-consuming process, meaning a lot of consultations, as fishing rights regimes have to be tailor-made to suit different fisheries, countries and cultures.

> Times have changed, and we can no longer ignore the questions of how to share our limited fisheries resources and how to determine who can catch what, however sensitive these questions may be. Indeed, the longer we avoid implementing allocation mechanisms, the more we risk making decisions that, ultimately, do not lead to fisheries that are as healthy as they could be.
>
> (Nomura, 2006)

Conclusion

The EAFM is a more sophisticated way of managing capture fisheries. As with any other advanced process management system, it requires systematic collection of more information about the process than hitherto.

However, just as Scandol *et al.* (2005) concluded, 'it would be nonsense to assume that the best way to improve a troubled simple management system is to replace it with a highly complex one', and Cochrane (1999) also pointed out the drawbacks of complex system of natural resource management.

Over recent decades the fish processing sector of the industry has gone through a significant change in philosophy on how to respond to ever more demanding product safety and quality regimes. Largely, the successful approach has been to move away from centralized government controls towards making the industry responsible for implementing 'self control' systems that are verified and audited by governments. Such systems require well-specified objectives and ample record-keeping for industry to be able to prove due diligence.

For the fishing industry, the framework provided by the EAFM is essentially yet another Quality System with which industry will have to comply. Indeed, many already call it the Environmental Management System (EMS).

When the EAFM/EMS becomes fully operational, it will require collecting far more information with regard to fishing activities than today. There will also be calls for transparency in presenting such information to the authorities and society at large in the form of 'social and environmental reporting'. Unfortunately, within 'Olympic' or 'Derby' fishery management regimes, whereby management imposes an increase in costs of fishing by limiting seasons and gears, such additional and costly information gathering will be difficult to achieve.

Fortunately, under management regimes where fishing rights are secure and foster long-term stewardship of the fishery ecosystem, fishermen will be pleased to participate in ecosystem-based approaches to fisheries management because they will able to realize the value of certified quality products as part of good business by being 'green'.

Acknowledgements

The author would like to thank colleagues for their consultations on this chapter, particularly Rebecca Metzner and William Emerson.

Mistakes and errors are the sole responsibility of the author.

References

Cochrane, K.L. (1999) Complexity in fisheries and limitations in the increasing complexity of fisheries management. *ICES Journal of Marine Science* 56, 917–926.

Cochrane, K.L. (2000) Reconciling sustainability, economic efficiency and equity in fisheries: the one that got away? *Fish and Fisheries* 1, 2–21.

Cochrane, K.L. and Doulman, D.J. (2005) The rising tide of fisheries instruments and the struggle to keep afloat. *Philosophical Transactions of the Royal Society of London B* 360, 77–94.

Dayton, P.K. (1998) Reversal of the burden of proof in fisheries management. *Science 6* 279(5352), 821–823.

Easterly, W. (2006) *White Man's Burden: Why the West's Efforts to Aid the Rest Have Done so Much Ill and So Little Good*. Penguin Group, New York, pp. 448.

FAO (2002) Report of the Expert Consultation on Catalysing the Transition away from Overcapacity in Marine Capture Fisheries – Rome, 15–18 October 2002 by Metzner, R and Ward, J.M. FAO Fisheries Report, No. 691. FAO, Rome, Italy, pp. 89.

FAO (2003a) Towards ecosystem-based fisheries management. In: Sinclair, M. and Valdimarsson, G. (eds) *Responsible Fisheries in the Marine Ecosystem*. FAO/CAB International, Rome, Italy/Wallingford, UK, pp. 393–403.

FAO (2003b) The ecosystem approach to fisheries. FAO Technical guidelines for responsible fisheries, No. 4, Suppl. 2. FAO, Rome, Italy, pp. 112.

FAO (2005) Responsible fish trade and food security. In: Kurien, J. (ed) *FAO Fisheries Technical Paper*. No.T456. FAO, Rome, Italy, pp. 93.

FAO (2007) State of World Fisheries and Aquaculture. FAO Fisheries and Aquaculture Department, 162 pp.

France, M. and Exel, M. (2000). No rights, no responsibility in use of property rights in fisheries management. *Proceedings of the Fish Rights '99 Conference*. Fremantle, Western Australia, 11–19 November 1999. Mini-course lectures and core conference presentations. FAO Fisheries Technical Paper 404/1, pp. 325–240.

Garcia, S.M. (2005) Fishery science and decision making: Dire straights to sustainability. *Bulletin of Marine Science* 76, 171–196.

Garcia S.M. and Boncoeur, J. (2004) Allocation and conservation of ocean fishery resources: connecting rights and responsibilities. Paper presented at the Fourth World Fisheries Congress. Vancouver, Canada, May 2004.

Greenlaw, L. (1999) *The Hungry Ocean: A Swordboat Captain's Journey*. Hyperion, New York, pp. 288.

Nomura, I. (2006) Allocation issues in fisheries management. Keynote delivered at Sharing the Fish 06 Conference, Fremantle, Western Australia, 27 February.

Scandol, J.P., Holloway, M.G., Gibbs, P.J. and Astles, L. (2005) Ecosystem-based fisheries management: an Australian perspective. *Aquatic Living Resources* 18, 261–273.

Sinclair, M. and Valdimarsson, G. (eds) (2003) *Responsible Fisheries in the Marine Ecosystem*. FAO/CAB International, Rome, Italy/Wallingford, UK, 426 pp.

Sinclair, M., Arnason, R., Csirke, J., Karnicki, Z., Sigurjonsson, J., Rune Skjoldal, H. and Valdimarsson G. (2002) Conference Report on Responsible fisheries in the marine ecosystem. *Fisheries Research* 58, 255–265.

Sissenwine, M.P. and Mace, P.M. (2003) Governance for responsible fisheries: an ecosystem approach. In: Sinclair, M. and Valdimarsson, G. (eds) *Responsible Fisheries in the Marine Ecosystem*. FAO/

CAB International, Rome, Italy/Wallingford, UK, pp. 363–390.

Stefansson, G. (2003) Multi-species and ecosystem models in a management context. In: Sinclair, M. and Valdimarsson, G. (eds) *Responsible Fisheries in the Marine Ecosystem*. FAO/CAB International, Rome, Italy/Wallingford, UK, pp. 171–188.

Valdimarsson, G. and Metzner, R. (2005) Aligning incentives for a successful ecosystem approach to fisheries management. In: Browman, H.J. and Stergiou, K.J. (eds) Politics and socio-economics of ecosystem based management of marine resources. Marine Ecology Progress Series 300, 286–291. Available at: http://www.int-res.com/articles/theme/m300 p241.pdf

The Economist (2006) Swimming against the tide: the battle against illegal fishing off east Africa's coast. August 5–11, 380, 8489, p. 37.

13 Ecosystem Approach to Management: Definitions, Principles and Experiences from Implementation in the North Sea

HEIN RUNE SKJOLDAL AND OLE ARVE MISUND

Institute of Marine Research, Bergen, Norway

Abstract

'Ecosystem' and 'Ecosystem approach' (EA) are intrinsically complex and difficult terms that are used in many different ways, often with lack of clear definitions. We provide here a brief summary of the concepts and definitions to facilitate the further convergence in thinking and to avoid confusion or disagreement on semantic grounds. We also explain the concepts of 'Ecological Quality Objectives' (EcoQOs) as used in OSPAR, and 'Environmental Status' as used in the new EU Marine Strategy Framework Directive (MSFD).

The North Sea Ministerial Conferences (NSC) has had a great influence on the ocean policy development in Europe. The attention given to fisheries at the 4th NSC in Esbjerg in 1995 and the follow-up Intermediate Ministerial Meeting on Integration of Fisheries and Environmental Issues in Bergen in 1997 led to the development of an EA as a guiding principle for integration of fisheries in a wider framework. Work on a system of EcoQOs was initiated after the 3rd NSC in The Hague in 1990, and EcoQOs were seen as an integral component of the EA as adopted in the Bergen Declaration from the 5th NSC in 2002. OSPAR and ICES were requested to continue work to further develop and evaluate the EcoQO system. Several of the EcoQOs relate to fisheries, such as the status of commercial fish populations, by-catch of harbour porpoise and proportion of large fish in bottom trawl surveys in the North Sea. Environmental status is the core concept of the EU MSFD that was adopted in 2008, and Good Environmental Status (GES) is the main objective to be achieved by 2020, at the latest, for defined geographical regions or subregions, equivalent to large marine ecosystems (LMEs). The MSFD is a legal and practical implementation of the EA to integrated management. Full integration of fisheries within a broader framework in the EU will remain a challenge, however, due to the institutional obstacles of working together across the boundaries of agencies and legal instruments.

Introduction

The ecosystem approach (EA) to management is now broadly accepted as a key management principle. The increased awareness and formalization of the EA have

 The Ecosystem Approach to Fisheries (eds G. Bianchi and H.R. Skjoldal)

emerged as a result of international environmental agreements within the frame of the United Nations (UN), and a fundamental description of the basis of an 'ecosystem approach' was first formalized in the Stockholm Declaration in 1972 (Turrell, 2004). The most authoritative account of the EA is that found in Decision V/6 from the meeting of the Conference of the Parties to the UN Convention on Biological Diversity (CBD) in Nairobi, Kenya, in 2000. This decision has an annex with a description, principles and operational guidance for application of the EA (see Vierros, Chapter 3, this volume).

The large marine ecosystem (LME) concept has been the basis for a practical development of EA to the management of marine resources and environment (Sherman, 1995; Sherman, this volume). Currently, 64 LMEs have been identified, dividing mainly the shelf regions of the globe into identified management units. Descriptions of these LMEs along with a range of general scientific and management issues have been considered in a large number of symposia and published books (http://www.edc.uri.edu/lme).

In Europe, the North Sea Ministerial Conferences (NSC) led to important political developments. At the 5th NSC in Bergen in March 2002, the Ministers agreed to a framework for an EA to the management of the North Sea (NSC, 2002). The work leading up to the 5th NSC informed the development of a white paper on integrated marine management by the Norwegian Government (Anon., 2002), which laid the basis for the subsequent integrated management plan for the Barents Sea (Anon., 2006; Olsen *et al.*, 2007; Winsnes and Skjoldal, this volume). The North Sea process also informed work in the European Union leading up to the Marine Strategies Framework Directive that was adopted in early 2008 (see http://www.europarl.europa.eu/oeil/FindByProcnum.do?lang=2&procnum=COD/2005/211).

In this chapter, the conceptual and political developments that have taken place will be further explored and some of the experiences from implementation of the EA in the North Sea and Norway will be described. While this contribution builds on from a previous chapter (Misund and Skjoldal, 2005), more emphasis will be given here to the issue of Ecological Quality Objectives (EcoQOs) as an integral component of the EA.

Ecosystems and the Ecosystem Approach

Concepts, definitions and terminology are important. The words we use to describe something reflect and influence how we think. Words such as 'ecosystem' and 'ecosystem approach' are intrinsically difficult terms and are used in different ways, often with lack of clear definitions. Unclear terminology has no doubt contributed to 'muddy' the debate about EA to management, and still continues to do so.

'Ecosystem' is defined in Article 2 of the UN CBD as 'a dynamic complex of plant, animal and micro-organism communities and their non-living environment interacting as a functional unit'. This definition makes it clear that 'ecosystem' includes not only the living part of nature (as it is often implied), but also the non-living part. The ecosystem is thus the abiotic living space or area for the living organisms, and the organisms of the variety of species that are resident inhabitants, or visitors that come to utilize the resources, within the defined physical space or area of the ecosystem. A simple way of saying this is that ecosystems comprise habitats and species. The distinction is in reality not so sharp since vegetation and structural biota like coral reefs are very much parts of the habitats. A forest without trees would not be a forest habitat.

The difficult part of the definition is the last, 'interacting as a functional unit'. It is intuitively clear what is meant, but how to determine where the interactions that give the character of a functional unit grade over into the next functional unit of the next neighbouring ecosystem? Both in theory and practice this is difficult if ecosystems are to be delineated.

The CBD definition of ecosystem does not specify any particular spatial unit or scale. In COP Decision V/6 it is made clear that it could be any scale from a grain of soil to the entire biosphere dependent on the problem being addressed, and that the application of the EA should be undertaken at the appropriate spatial and temporal scales (principle 7; Vierros, this volume). The concept of LMEs provides guidance to what would be the appropriate scales for integrated ocean management. The L stands for Large, and an LME is defined as 'a relatively large ocean area, typically 200,000 km^2 or larger, with characteristic bottom topography, hydrography and productivity, and trophically coupled populations'.

The LMEs are delineated by applying ecological criteria, and these criteria are those contained in the definition. The 64 or so LMEs that have been delineated globally are located mainly on the shelves surrounding the continents. Here the bottom topography has a strong steering of currents and water mass distribution. The physical conditions again determine the characteristics of plankton production.

The last criterion - having trophically coupled populations - distinguishes LMEs from other classification systems such as biogeographical partitioning. Commercial fish populations are usually important ecological components as prey and predators for other marine biota. Because of their large size, such fish populations require a large living space as they need to feed on the production of prey organisms over a large area. The populations at the same time need to achieve geographical life cycle closure, where spawning areas, larval drift routes, juvenile nursery areas, feeding areas and spawning migrations form a spatial life cycle context in relation to ocean currents and circulation patterns (Skjoldal, 2004b). The distributions of commercial fish populations are therefore an important element to consider when delineating LMEs. Since their distributions reflect circulation and water mass distributions, this criterion is related to the other criteria of characteristic bottom topography, hydrography and productivity.

The LMEs are open systems with flux of water and plankton and migration of fish, birds and marine mammals across their boundaries, making them fuzzy. Nevertheless, there are more or less sharp discontinuities in physical features, such as capes, ridges and fronts, which are reflected in distribution patterns of organisms and can be used when drawing the boundaries of LMEs based on ecological criteria (Skjoldal, 2004a,b). The fact that commercial fish populations play an important role when ecological criteria are applied to delineate LMEs means that the LMEs are very appropriate units for integrated management including the sector of fisheries.

Ecosystem approach is a management principle. As such it builds on the recognition that the nature of nature is integrated and that we must take a holistic approach to nature management. The science to support the EA to management must also be integrated and holistic. A core element of this science is *ecology* with a focus on the properties and dynamics of ecosystems (Fenchel, 1987). Many scientists and managers have recognized the need for an EA for a long time (Likens, 1992), although it is only during the last 10–15 years that a broad awareness of the need for such an approach has developed.

The CBD does not provide a definition of the EA, but characterizes it as 'a strategy for the integrated management of land, water and living resources that promotes conservation and sustainable use in an equitable way'.

A definition of EA was proposed by the ICES Study Group on Ecosystem Monitoring and Assessment (ICES, 2000b): 'Integrated management of human activities based on knowledge of ecosystem dynamics to achieve sustainable use of ecosystem goods and services, and maintenance of ecosystem integrity.'

This formed the basis for the technical definition of EA used in a statement from the First Joint Ministerial Meeting of the Helsinki and OSPAR Commissions (JMM) in Bremen in June 2003 (http://www.ospar.org), and in the work on developing the thematic Marine Strategy within the EU:

> The comprehensive integrated management of human activities based on the best available scientific knowledge about the ecosystem and its dynamics, in order to identify and take action on influences which are critical to the health of marine ecosystems, thereby achieving sustainable use of ecosystem goods and services and maintenance of ecosystem integrity.

It is worth stressing the emphasis on integrated management of human activities in this definition. Sector integration is a key element of the EA and this has scientific and institutional implications. Scientifically we need the ability to assess the combined impacts from different sectors on the marine ecosystems, and institutionally the sectors need to work closely together. This means, for instance, that close collaboration between the fisheries and environmental conservation sectors is a prerequisite for an effective EA to management.

The North Sea Process

The shallow and productive North Sea is an arena for diverse human activities including fishing, dredging, oil and gas exploration, and shipping, and it is the recipient for discharges of nutrients and contaminants from sources on land or offshore. During the last decades, there has been an increasing awareness of the need for measures to protect the environment of the North Sea. Several International Conferences on the Protection of the North Sea have been held, the first in Bremen in Germany in 1987. The Ministers at the 3rd Conference in The Hague in 1990 requested that OSPAR and ICES should establish a North Sea Task Force (NSTF), with one of the tasks being to produce a Quality Status Report (QSR) for the North Sea. This was completed in 1993 (NSTF, 1993) and identified fisheries as having major impacts on the North Sea ecosystem.

Norway hosted an Intermediate Ministerial Meeting on the Integration of Fisheries and Environmental Issues in Bergen in March 1997. In their Statement of Conclusions (IMM, 1997), it was agreed that an EA should be developed and implemented as a guiding principle for the further integration of fisheries and envi-

Fig. 13.1. Reports from Ministerial meetings and workshops that were central in developing the framework for the ecosystem approach (EA) to the management of the North Sea and associated Ecological Quality Objectives (EcoQOs).

ronmental management measures. In the following years, a number of technical meetings were held that elaborated on the concept of the EA and on a number of related issues such as EcoQOs, monitoring and priority research needs (Fig. 13.1).

The ministers at the 5th NSC in Bergen in 2002, agreed to a framework for the EA contained in Annex I to the Bergen Declaration (http://www.regjeringen.no/nb/dep/md/dok/rapporter_planer/rapporter/2002/T-1410-Bergen-Declaration.html?id=420161). This framework has five main components that are linked in a decision cycle as shown in a simplified representation in Fig. 13.2. The five components are:

- *Objectives*, set for the overall condition in the ecosystem and translated into operational objectives or targets.
- *Monitoring* and *research*, to provide updated information on the status and trends and insight into the relationships and mechanisms in the ecosystem.
- *Assessment*, building on new information from monitoring and research, of the current situation, including the degree of impacts from human activities.
- *Advice*, translating the complexities of nature into a clear and transparent basis for decision makers and the public.
- *Adaptive management*, where measures are tailored to the current situation in order to achieve the agreed objectives.

Ecological Quality Objectives

The Ministers at the 3rd Conference in The Hague in 1990 requested that methodology for setting ecological objectives should be developed. This turned out

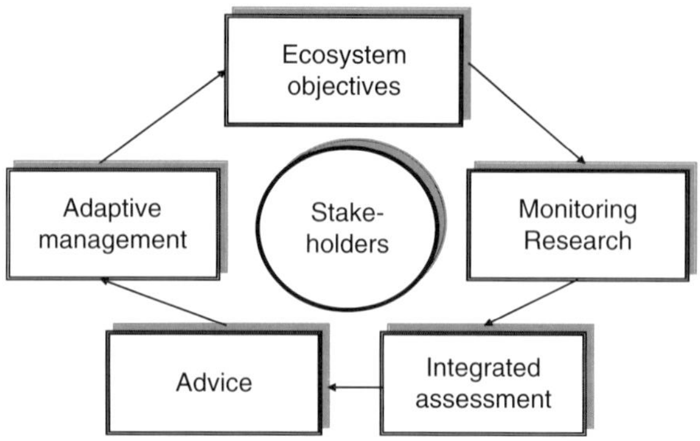

Fig. 13.2. A framework for ecosystem approach (EA) to ocean management with main components or modules shown in an iterative management decision cycle. This is a slightly simplified version of the framework in the Bergen Declaration (NSC 2002). Stakeholders should be included in the process to promote openness and transparency.

to be a long and rather complicated process with many meetings and institutions involved (Fig. 13.3). The NSTF started on the development of ecological objectives, and OSPAR continued the process. A general approach was agreed within the OSPAR system in 1997 using the North Sea as a test case (Skjoldal, 1999). After considerable input of advice from ICES (http://www.ices.dk/products/cooperative.asp), a set of 21 ecological quality elements with EcoQOs set for ten of them, were agreed by the ministers at the 5th NSC (NSC, 2002, Annex 3; Table 13.1).

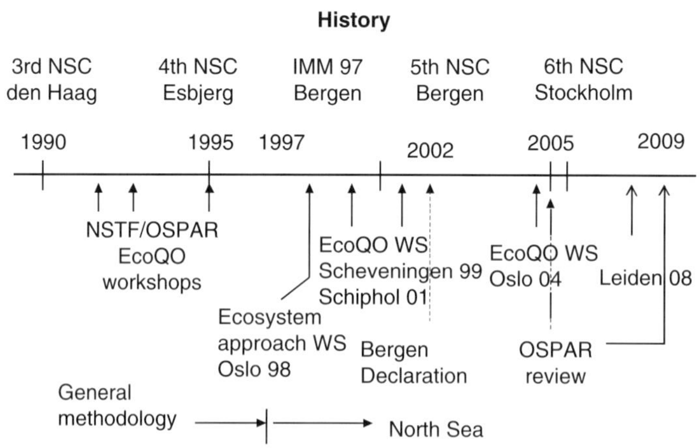

Fig. 13.3. Milestones in the development of Ecological Quality Objectives (EcoQOs) and the ecosystem approach (EA) for the North Sea. North Sea Ministerial Conferences (NSC), EcoQO workshops and OSPAR reviews in 2005 and 2009.

Table 13.1. Ecological quality issues, related ecological quality elements and corresponding ecological quality objectives (EcoQOs), as developed by the Fifth North Sea Conference. EcoQOs are shown in *italics* and advanced ecological quality elements are shown in **bold**. (From OSPAR, 2006b.)

Issue	Ecological quality element and related ecological quality objective (EcoQO)
1. Commercial fish species	(a) **Spawning stock biomass (SSB) of commercial fish species in the North Sea**. *Above precautionary reference points*[a] *for commercial fish species where those have been agreed by the competent authority for fisheries management*
2. Threatened and declining species	(b) Presence and extent of threatened and declining species in the North Sea
3. Sea mammals	(c) **Seal population trends in the North Sea.** *No decline in population size or pup production of ≥10% over a period of up to 10 years*
	(d) Utilization of seal-breeding sites in the North Sea
	(e) **By-catch of harbour porpoises.** *Annual by-catch levels should be reduced to below 1.7% of the best population estimate*
4. Seabirds	(f) **Proportion of oiled common guillemots among those found dead or dying on beaches.** *The proportion of such birds should be 10% or less of the total found dead or dying, in all areas of the North Sea*
	(g) Mercury concentrations in seabird eggs and feathers
	(h) Organochlorine concentrations in seabird eggs
	(i) Plastic particles in stomachs of seabirds
	(j) Local sand eel availability to black-legged Kittiwakes
	(k) Seabird population trends as an index of seabird community health
5. Fish communities	(l) Changes in the proportion of large fish and hence the average weight and average maximum length of the fish community
6. Benthic communities	(m) **Changes/kills in zoobenthos in relation to eutrophication.**[b] *There should be no kills in benthic animal species as a result of oxygen deficiency and/or toxic phytoplankton species*
	(n) **Imposex in dog whelks (*Nucella lapillus*)**. *A low (<2) level of imposex in female dog whelks, as measured by the Vas Deferens Sequence Index*
	(o) Density of sensitive (e.g. fragile) species
	(p) Density of opportunistic species
7. Plankton communities	(q) **Phytoplankton chlorophyll a**[b]**.** *Maximum and mean chlorophyll a concentrations during the growing season should remain*

Continued

Table 13.1. Continued

Issue	Ecological quality element and related ecological quality objective (EcoQO)
	below elevated levels, defined as concentrations >50% above the spatial (offshore) and/or historical background concentration
	(r) **Phytoplankton indicator species for eutrophication.**[b] *Region/area-specific phytoplankton eutrophication indicator species should remain below respective nuisance and/or toxic elevated levels (and increased duration)*
8. Habitats	(s) Restore and/or maintain habitat quality
9. Nutrient budgets and production	(t) **Winter nutrient (DIN and DIP) concentrations.**[b] *Winter DIN and/or DIP should remain below elevated levels, defined as concentrations >50% above salinity-related and/or region-specific natural background concentrations*
10. Oxygen consumption	(u) **Oxygen.**[b] *Oxygen concentration, decreased as an indirect effect of nutrient enrichment, should remain above region-specific oxygen deficiency levels, ranging from 4 to 6 mg oxygen per litre*

[a]In this context 'reference points' are those for spawning stock biomass (B_{pa}), also taking into account fishing mortality (F_{pa}), used in advice given by ICES in relation to fisheries management.
[b]The ecological quality objectives for elements (m), (q), (r), (t) and (u) are an integrated set and cannot be considered in isolation.

Ecological quality (EcoQ) is defined as:

> an overall expression of the structure and function of the marine ecosystem taking into account the biological community and natural physiographic, geographic and climatic factors as well as physical and chemical conditions including those resulting from human activities.

It is expressed by a number of ecological quality elements or variables, reflecting the different parts of the ecosystem, to which objectives or targets (EcoQOs) can be set. In developing the system of EcoQOs (OSPAR, 2006b), the first step was to select ten ecological quality issues which are broad fields or compartments of the North Sea ecosystem. Under each of these issues, one or more EcoQ elements have been identified as individual aspects of ecological quality on which it is appropriate to focus. Finally, the possibility to set EcoQOs for the various EcoQ elements has been explored, where an EcoQO 'is the desired level of an ecological quality. Such a level may be set in relation to a reference level'.

Reference level is the level where the anthropogenic influence on the ecosystem is minimal. This may be difficult to establish in many cases, and a pragmatic choice of reference may be done, for instance as the value at the start of an existing time series of the EcoQ element.

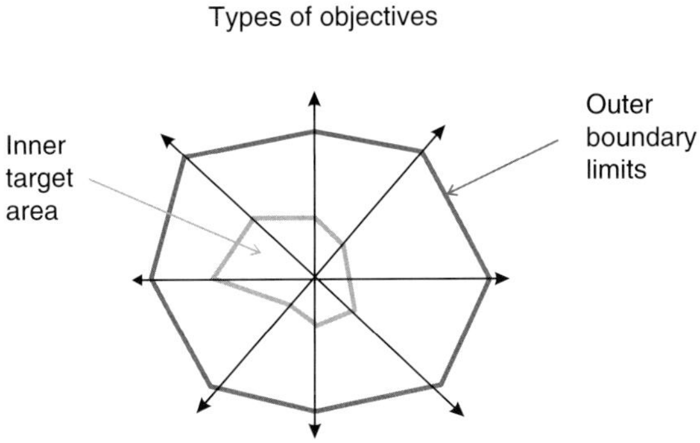

Fig. 13.4. Illustration of two different types of Ecological Quality Objectives (EcoQOs). Ecological quality is the overall state of the ecosystem and can only be expressed by a number of different variables or indicators for different components or aspects of the ecosystem. Operational objectives set for each of these variables or indicators can either be an outer envelope of limits or an inner envelope defining a target area for the state of the ecosystem.

Taken together, the suite of EcoQOs can be seen as an envelope defining the acceptable state of the ecosystem compatible with sustainability. This can either be a wide outer envelope of *limits* which should not be exceeded due to risk of serious or irreversible damage to the ecosystem, or a more restricted inner envelope defined by *targets* based on some considerations of optimum use of ecosystem goods and services (Fig. 13.4). The envelope could also be a combination of the two, with outer boundary limits in some parts and optimum target zones in others.

The North Sea Ministers in 2002 invited OSPAR, in collaboration with ICES and other relevant bodies, to review progress in 2005 with the aim of adopting a comprehensive and consistent scheme of EcoQOs (NSC, 2002). In its review, OSPAR concluded that the system of EcoQOs is a workable and scientifically valid system and a suitable operational tool for implementing the EA to the management of human activities (OSPAR, 2006b). However, it was noted that it was not yet a comprehensive system of EcoQOs, and that additional steps were needed to ensure a successful implementation and to bring the EcoQO system to completion. In 2006 OSPAR adopted an agreement on the application of the EcoQO system describing further actions to be taken, including further evaluation of the EcoQO system in 2008 and 2009 (OSPAR, 2006a). The evaluation aims to produce, in 2008, an assessment of how to achieve a full suite of EcoQOs, including a timetable for its implementation, and, in 2009, an assessment of the results of the EcoQO system as a contribution to the OSPAR Quality Status Report 2010.

Table 13.1 gives an overview of the 10 EcoQ issues, 21 EcoQ elements and 10 EcoQOs agreed in the Bergen Declaration as the basis for the follow-up work by OSPAR and ICES. The set of ten EcoQOs has been used in a pilot project and they are considered the more advanced of the EcoQOs in terms of development and possible application. For each of them, a background document has been prepared and published as an OSPAR Report in the Biodiversity Series

	Pollution	Eutrophi-cation	Litter	Fisheries	Mari-culture	Ecosystem /habitats
1. Reference points for commercial fish species				a		
2. Threatened or declining species				(b)		(b)
3. Sea mammals				e		c,d
4. Seabirds	f,g,h, (k)		i	j, (k)		k
5. Fish communities				l		l
6. Benthic communities	n, (o,p)	m, (o,p)		(o,p)		
7. Plankton communities		q,r				
8. Habitats						s
9. Nutrient budgets and production		t				
10. Oxygen consumption		u				

Fig. 13.5. Distribution of EcoQ elements and EcoQOs in a matrix of the ten EcoQO issues (ecosystem compartments) and human pressures. The letters refer to the identification of the EcoQ elements and EcoQOs in Table 13.1. Letters in parentheses denote elements that potentially could link issues and pressures.

(http://www.ospar.org). The remaining 11 EcoQ elements are considered less advanced and work is progressing on many of them with the aim to include them in the EcoQO system.

The 21 EcoQ elements including the 10 advanced EcoQOs are distributed across various human pressures as illustrated in Fig. 13.5. Four of the EcoQ elements relate to pollution; three for seabirds and one for benthic communities. The EcoQO for proportion of oiled common guillemots is used as an indicator of the level of oil pollution (OSPAR, 2005f), while that for imposex in dogwhelks is an indicator of the toxic effect of TBT (tri-butyl-tin), used previously as an antifouling agent in ship paints (OSPAR, 2005e). Five of the EcoQOs relate to eutrophication (changes/kills in zoobenthos, phytoplankton chlorophyll *a*, phytoplankton indicator species, winter nutrient concentrations and oxygen). They are to be used as an integrated set as part of the application of the OSPAR Common Procedure for the Identification of the Eutrophication Status of the OSPAR Maritime Area (OSPAR, 2005a,2006b). One EcoQ element relates to litter (plastic particles in stomachs of seabirds), while none relate to mariculture.

The EcoQO on seal population trends in the North Sea is developed for two species (harbour seal and grey seal) and is intended to serve as an early warning to trigger investigations into causes and what measures should be taken in case of population declines (OSPAR, 2005d). A similar EcoQ element is under development for seabird population trends. For the EcoQ element for habitats the focus is now on threatened and/or declining habitats, which have been identified on an OSPAR List of threatened and declining species and habitats (OSPAR, 2004, 2006b). There may also be future work on EcoQOs for the species on this list.

OSPAR concluded in the 2005 review that the EcoQO system (as shown in Table 13.1) had some gaps identified through internal (food web) and external (human impacts) analysis. The areas needing further attention were: water and sediment quality, macrophytes, radioactive substances, persistent organic substances other than the classic chlorinated compounds, noise, non-indigenous species, use of marine space and marine litter (OSPAR, 2006b).

Fisheries-related EcoQOs

Four of the EcoQO elements including two EcoQOs relate directly to fisheries while a few more may do so indirectly (e.g. seabird population trends). OSPAR has no legal competence in fisheries management (in Article 4 of Annex V to the Convention it is stated: 'In accordance with the penultimate recital of the Convention, no programme or measure concerning a question relating to the management of fisheries shall be adopted under this Annex' (http://www.ospar.org/eng/html/welcome.html). The EcoQOs related to fisheries are objectives to be achieved through fisheries management and are thus outside OSPAR competence. However, the framework for EA to the management of the North Sea, including the EcoQO system, is a response to the request of the North Sea Ministers (and EU Commissioners) in 1997 to use the EA as a guiding principle to integrate fisheries and environmental issues (IMM, 1997).

The EcoQO for commercial fish species builds on the system used by ICES to evaluate and advise on the status and exploitation of fish stocks. ICES has developed a system with precautionary reference points for spawning stock biomass (SSB; B_{pa}) and fishing mortality (F_{pa}), and the EcoQO is to stay above B_{pa} for the assessed stock with fishing mortality kept below F_{pa} (OSPAR, 2005b). The ICES system is simple in principle, but somewhat complex in practice in that it takes into account assessment uncertainty and probability to avoid limits. With regard to SSB (or B), B_{pa} is set with a precautionary 'safety zone' above B_{lim} so that when the stock is assessed to be above B_{pa}, the probability that the stock in reality (because of assessment uncertainty) is below B_{lim}, is low (Fig. 13.6). B_{lim} is a limit reference point, below which recruitment is impaired and

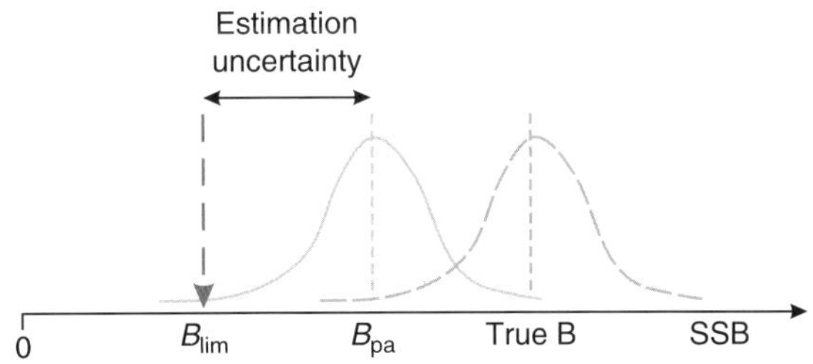

Fig. 13.6. Schematic illustration of the positions of B_{lim} and B_{pa} on the spawning stock biomass (SSB) axis. B_{pa} is positioned higher than B_{lim} with a distance related to the estimation uncertainty. This is determined by the probability distribution around the median estimate and the level set to ensure a low probability of estimates falling below B_{lim}. If the point estimates of SSB are to have a high probability of falling above B_{pa}, then the true biomass value must be some distance above B_{pa}, corresponding to the estimation uncertainty. (From OSPAR, 2005b.)

there may be a danger of stock collapse, or of getting into a zone of low population size with unknown dynamic properties. When the stock is assessed to be at or just above B_{pa}, there is a fairly high probability that it, in reality, could be below B_{pa} due to the assessment uncertainty. If the requirement was that it should also have a high probability to be above B_{pa}, this would require an extra 'safety zone' that would correspond to double precaution (Fig. 13.6; OSPAR, 2005b).

In the OSPAR system, the EcoQO for commercial fish species is reported as the proportion of the total number of stocks in the North Sea for which the assessments indicate that the stocks are below B_{pa}. In addition, the stocks which fail to meet this criterion are listed. ICES provides advice on the status and fisheries of 26 fish stocks in the North Sea (ICES, 2007a). For 2006, eight stocks were assessed to have stock levels below B_{pa} (four of them also below B_{lim}), two stocks were fished outside F_{pa}, while five stocks were assessed to be in good status (stock above B_{pa} and fished below F_{pa}). For 11 stocks, the status was unknown or uncertain, either because reference points had not been set or because data were not of sufficient quality.

A time series of cumulative assessment results for 14 North Sea fish stocks from 1980 to 2006 reveals that the proportion of stocks below B_{pa} increased from 30% to 40% in the 1980s to around 60% in the late 1990s (Fig. 13.7). At the same time the proportion of stocks fished above F_{pa} have decreased from around 40% to 50% in the 1980s to around 20% in the 2000s. This indicates a pattern where too high fishing pressure in the 1980s resulted in an increased number of stocks below B_{pa} in the 1990s. Over the last 10 years or so there has been an increasing 'polarization', with more stocks falling below B_{lim}, but also more stocks coming within safe limits. Stocks below B_{lim} now include North

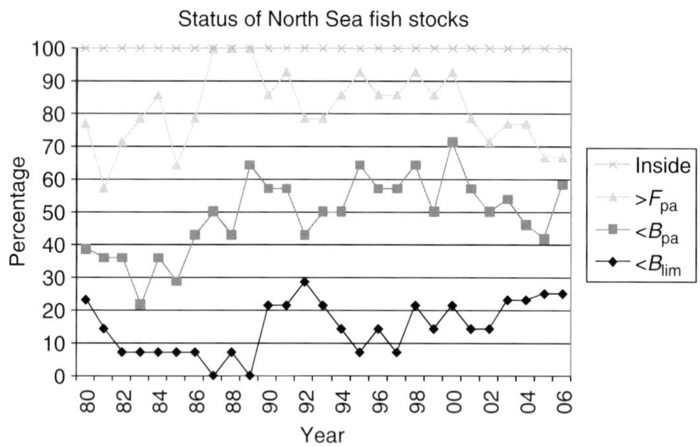

Fig. 13.7. Proportions (cumulative) of 15 North Sea fish stocks assessed by ICES to have spawning stock biomass (SSB) $<B_{lim}$, $<B_{pa}$ (but $>B_{lim}$), fishing mortality higher than F_{pa}, and stocks being within safe limits (biomass $>B_{pa}$, fishing mortality $<F_{pa}$.) (From OSPAR Document MASH 07/2/5 presented at the meeting of the Working Group on Marine Protected Areas, Species and Habitats (MASH), Brest, France, 5–8 November 2007.)

Sea cod, cod in Kattegat, Norway pout and North Sea mackerel. The stocks in good condition comprise haddock, saithe, sole in Skagerrak-Kattegat, sole in the English Channel and hake.

The EcoQO for commercial fish species has not been met. The proportion of stocks below B_{lim} (about 25% representing 4 stocks out of 15) is higher than expected if the objective is to keep all stocks above B_{lim} with high probability. This reflects partly the difficult situation for the North Sea cod stocks (Beaugrand *et al.*, 2003; Beaugrand, 2004). The proportion of stocks below B_{pa} has fluctuated around 50% in the recent years (Fig. 13.7). This is not so far from what should be expected if fisheries managers aim to keep the stocks at or above B_{pa} with just a small margin (OSPAR, 2005b).

The EcoQO on by-catch of harbour porpoise is to have a level of by-catch below 1.7% of the best population estimate (OSPAR, 2005c). There has been a concern that high by-catch of harbour porpoises in some gill net fisheries in the North Sea may represent a threat to the population. The level of 1.7% has been set with a precautionary margin based on the best scientific information available (ICES, 2001). This EcoQO is probably not met in several parts of the North Sea. A practical difficulty is in having information on by-catch of sufficient quality for all relevant fisheries and in updating estimates of the population size of harbour porpoise (OSPAR, 2005c).

The EcoQ element on changes in the proportion of large fish is one of the less-advanced EcoQOs, which is currently under development. Scientific bottom trawl surveys have shown decrease in the mean size and proportion of large fish caught in the surveys in recent decades. Strong year classes of fish, especially of North Sea haddock, may cause the proportion of large fish to decrease for natural reasons. To reduce the influence of such events, ICES (2007b) suggested using the proportion of fish larger than 40 cm caught in the International Bottom Trawl Survey (IBTS) as the metric for this EcoQO. Based on historical data, a value of 0.3 for the proportion of large fish was suggested as the EcoQO (Fig. 13.8). This proposal is now being considered by OSPAR together with competent fisheries management authorities in Norway and the EU.

The EcoQ element on local sand eel availability to black-legged kittiwakes uses breeding success of kittiwakes as a metric to indicate the availability of sand eels as a predominant food source in the vicinity of the breeding colonies. Low breeding success over 3 successive years could be used as a trigger for closing sand eel fisheries to prevent further local depletion of the sand eel stock. OSPAR (2006b) concluded that the proposed EcoQO is generally sound as a strategy to protect seabirds from local depletion of sand eels by fishing, but that more work was needed to understand the performance of the EcoQO. This EcoQO is related to previous advice from ICES to the European Commission regarding regulation of local sand eel fisheries in the northwestern North Sea (ICES, 2000a).

The EU Marine Strategy Framework Directive

The European Commission presented in 2005 a Communication on a Thematic Environmental Strategy and a proposal for a new MSFD. The directive builds

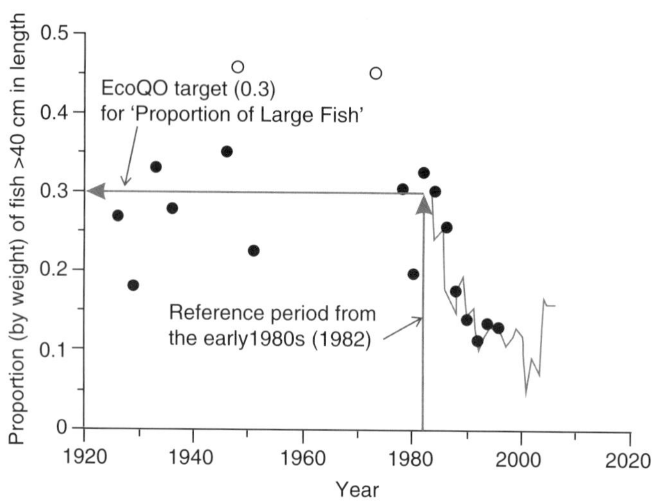

Fig. 13.8. Plot showing the proportion of large fish (>40cm) in time series of scientific bottom trawl surveys in the North Sea. Circles are data from the Scottish Autumn Ground Fish Survey (unfilled circles indicating two outliers related to strong year classes of gadoids). The solid line shows variation in the ICES International Bottom Trawl Survey (IBTS) data set between 1982 and 2006. 1982 was considered to represent the 'early 1980s' reference period and derivation of 0.3 as the proposed EcoQO is illustrated. (From ICES, 2007b.)

on the EA as one of its key elements, and the marine strategy can be seen very much as a legal and technical implementation of the EA. The MSFD has been through political negotiations and was adopted with amendments in the second reading by the European Parliament in December 2007 (http://www.europarl.europa.eu/oeil/FindByProcnum.do?lang=2&procnum=COD/2005/211). It is intended to be the environmental sustainability pillar in the European maritime policy (EC, 2006).

The core of the proposed directive is the concept of good environmental status (GES), which is also the overall objective to be achieved by the year 2020 at the latest. Environmental status is defined as:

> the overall state of the environment in marine waters, taking into account the structure, function and processes of the constituent marine ecosystems together with natural physiographic, geographic, biological, geological and climatic factors, as well as physical, acoustic and chemical conditions, including those resulting from human activities inside or outside the area concerned.

This definition is very similar to the definition of ecological quality in the Bergen Declaration (NSC, 2002) and OSPAR, reflecting a common origin (the definition of ecological quality was based on a definition in a preparatory document in the process leading up to the EU Water Framework Directive). Environmental status and ecological quality are thus basically the same thing, and the work in OSPAR on the EcoQO system is expected to be used to also inform the implementation of the MSFD.

The MSFD has two main parts. The first, called Preparation, prescribes how environmental status is to be expressed and made operational. The second prescribes how programmes of measures are to be developed to achieve the goal of GES. The Preparation contains four elements: (i) initial assessment; (ii) determination of environmental status; (iii) environmental targets; and (iv) monitoring programme. These elements are to be prepared by Member States within an indicated time frame of 4–6 years after the directive enters into force (by 2012–2014). The European Commission is to prepare generic qualitative descriptors, detailed criteria and standards for the recognition of GES within 2 years after entry into force (by 2010).

European marine regions and sub-regions as management units for implementation will be established by the directive. According to Article 4 there are four regions (Baltic Sea, Northeast Atlantic Ocean, Mediterranean Sea and Black Sea), with four sub-regions recognized for the Northeast Atlantic Ocean (North Sea, Celtic Seas, Bay of Biscay and Iberian coast, and Macaronesian biogeographical region) and for the Mediterranean Sea (Western Mediterranean Sea, Adriatic Sea, Ionian Sea and the Central Mediterranean Sea, and Aegean-Levantine Sea). This division into regions and sub-regions is based upon advice from ICES, who reviewed the LME divisions of European waters (Sherman and Skjoldal, 2002) as well as several different biogeographical classification systems (ICES, 2005). While none of these systems were used directly, the proposed division into ecological regions followed closely the LME divisions, with adjustments of some of the borders. The criteria used to define the ecological regions were also similar to those for identifying LMEs. The regions or sub-regions used in the MSFD are therefore equivalent to LMEs. For consistency with terminology used in other international contexts, this should be recognized.

For their marine waters within each marine region, EU member states will be required to develop Marine Strategies. By 2015, at the latest, a programme of measures designed to achieve or maintain GES shall be developed. Member States sharing a marine region or sub-region are obliged to cooperate to ensure that the different elements of the marine strategy are coherent and coordinated. GES is to be documented through monitoring and reporting, and to be achieved by 2020 at the latest.

GES is defined as (MSFD Article 3):

> the environmental status of marine waters where these provide ecologically diverse and dynamic oceans and seas which are clean, healthy and productive within their intrinsic conditions, and the use of the marine environment is at a level that is sustainable, thus safeguarding the potential for uses and activities by current and future generations.

This is further specified with a list of qualitative descriptors of GES in Annex I, and indicative lists of characteristics, pressures and impacts to be considered when expressing environmental status, in Annex III of the MSFD. Two of the qualitative descriptors of GES from Annex I that are relevant for fisheries, are:

> (3) Populations of all commercially exploited fish and shellfish are within safe biological limits, exhibiting a population age and size distribution that is indicative of a healthy stock.

(4) All elements of the marine food webs, to the extent that they are known, occur at normal abundance and diversity and levels capable of ensuring the long-term abundance of the species and the retention of their full reproductive capacity.

The EU Member States have transferred the competence for fisheries management to the European Commission. Therefore the MSFD, which is for measures to be taken by Member States, does not include fisheries directly. However, the directive makes it clear that: 'Measures regulating fisheries management can be taken in the context of the Common Fisheries Policy, as set out in Council Regulation (EC) No 2371/2002 of 20 December 2002 on the conservation and sustainable exploitation of fisheries resources under the Common Fisheries Policy, based on scientific advice with a view to supporting the achievement of the objectives addressed by this Directive, including the full closure to fisheries of certain areas, to enable the integrity, structure and functioning of ecosystems to be maintained or restored and, where appropriate, in order to safeguard, *inter alia*, spawning, nursery and feeding grounds' (MSFD, preambular paragraph 39). Coordination and integration of measures taken by Member States under the MSFD and by the European Commission under the Common Fisheries Policy will be a challenge that will influence the successful establishment of the sustainability pillar of the EU Maritime Policy.

Concluding Remarks

The North Sea process with the NSC has had a great influence on ocean policy development in Europe. The attention given to fisheries at the 4th NSC in Esbjerg in 1995 and the follow-up Intermediate Ministerial Meeting on Integration of Fisheries and Environmental Issues in Bergen in 1997 led to a development of the EA as a guiding principle for integration of fisheries in a wider framework. Thus, the policy development in the EU, as well as in Norway (Anon., 2002; Winsnes and Skjoldal, this volume), has focused on a broad framework with integration across sectors, with fisheries as one of them. The MSFD is a key component of this framework, being the environmental pillar of the European maritime policies. Full integration of fisheries within a broader framework in the EU will remain a challenge, however, due to the institutional obstacles of working together across the boundaries of agencies and legal instruments, and the political realities of different perceptions among member states and stakeholders about the need for reform.

The concept of 'good environmental status' of the MSFD and that of 'Ecological Quality Objectives' in OSPAR are very similar, and the work in OSPAR is likely to be used as a basis to inform the implementation of the MSFD in the North Sea and other regions/sub-regions (or LMEs) of the European waters. OSPAR has no competence in questions of fisheries management and its work will depend on constructive cooperation with the Northeast Atlantic Fisheries Commission (NEAFC) as a regional fisheries organization, and with fisheries agencies in the different countries and the European Commission that are Contracting parties to OSPAR. Working together is always a challenge and the situation is still characterized by defensive attitudes and 'watching ones turf'.

The division of European marine waters into regions and sub-regions, equivalent to LMEs, for implementing the MSFD is important. This builds on the recognition that ecosystems are geographical units with boundaries where member states are to cooperate when drawing up their programmes of measures to achieve GES. There is an increasing awareness of this internationally. Thus, the meeting of UNICPOLOS in New York in 2006 had as one of the conclusions that the EA 'should be applied within geographically specific areas based on ecological criteria' (Ridgeway and Maquieira, 2006). Norway has taken a similar approach when implementing the EA by developing management plans for the Barents Sea, Norwegian Sea and North Sea LMEs (Winsnes and Skjoldal, this volume).

The EA has requirements to science to deliver the knowledge and information required for adaptive management to achieve the objectives agreed for the acceptable degree of impacts and state of the ecosystem. The framework shown in Fig. 13.2 emphasizes the scientific components of monitoring and research, assessment and scientific advice for management. Integrated assessment remains a main challenge for the scientific community to better support the practical performance of the EA to management. ICES had a 3-year study group (Regional Ecosystem Group for the North Sea (REGNS)) in 2004–2006 that carried out an integrated assessment of the North Sea as a demonstration of how this could be done (Kenny *et al.*, 2008). In Norway, the Institute of Marine Research (IMR) was reorganized to focus on the three LMEs (Barents, Norwegian and North Seas) (Misund and Skjoldal, 2005). However, integration of traditionally narrow scientific disciplines into a broader and coordinated ecological approach and successful provision of effective scientific support for the EA to management remain key challenges.

References

Anon. (2002) Rent og rikt hav, St.meld. nr. 12 (2001–2002) (Clean and rich sea, Government White Paper; in Norwegian), Ministry of Environment, Oslo. Available at: http://www.regjeringen.no/nb.html?id=4

Anon. (2006) Report No. 8 to the Storting (2005–2006) Integrated management of the marine environment of the Barents Sea and the sea areas off the Lofoten Islands (Translation in English), Ministry of Environment, Oslo. Available at: http://www.regjeringen.no/nb.html?id=4

Beaugrand, G. (2004) The North Sea regime shift: evidence, causes, mechanisms and consequences. Progress in Oceanography, 60: 245–262.

Beaugrand, G., Brander, K.M., Lindley, J.A., Souissi, S., and Reid, P.C. (2003) Plankton effect on cod recruitment in the North Sea. Nature, 426: 661–664.

EC (2006) Towards a future Maritime Policy for the Union: a European vision for the oceans and seas. European Communities, Office for Official Publications of the European Communities, Luxembourg, 58 pp. (ISBN 92-79-01825-6). Available at: http://ec.europa.eu/maritimeaffairs/pdf/greenpaper_brochure_en.pdf

Fenchel, T. (1987) Ecology – potentials and limitations. In: Kinne, O. (ed) *Excellence in Ecology*, Book 1. International Ecology Institute, Oldendorf/Luhe.

ICES (2000a) Report of the ICES Advisory Committee on the Marine Environment, 1999. ICES Cooperative Research Report No. 239.

ICES (2000b) Report of the Study Group on Ecosystem Assessment and Monitoring. ICES

CM 2000/E:09, International Council for the Exploration of the Sea, Copenhagen.

ICES (2001) Report of the ICES Advisory Committee on Ecosystems, 2001. ICES Cooperative Research Report NO. 249, International Council for the Exploration of the Sea, Copenhagen.

ICES (2005) Report of the ICES Advisory Committee on Fishery Management, Advisory Committee on the Marine Environment and Advisory Committee on Ecosystems, 2005 (2.2.2.2 Information and advice about appropriate eco-regions for the implementation of an ecosystem approach in European waters, pp. 31–53). ICES Advice 2005, Volume 1. International Council for the Exploration of the Sea, Copenhagen.

ICES (2007a) Report of the ICES Advisory Committee on Fishery Management, Advisory Committee on the Marine Environment and Advisory Committee on Ecosystems, 2007. ICES Advice 2007, Books 6 and 9. International Council for the Exploration of the Sea, Copenhagen.

ICES (2007b) Report of the Working Group on Ecosystem Effects of Fishing Activities (WGECO). ICES CM 2007/ACE:04. International Council for the Exploration of the Sea, Copenhagen.

IMM (1997) Statement of conclusions. Intermediate Ministerial Meeting on the Integration of Fisheries and Environmental Issues, 13–14 March 1997, Bergen, Norway. Ministry of Environment, Oslo. Available at: http://odin.dep.no/nsc

Kenny, A.J., Skjoldal, H.R., Engelhard, G.H. and Kershaw, P. (2008) An integrated approach for assessing the relative significance of human pressures and environmental forcing on the status of a Large Marine Ecosystem. Manuscript in review.

Likens, G. (1992) The ecosystem approach: its use and abuse. In: Kinne, O. (ed) *Excellence in Ecology*, Book 3. International Ecology Institute, Oldendorf/Luhe.

Misund, O.A. and Skjoldal, H.R. (2005) Implementing the ecosystem approach: experiences from the North Sea, ICES, and the Institute of Marine Research, Norway. Marine Ecology Progress Series, 300: 260–265.

NSC (2002) Bergen Declaration. Fifth International Conference on the Protection of the North Sea, 20–21 March 2002, Bergen, Norway. Ministry of Environment, Oslo. ISBN 82-457-0361-3.

NSTF (1993) North Sea Quality Status Report 1993. North Sea Task Force, Oslo and Paris Commissions, London.

Olsen, E., Gjøsæter, H., Røttingen, I., Dommasnes, A., Fossum, P. and Sandberg, P. (2007) The Norwegian ecosystem-based management plan for the Barents Sea. ICES Journal of Marine Science, 64: 599–602.

OSPAR (2004) 2004 Initial OSPAR list of threatened and/or declining species and habitats. OSPAR Agreement Reference Number: 2004-06. OSPAR Commission, London.

OSPAR (2005a) Common procedure for the identification of the eutrophication status of the OSPAR Maritime Area. OSPAR Agreement Reference number: 2005-3. OSPAR Commission, London.

OSPAR (2005b) Background document on the Ecological Quality Objective for spawning stock biomass of commercial species in the North Sea. Biodiversity Series, Publication Number: 2005/242. OSPAR Commission, London.

OSPAR (2005c) Background document on the Ecological Quality Objective on bycatch of harbour porpoises in the North Sea. Biodiversity Series, Publication Number: 2005/244. OSPAR Commission, London.

OSPAR (2005d) Background document on the Ecological Quality Objective for seal population trends in the North Sea. Biodiversity Series, Publication Number: 2005/245. OSPAR Commission, London.

OSPAR (2005e) Background document on the Ecological Quality Objective on imposex in dog whelks *Nucella lapillus*. Biodiversity Series, Publication Number: 2005/247. OSPAR Commission, London.

OSPAR (2005f) Background document on the Ecological Quality Objective on oiled guillemots. Biodiversity Series, Publication Number: 2005/252. OSPAR Commission, London.

OSPAR (2006a) Agreement on the application of the EcoQO system. OSPAR Agreement Reference number: 2006-4. OSPAR Commission, London.

OSPAR (2006b) Report on North Sea Pilot Project on Ecological Quality Objectives.

Biodiversity Series, Publication Number: 2006/239. OSPAR Commission, London.

Sherman, K. (1995) Achieving regional cooperation in the management of marine ecosystems: the use of the large marine ecosystem approach. Ocean and Coastal Management, 29, 165–185.

Sherman, K. and Skjoldal, H.R. (eds) (2002) *Large Marine Ecosystems of the North Atlantic. Changing States and Sustainability*. Elsevier, Amsterdam, 449 pp.

Skjoldal, H.R. (1999) Overview report on Ecological Quality (EcoQ) and Ecological Quality Objectives (EcoQOs). Institute of Marine Research. ISBN 82-7461-050-4.

Skjoldal, H.R. (2004a) An introduction to the Norwegian Sea ecosystem. In: Skjoldal, H.R., Sætre, R., Fernø, A., Misund, O. A. and Røttingen, I. (eds) *The Norwegian Sea Ecosystem*. Tapir Academic Press, Trondheim, Norway, pp. 15–32.

Skjoldal, H.R. (2004b) Fish stocks and fisheries in relation to climate variability and exploitation. In: Wolanski, E. (ed) *Natural Resource System Challenge: Oceans and Aquatic Ecosystems*. In: Encyclopedia of Life Supporting Systems (EOLSS), Developed under the auspices of the UNESCO. Eolss Publishers, Oxford, UK. Available at: http://www.eolss.net

Turrell, W. (2004) The policy basis of the 'ecosystem approach' to fisheries management. EuroGOOS Publication No. 21, EuroGOOS Office, SMHI, 601 76 Norrköping, Sweden. ISBN 91-974828-1.

Ridgeway, L. and Maquieira, C. (2006) Report on the work of the United Nations Open-ended Informal Consultative Process on Oceans and the Law of the Sea at its 7th meeting. 28 pp.

14 Management Plan for the Norwegian Part of the Barents Sea Ecosystem

INGER WINSNES[a] AND HEIN RUNE SKJOLDAL[b]

[a]*Norwegian Ministry of the Environment;* [b]*Institute of Marine Research, Bergen, Norway*

Abstract

Norway adopted the ecosystem approach to ocean management in 2002. A management plan for its implementation in the Barents Sea and the sea areas off the Lofoten Islands was presented in 2006. The preparation of the management plan was overseen by a steering group with representatives of many ministries, and the work was carried out by a large number of government agencies. Several reports were prepared that summarized the scientific knowledge on the area and the results from environmental impact assessments (EIAs) for key sectors (fisheries, petroleum development, shipping and external pressures), as well as overall. There were also reports on identification of particularly valuable and vulnerable areas, and indicators or variables that could be used as a basis for setting environmental quality objectives. Three coordination groups with broad representation of relevant agencies were established as part of a new management regime: an advisory group on monitoring, a forum on risk management and a management forum that reports on need for measures to a steering group of ministries. A monitoring system based on indicators or variables with reference levels and action thresholds for management intervention has been put in place and is operated by the advisory management group. The management regime also includes a reference group with participation of relevant stakeholders. The management plan is to be revised on a regular basis, the first time in 2010. A similar management plan for the Norwegian Sea is now under development (to be finished in 2009) and one for the Norwegian part of the North Sea is being planned.

Clean and Rich Sea

As a follow-up of the World Summit on Sustainable Development in Rio de Janeiro in 1992, Norway established a new and more comprehensive national environmental policy. This policy is regularly updated with a White Paper on the government's environmental policy and the state of the environment, presented to the Norwegian Parliament every second year. A core element is national goals within the main areas of the environmental policy, with key variables or indicators identified to measure progress and achievements in relation to the goals.

 The Ecosystem Approach to Fisheries (eds G. Bianchi and H.R. Skjoldal)

As regards the use of living resources, including by fisheries, the main goal is that no species or stocks should be made extinct or threatened by exploitation. Threatened species and species for which Norway has a special conservation responsibility should be maintained or restored to viable population levels (Anon., 2007).

Norway held the chairmanship of the North Sea Ministerial Conferences (NSC) in the period from the 4th NSC in Esbjerg in Denmark in 1995 to the 5th NSC in Bergen in March 2002. In Esbjerg, the Ministers focused on fisheries as an important topic for the North Sea. This was followed up by an Intermediate Ministerial Meeting on Integration of Fisheries and Environmental Issues held in Bergen in 1997. In the Statement of Conclusions from that meeting, the Ministers agreed to develop and use an ecosystem approach as a guiding principle for the needed integration (IMM, 1997). A framework for an ecosystem approach was agreed by the Ministers in the Bergen Declaration from the 5th NSC (NSC, 2002; Skjoldal and Misund, this volume).

The work on developing the framework of the ecosystem approach in the Bergen Declaration influenced the ocean policy development in Norway. Just before the 5th NSC in Bergen, the Norwegian Government presented the White Paper 'Clean and rich sea' to the Norwegian Parliament where an ecosystem approach was adopted (Anon., 2002). The White Paper announced that the ecosystem approach would be implemented by preparing integrated management plans for the three large marine ecosystems (LMEs) that are included in the Norwegian Exclusive Economic Zone, starting with the Barents Sea area. We describe here the process of planning and implementing the management plan for the Norwegian part of the Barents Sea. A similar plan is now being prepared for the Norwegian Sea ecosystem, and preparatory steps have been taken to start work on one for the North Sea. The goal is that all these plans should be in operation by 2015 as practical implementation of the ecosystem approach.

Preparing the Management Plan (2002–2006)

The purpose of the management plan was to provide a framework for the sustainable use of natural resources and goods derived from the Barents Sea ecosystem, while at the same time maintaining its structure, functioning and productivity. The plan is intended to clarify the overall framework for both existing and new activities in these waters. Of particular importance were the future development of the oil and gas industry in the area and the possible coexistence of petroleum developments along with fisheries. Increased shipping in the area associated with export of oil from northwestern Russia and cruise traffic along the coast and around Svalbard was also an important issue.

The geographical area for the plan encompassed the Norwegian part of the Barents Sea including the fishery protection zone around the Svalbard archipelago. It also included the waters off the Lofoten Islands and the adjacent part of the Norwegian Sea and Greenland Sea to the west of the Barents Sea and Svalbard (Fig. 14.1). This differs from the delineation of large marine ecosystems (LMEs) where the boundary between the Barents Sea and Norwegian Sea LMEs

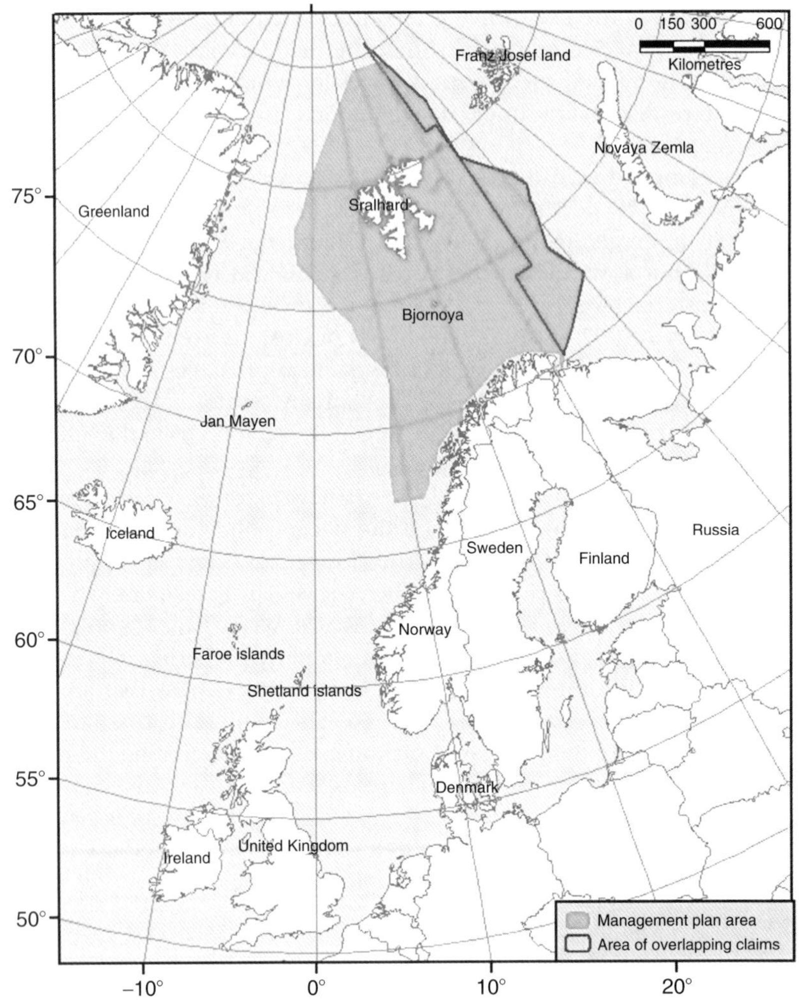

Fig. 14.1. Map with the area for the management plan for the Barents Sea and the sea areas off Lofoten Islands, including adjacent parts of the Norwegian and Greenland Seas.

follows the continental slope (Arctic Council, PAME Working Group; http://arcticportal.org/en/pame/ecosystem-approche). The area for the management plan was chosen for administrative reasons, and it was recognized that the area comprised parts of different, although ecologically linked, ecosystems.

The management plan was presented in a Government White Paper in 2006 (Anon., 2006). The work of preparing the plan was overseen by a steering group lead by the Ministry of Environment and with representatives from several other ministries (Petroleum and Energy, Fisheries and Coastal Affairs, Foreign Affairs, Labour and Social Inclusion, and Trade and Industry). The work was carried out by a large number of government agencies and institutions in an open and transparent process (Fig. 14.2).

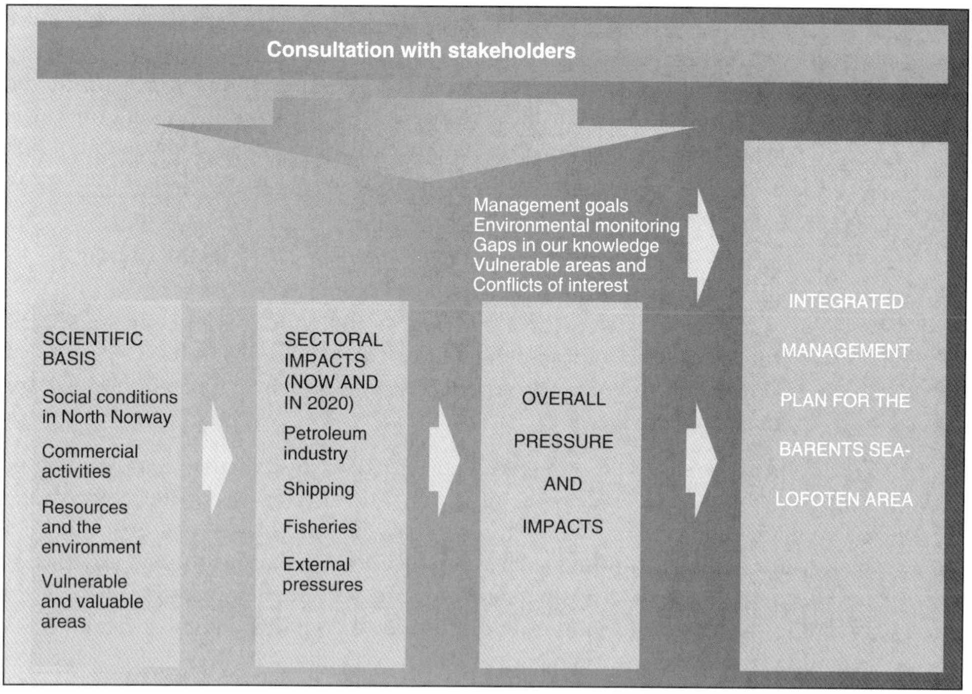

Fig. 14.2. The process of consultations and preparations of scientific reports leading to the integrated management plan for the Barents Sea–Lofoten area. (From Anon., 2006.)

The first step was to prepare the scientific basis, summarized in four reports on social conditions in North Norway, commercial activities in the area, living resources and the environment, and identification of vulnerable and valuable areas. This was followed by preparation of four thematic environmental impact assessments (EIAs) of the petroleum industry, shipping, fisheries and external pressures. The external pressures that were examined included long-range transport of pollutants, impacts from climate change and introduction of alien species. In the third step, the information in the thematic assessments was brought together in an assessment of the overall pressure and impacts in the area.

In parallel there was work on developing operational management goals. A report was prepared on indicators or variables that could be used as basis for setting environmental quality objectives (von Quillfeldt and Dommasnes, 2005). Monitoring requirements associated with these indicators or variables were also considered. Gaps in knowledge were identified when preparing the reports on the scientific basis and impact assessments, and a key principle was to use caution where there was uncertainty. Special attention was also given to spatial planning aspects, such as vulnerable areas and areas of potential conflicts between different activities and industries. These steps were then used when drawing up the management plan for the Barents Sea and the Lofoten area (Fig. 14.2; Anon., 2006; Olsen *et al.*, 2007).

Particularly Valuable and Vulnerable Areas

Particularly valuable and vulnerable areas were identified that were considered to be of great importance for biodiversity and for biological production in the entire Barents Sea–Lofoten area, and where possible adverse impacts might persist for many years. Vulnerability of an area was assessed based on a number of criteria:

- High level of production and concentrations of animals.
- Large proportion of endangered or vulnerable habitats.
- Whether it is a key area for species for which Norway has a special responsibility or for endangered or vulnerable species.
- Whether it supports internationally or nationally important populations of certain species.

Seven broad areas were identified as particularly valuable and vulnerable (Fig. 14.3; Olsen and von Quillfeldt, 2003). The shelf area outside Lofoten and Versterålen is narrow and constitutes a spawning area and larval drift route for major commercial fish stocks such as the Barents Sea cod (northeast Arctic stock) and herring (Norwegian spring spawning stock). The area is also important for breeding, moulting and wintering seabirds and contains cold-water coral reefs, including the largest known reef (Røstrevet; Fosså *et al.*, 2005). Tromsøflaket is a large bank area in the southwestern Barents Sea that acts as a retention area for fish larvae, which is important for breeding and wintering seabirds, and contains very rich sponge communities and possibly coral reefs. The shelf edge, northwest from Tromsøflaket, is a particularly rich zone in biological and ecological terms. The coast along northern Norway in the Barents Sea east of Tromsøflaket is also rich with large colonies of seabirds and important wintering and moulting areas for eiders and other seabirds.

The marginal ice zone is an important area for seasonal plankton production and contains concentrated animal life with seabirds and marine mammals feeding and resting in the area. The marginal ice zone withdraws northwards in summer, but in winter or early spring, when ice distribution is at its maximum, is located in the area of the physical polar front. This is the area where relatively warm Atlantic water meets cold Arctic water. The front is sharp and topographically positioned (over about 150 m depth) in the western and central Barents Sea and is also an important area ecologically. The last area identified as particularly valuable and vulnerable were the coastal areas around the Svalbard archipelago, including around Bear Island. This area is important for walrus, as a breeding and feeding area for many types of seabirds, and as a moulting area for eiders, geese and auks.

Environmental Status and Impact Assessments

The general state of the ecosystems in the area covered by the management plan is considered to be fairly good today, and can be characterized as being clean

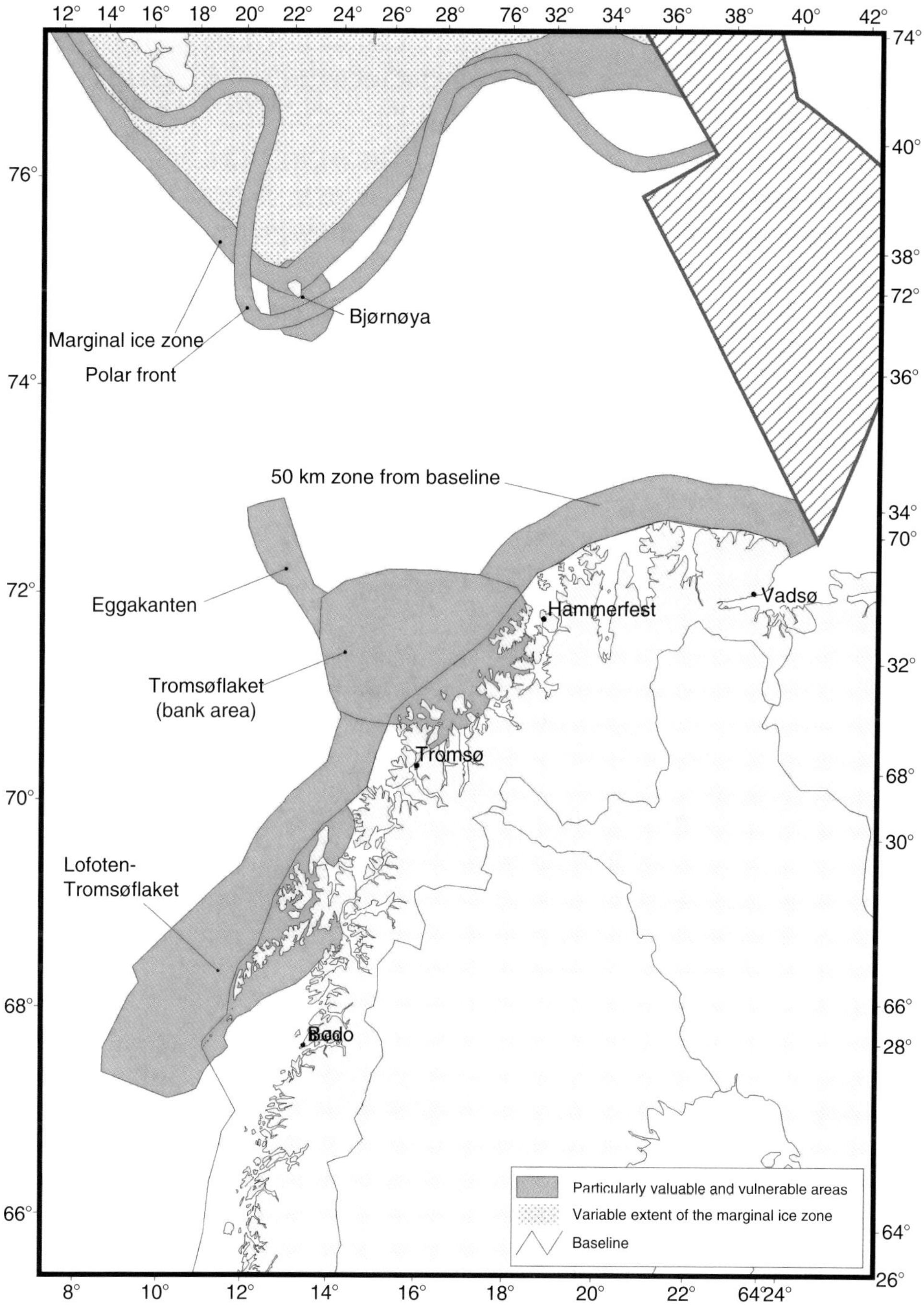

Fig. 14.3. Particularly valuable and vulnerable areas identified in the area covered by the management plan. (From Olsen and von Quillfeldt, 2003; Anon., 2006.)

and rich in resources. The government considers it very important to safeguard the basic structure and functioning of the ecosystems in the long term, so that they continue to be clean, rich and productive.

The EIA of fisheries considered a range of effects including direct effects on the targeted stocks, impacts on benthic habitats and by-catch of seabirds and marine mammals (Anon., 2004a). The most important commercial fish stocks in the Barents Sea–Lofoten area are cod, haddock, saithe, Greenland halibut, herring and capelin. The shrimp fishery is also relatively important in economic terms. In addition, there are large stocks of polar cod and long-rough dab and also a wide range of other non-commercial Arctic fish species (Føyn *et al.*, 2002).

The status of the commercial fish species is a mixed picture. Cod, haddock and saithe are in relatively good condition with stock levels above precautionary reference points. Haddock and saithe are harvested sustainably (with fishing mortality lower than a precautionary reference level). A major concern for the cod fishery has been illegal and unreported landings, and hence the harvest rate (fishing mortality) of cod has been higher than that intended under the management plan. The Greenland halibut stock has been low and is rebuilding slowly. Stocks of redfish (*Sebastes marinus* and *S. mentella*) have been low for many years due to high fishing pressure and poor recruitment (ICES, 2007) and both species are red-listed as Vulnerable in Norway (Kålås *et al.*, 2006). The capelin stock is a key component of the Barents Sea ecosystem, and it has been fluctuating very much due to variable predation from herring (on capelin larvae), cod and other consumers (Skjoldal and Rey, 1989; Gjøsæter, 1998; Dalpadado *et al.*, 2002). The capelin fishery is strictly regulated and opened only when there is a surplus above what is needed as a minimum spawning stock biomass and as food for cod. The capelin stock was down to a low level in 2005 after a maximum of about 4 million t in 2000, but increased again in 2007 (Gjøsæter *et al.*, 2008). The Barents Sea is the main nursery area for juvenile herring of the Norwegian spring spawning stock, but there is no fishery for it in this area.

The fisheries can also have major impacts on other parts of the ecosystem. If the size of commercial fish stocks is reduced by an increase in the harvest or recruitment failure, this has repercussions for the whole ecosystem, regardless of whether the reduction is a result of human activity or is caused by natural events. A reduction in fish stocks can result in poor food supplies for both seabirds and marine mammals. This was considered to be an important reason behind the serious decline in populations of some seabirds (e.g. Atlantic puffin and black-legged kittiwake) in the Barents Sea–Lofoten area.

By-catches of seabirds and marine mammals in fishing gear can be a problem in certain areas and at certain times of the year. The species caught as by-catches in gill nets are mainly diving birds, which become entangled and drown. This is one explanation for the decline of several species of auk in north Norway. The prohibition on drift netting for salmon introduced in 1989 has substantially reduced by-catches in gill nets.

By-catches of fish by various types of fishing gear, particularly below the minimum size in shrimp trawls, represent another important pressure on the ecosystem. Considerable efforts are being made to reduce by-catches, for example, by developing selective gear, using sorting grids in trawls and temporarily closing areas to fishing when by-catches exceed specified limits.

Bottom trawling can have direct effects on the seabed habitats and bottom communities. Both trawls and other gear types that are towed along the seabed can seriously damage and disturb benthic communities, and also resuspend particles and shift sediments. Near coral reefs, the use of these types of gear can smother corals with sediment in addition to causing mechanical damage. Coral reefs and sponge communities in northern waters may be very sensitive to mechanical disturbance due to slow growth rate. To protect coral reefs from damage, a provision has been laid down in the Regulations relating to sea-water fisheries, requiring special care to be exercised near known coral reefs. In addition, the use of bottom gear is prohibited on and near certain large coral reefs, such as Røstrevet.

The spatial distribution of fishing activities based on satellite tracking data from fishing vessels over 24 m in length for the year 2004 was used to illustrate the level of fishing activity in different areas (Fig. 14.4). In addition, there is a substantial coastal fishery using smaller vessels. Vessels fish in different areas depending on the type of fishing gear they use. While the data in Fig. 14.4 is for one selected year, it illustrates the concentrated occurrence of the fishing activities in some areas such as along the shelf edge (long-lining, bottom trawling and gill-netting), in the Hopen Depth southeast of Bear Island (bottom trawling), and in the area east of Tromsøflaket (bottom trawling and long-lining). It should be noted that there is a general prohibition on bottom trawling except for shrimp inside 4 nautical miles off the baseline. The impacts of bottom trawling on benthic habitats and communities are not well known, but are being further investigated in the bottom mapping programme MAREANO (http://www.mareano.no/).

Seismic surveys and exploration drilling for oil and gas began in the Barents Sea–Lofoten area in 1980. Some discoveries have been made, mainly of gas and also some oil, but prior to the management plan there has been no year-round petroleum activity in this area. However, the plan for development and operation for the gas and condensate field Snøhvit northwest of Hammerfest had been approved, and Snøhvit came on stream in 2007. There were plans to drill two more appraisal wells on the nearby Goliat oil field before a decision would be taken on whether it could be developed. Sixty-five exploration and appraisal wells had been drilled in the Barents Sea–Lofoten area up to 2005. The petroleum activities in the Barents Sea were not believed to have had significant environmental impacts up to the present (Anon., 2003).

The management plan included area and seasonal restrictions of petroleum activities (Fig. 14.5). The areas identified as particularly valuable and vulnerable (Fig. 14.3) would remain closed for petroleum activities or with seasonal restrictions to avoid accidents in the environmentally sensitive spring and summer period. Strict regulations were also imposed according to a zero-discharge policy, requiring that there would be no regular discharges to the sea. Produced water was to be reinjected, and a maximum of 5% discharged during operational deviations provided that it was treated before discharge. Drill cuttings and drilling mud must also be reinjected or taken ashore for treatment, with exception for

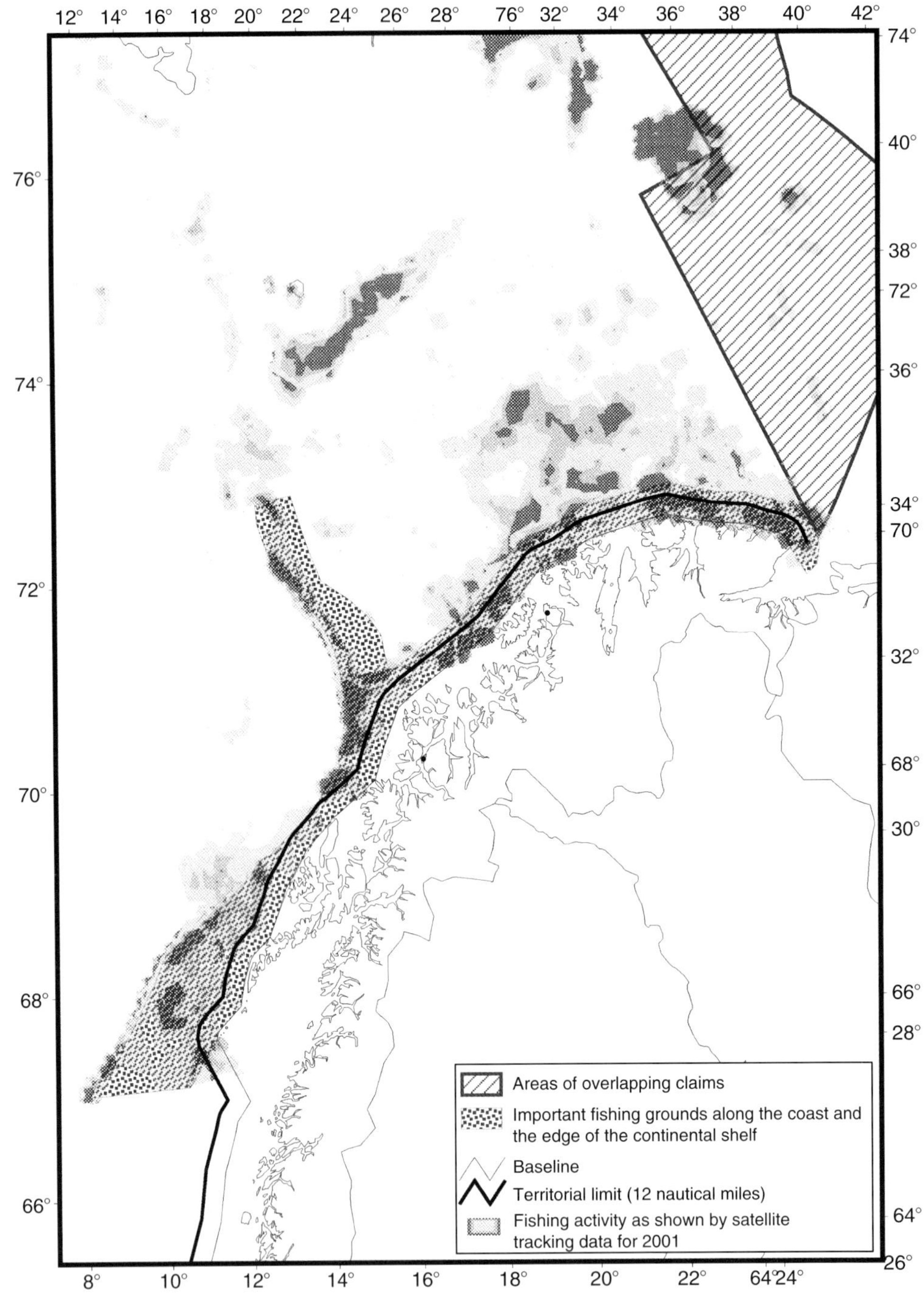

Fig. 14.4. The most important fishing grounds and fishing activity as indicated by satellite tracking data from vessels over 24 m in length in 2001. (From Anon., 2004a, 2006.)

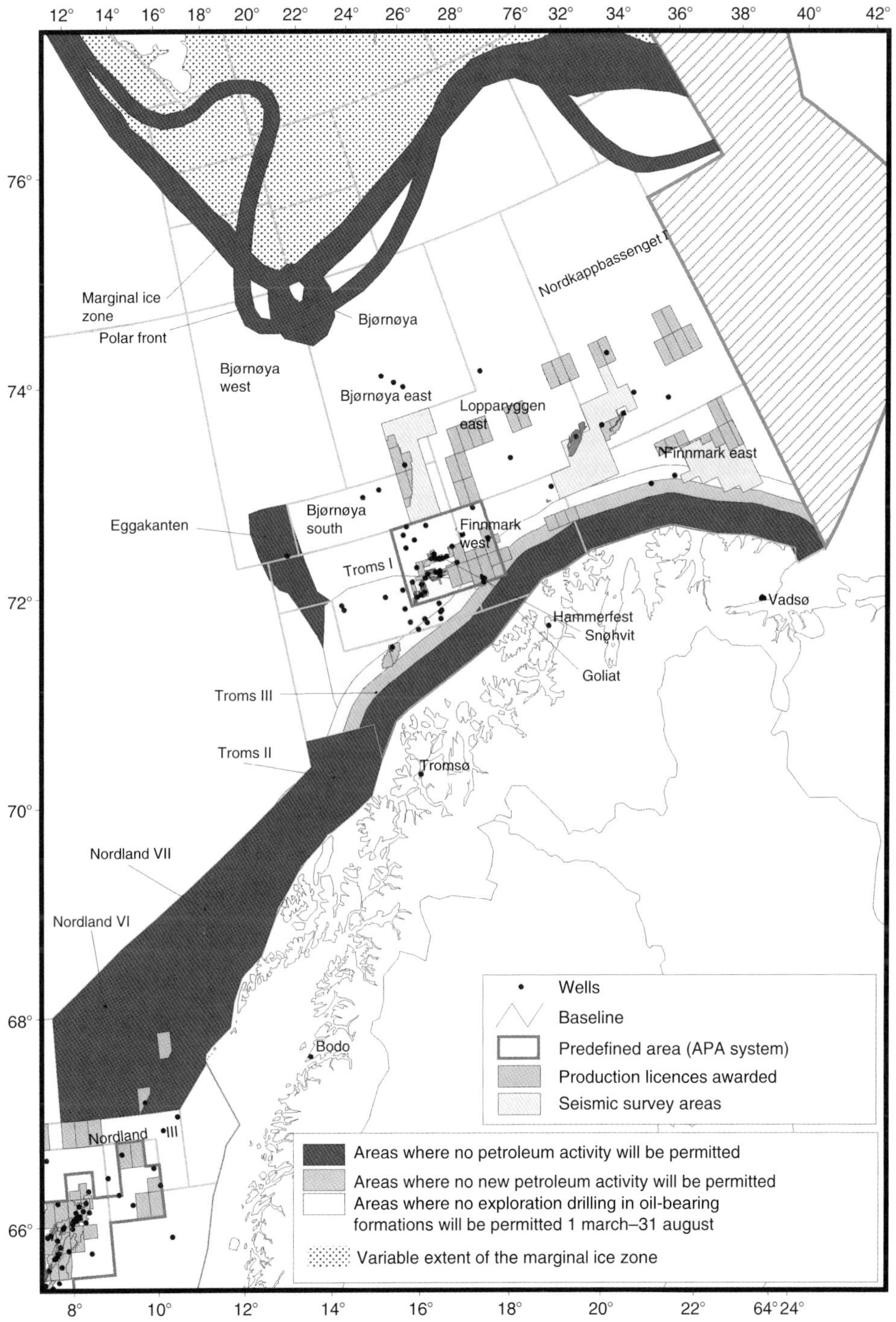

Fig. 14.5. Framework for petroleum activities as established in the management plan. The map shows areas that will be closed for all or new petroleum activities and areas with seasonal restrictions. Also shown are positions for drilled exploration wells, areas where licenses have been awarded, and areas opened for seismic surveys. (From Anon., 2006.)

cuttings and mud from the top-hole section that may be discharged, provided that they do not contain substances with unacceptable properties, and only in areas where assessments indicate that damage to vulnerable components of the environment is unlikely.

With such restrictions, regular petroleum operations are not expected to have significant negative impacts on the marine environment. What remains of concern is the risk for accidental release of oil, which could have serious environmental effects. The probability of a major oil spill during drilling is very low, and the area and seasonal restrictions on petroleum activities imposed in the management plan would further reduce the likelihood of serious environmental impact should an accident occur.

The EIA of shipping examined adverse impacts on the environment through operational discharges to water and air, releases of pollutants from antifouling systems, noise, introduction of alien species via ballast water or attached to hulls and local discharges from zinc anodes in ballast tanks (Anon., 2004b). It also considered the risk of spills of oil and chemicals in case of accidents. The volume of shipping in the management plan area is expected to increase for cargo and passenger vessels, and also due to transit of oil from northwestern Russia and transport of gas from the LNG plant at Melkøya near Hammerfest.

The MARPOL Convention permits a certain level of discharges of oily bilge water and oily mixtures from tank washings. However, all ships are required to have segregated ballast tanks by 2010, and this will eliminate discharges of oily ballast water. Oil slicks on the sea are reported every year, and most of these are believed to be from illegal discharges from ships. The steady pressure on the marine environment caused by oil pollution will have negative impacts, particularly on seabird populations. However, it has not been possible to quantify the impacts on the management plan area. There is a risk of spread of alien species to the management plan area, either in ballast water or attached to ships' hulls. A particular concern is the possible introduction of species from the north Pacific with increased future traffic along the Northeast Passage.

The most serious impact factors and pressures today are considered to be fisheries and long-range transport of pollutants (Anon., 2005). Levels of persistent organic pollutants (such as the classical PCBs and DDT) are relatively high in some compartments of the ecosystem and their effects on predators on the top of Arctic food chains (e.g. glaucous gull and polar bear) are of concern (AMAP, 2004; Anon., 2004c).

The EIAs considered scenarios up to 2020 (Anon., 2005). Petroleum activities were expected to constitute the main change of activity in years to come. The main concern for impacts is related to possible accidental discharges. For fisheries, the present management is based on precautionary principles. Increased harvest and insufficient control in the future could have serious consequences for resources, environment, trade and communities. With regard to long-range transport of pollutants, the situation in 2020 would probably be somewhat similar to the one today. The flow of 'old' hazardous substances (e.g. PCBs and DDT) is expected to decrease, but these substances are already widely dispersed and their degradation is slow. The flow of 'new' hazardous substances is expected to increase. Introduced species represent a potentially large threat, which increases as traffic with tankers increases. The consequences may be substantial changes in the ecosystem.

Implementing the Management Plan

Responsibility for the management of the Barents Sea–Lofoten area is currently divided between several different sectors, and implementation of the management plan will require better coordination between them. To provide a sound foundation for an improved management regime, the government has established several new coordination groups (Fig. 14.6).

An *advisory group on monitoring* of the Barents Sea is tasked with the coordination and implementation of the monitoring system for the Barents Sea–Lofoten area under the management plan. This group is headed by the Institute of Marine Research and has a broad membership of relevant public institutions with responsibility for research and monitoring in the area. This group will produce annual reports summarizing and evaluating the results from the monitoring of a set of indicators.

A *forum on environmental risk management* is tasked with providing better information on risk trends in the area, especially as regards acute oil

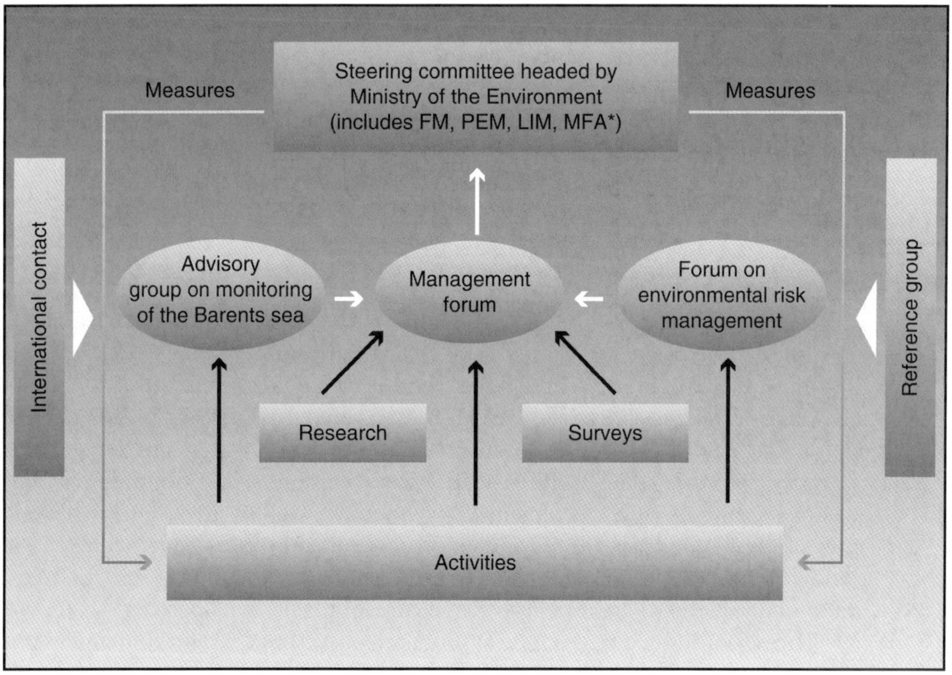

Fig. 14.6. Overview of the elements of the system for implementing the plan. Abbreviations: FM = Ministry of Fisheries and Coastal Affairs, PEM = Ministry of Petroleum and Energy, LIM = Ministry of Labour and Social Inclusion, MFA = Ministry of Foreign Affairs. (From Anon., 2006.)

pollution, to improve risk management both across and within sectors. This group is headed by the Coastal Administration and has a broad membership of several other agencies.

A *management forum* is responsible for the coordination and overall implementation of the scientific aspects of the management plan. The forum is headed by the Norwegian Polar Institute and is to prepare status reports on the results obtained through research, monitoring, surveys and other scientific activities relevant to the goals of the management plan.

The management forum reports to a steering group headed by the Ministry of Environment and with participation of many other ministries (e.g. Fisheries and Coastal Affairs, Petroleum and Energy, and others). The steering group will consider measures that need to be taken, and the relevant agencies will carry them out as specific activities. The Government has also established a reference group with representatives of the various stakeholders involved, including business and industry, environmental organizations and Sami interest groups.

The Government will consider the need for further measures to achieve the goals and for any update of the management plan on the basis of status reports to be prepared in 2010.

Goals, Indicators and Action Thresholds

The new management regime for the Barents Sea–Lofoten area is performance-oriented, and the Government has decided on a set of goals against which performance can be measured. These consist of general objectives concerned with value creation and coexistence between industries, and more specific targets for managing biodiversity, combating pollution and ensuring safe seafood (Table 14.1). The goals are in line with the goals and guidelines of the national environmental policy. An overall goal is that the management shall ensure that activities in the area do not threaten the environment and living resources and thus future opportunities for continued value creation.

In order to measure progress systematically, a system was established for monitoring the state of the environment, or the ecological quality, by means of indicators, reference values and action thresholds (von Quillfeldt and Dommasnes, 2005). For fish stocks, benthic organisms, seabirds and marine mammals, indicators have been selected with reference values ascribed to them (Table 14.2). Reference values correspond to the ecological quality expected in a similar, but more or less undisturbed, ecosystem, adjusted for natural variation and development trends. Precautionary reference values are used for harvestable stocks. For many of these indicators, action thresholds have been set. The action threshold is the point at which a change in an indicator in relation to the reference value is so great that new measures must be considered.

The set of indicators includes a number of factors that are fundamental to the state and functioning of the ecosystem, such as temperature, salinity, water transport, extent of the sea ice, nutrient distribution, and the occurrence and production of phytoplankton and zooplankton. There are naturally no action thresholds for these indicators.

Table 14.1. Management goals established for the Barents Sea and the area off Lofoten.

Issue	Goal
Pollution	Releases and inputs of pollutants to the Barents Sea–Lofoten area will not result in injury to health or damage the productivity of the natural environment and its capacity for self-renewal. Activities in the area will not result in higher levels of pollutants
	The environmental concentrations of hazardous and radioactive substances will not exceed the background levels for naturally occurring substances and will be close to zero for man-made synthetic substances. Releases and inputs of hazardous or radioactive substances from activity in the area will not cause these levels to be exceeded
	Operational discharges from activities in the area will not result in damage to the environment or elevated background levels of oil or other environmentally hazardous substances over the long term
	Litter and other environmental damage caused by waste from activities in the Barents Sea–Lofoten area will be avoided
Safe seafood	Fish and other seafood will be safe and will be perceived as safe by consumers in the various markets
Accidents and acute pollution	The risk of damage to the environment and living marine resources from acute pollution will be kept at a low level and continuous efforts will be made to reduce it further. Activity that involves a risk of acute pollution will be managed with this objective in mind
	Maritime safety measures and the oil spill response system will be designed and dimensioned to effectively keep the risk of damage to the environment and living marine resources at a low level
Biodiversity	*General*
	Management of the Barents Sea–Lofoten area will ensure that diversity at ecosystem, habitat, species and genetic levels, and the productivity of ecosystems are maintained. Human activity in the area will not damage the structure, functioning, productivity or dynamics of ecosystems
	Valuable and vulnerable areas
	Activities in particularly valuable and vulnerable areas will be conducted in such a way that the ecological functioning and biodiversity of such areas are not threatened
	Damage to marine habitats that are considered to be threatened or vulnerable will be avoided
	In marine habitats that are particularly important for the structure, functioning, productivity and dynamics of ecosystems, activities will be conducted in such a way that all ecological functions are maintained
	Species
	Naturally occurring species will exist in viable populations and genetic diversity will be maintained

Continued

Table 14.1. Continued

Issue	Goal
	Harvested species will be managed within safe biological limits so that their spawning stocks have good reproductive capacity Species that are essential to the structure, functioning, productivity and dynamics of ecosystems will be managed in such a way that they are able to maintain their role as key species in the ecosystem concerned Populations of endangered and vulnerable species and species for which Norway has a special responsibility will be maintained or restored to viable levels. Unintentional negative pressures on such species as a result of activity in the Barents Sea–Lofoten area will be reduced as much as possible by 2010 The introduction of alien species through human activity will be avoided A representative network of protected marine areas will be established in Norwegian waters, at the latest by 2012. This will include the southern parts of the Barents Sea–Lofoten area

Table 14.2. Elements of the monitoring system for environmental quality of the Barents Sea. Proposed set of indicators for fish stocks and other biological compartments.

Indicator	Reference values	Action threshold
Fish stocks that are not harvested		
Biomass and distribution of juvenile herring	Historical data	
Biomass and distribution of blue whiting	Historical data	
Fish stocks that are harvested		
Spawning stock of cod	Precautionary reference point	Estimated spawning stock is below the precautionary reference point
Spawning stock of capelin	Precautionary reference point	Estimated spawning stock is below the precautionary reference point
Spawning stocks of fish stocks that are being rebuilt to sustainable levels (indicator under development)	Precautionary reference point	Estimated spawning stock is below the precautionary reference point
Benthic organisms		
Species composition and quantity of benthic organisms and fish taken during research bottom trawling	Historical data	

Continued

Table 14.2. Continued

Indicator	Reference values	Action threshold
Distribution of coral reefs and sponge communities	Distribution and state of known sites	Significant rise in the extent of damage or reduction in distribution in areas that are monitored
Occurrence of red king crab	Distribution of red king crab	Spread of red king crab to new areas
Seabirds and marine mammals		
Spatial distribution of seabird and marine mammal communities (indicator under development)	Average population numbers, last 10 years, and historical data	
Population trend for common guillemot	Average population numbers, last 10 years, and historical data	Viable population level when population is below this: or a population decrease of 20% or more in 5 years, or failed breeding 5 years in a row
Population trend for Atlantic puffin	Average population numbers, last 10 years, and historical data	Viable population level when population is below this: or a population decrease of 20% or more in 5 years, or failed breeding 5 years in a row
Population trend for Brünnich's guillemot	Average population numbers, last 10 years, and historical data	Viable population level when population is below this: or a population decrease of 20% or more in 5 years, or failed breeding 5 years in a row
By-catch of common porpoise (indicator under development)	Average for the past 5 years	
Alien species		
Records of alien species	Historical data	Alien species recorded during monitoring
Vulnerable and endangered species		
Vulnerable and endangered species (indicator under development)	Viable population level and historical data on population levels	Population of selected species is below the level considered to be viable

The advisory group on monitoring (see Fig. 14.6) has worked on further development and implementation of the indicators and the coordinated monitoring system. The group presented results from the first year (2007) of using the indicators for monitoring in a recent report (Sunnanå and Fossheim, 2008). The evaluation based on this first year of use is that the system appears to provide useful information for management. However, the group notes that there is scope for improvement in some of the indicators and that others are still under development. Additional information beyond the system of indicators was also used when evaluating the status and trend in the ecosystems of the area of the management plan.

Concluding Remarks

The integrated management plan for the Barents Sea and the sea areas off the Lofoten Islands is an overall framework for sector integration and coordination. As such it is a practical example of the implementation of the ecosystem approach as adopted by UN CBD (CBD, 2000; Vierros, this volume). It builds on the framework for ecosystem approach developed under the Norwegian chairmanship of the NSC process (NSC, 2002) and is quite similar in many respects to that of the EU Marine Strategy Framework Directive (http://www.europarl.europa.eu/oeil/FindByProcnum.do?lang=2&procnum=COD/2005/0211; Skjoldal and Misund, this volume).

The area for the management plan includes the Norwegian part of the Barents Sea ecosystem and the adjacent part of the Norwegian Sea ecosystem including the sea area off Lofoten (Fig. 14.1). It thus contains parts of two LMEs. The Barents Sea LME includes the Russian part, and the Norwegian Government wishes to strengthen the cooperation with Russia. This includes enhanced cooperation with Russia on using the ecosystem approach to the management of the Barents Sea, aiming to establish common management principles and environmental standards.

Preparation of a similar management plan for the Norwegian Sea is now underway and a White Paper is planned to be presented in 2009. The boundary issue between the areas of management plans for the Barents Sea and Norwegian Sea ecosystems may have to be addressed in this context. The working group Protection of the Arctic Marine Environment (PAME) of the Arctic Council has made a working map of the delineation of LMEs of the Arctic and sub-Arctic marine areas, including the Barents Sea and Norwegian Sea LMEs (http://arcticportal.org/uploads/zT/PI/zTPIu_976jm7f6E3yrMRVA/17-Arctic-LMEs-2006-new-version.jpg). There may be a need to consider realigning the areas for the Norwegian management plans with the boundaries of the LMEs. In any case, the boundaries are open and there are important interactions (water transport, migration of fish and marine mammals, etc.) across them that must be taken into account in the management plans.

The use of indicators to inform management decisions is an attractive idea, but not without its challenges. The further evaluation of the indicator-based monitoring system by the advisory group on monitoring will be an important basis for any adjustments in the design or use of the system. One aspect that should be given attention is the role of integrated ecosystem assessments as a complementary component to the use of indicators. Integrated assessment is seen as a key component of the framework of ecosystem approach to management as contained in the Bergen Declaration from the 5th NSC (NSC, 2002; Skjoldal and Misund, this volume; Garcia, this volume). ICES had a study group (Regional Ecosystem Group for the North Sea – REGNS) that carried out an integrated assessment of the North Sea ecosystem based on data for over 100 variables over the time period 1983–2003 (Kenny *et al.*, 2008). The experiences from REGNS should be examined and a similar approach to integrated assessment of the Barents Sea ecosystem could possibly be attempted.

The management plan for the Barents Sea and the area off Lofoten draws up an ambitious set of goals for the sustainable use and conservation of the biodiversity of the ecosystems in this area. Assessment of whether or not the goals and targets are met is an important step to inform management measures. Therefore, it is of the utmost importance that the assessment is as accurate as possible. Measures to maintain or restore good environmental status may have large socio-economic consequences in the form of short-term costs, while securing long-term benefits. Assessments form the basis for evaluations of the effectiveness of existing measures and considerations of whether changes in policies and measures are required. Since the assessments are largely based on scientific knowledge, scientific objectivity and independence must be secured. This requires clear delegation of tasks and responsibilities to scientific institutions with independent roles, but secured funding to carry out their independent scientific advisory tasks within the institutional framework of ecosystem approach to management.

This is one aspect of the management plan that may require further attention to secure the provision of independent scientific advice in a transparent manner. The system with three groups (monitoring, risk management and forum on management) is designed to secure the needed coordination among the many agencies and institutions involved in management in the geographical area of the plan. At the same time the mix of management agencies and scientific institutions in these groups may obscure where the responsibility for scientific advice lies, and blur the interface between scientific advice and management actions.

References

AMAP (2004) AMAP Assessment 2002: Persistent Organic Pollutants in the Arctic. Arctic Monitoring and Assessment Programme (AMAP), Oslo, Norway. xvi+310 pp.

Anon. (2002) Rent og rikt hav, St.meld. nr. 12 (2001–2002) (Clean and rich sea, Government White Paper; in Norwegian), Ministry of Environment, Oslo. Available at: http://www.regjeringen.no

Anon. (2003) Utredning av konsekvenser av helårig petroleumsvirksomhet i området Lofoten - Barentshavet (Impact assessment of year-round petroleum activities in the Lofoten- Barents Sea area), 120 pp. Ministry of Petroleum and Energy (In Norwegian).

Anon. (2004a) Utredning av konsekvenser av fiskeri i området Lofoten - Barentshavet. Directorate of Fisheries, 177 pp. (In Norwegian).

Anon. (2004b) Utredning av konsekvenser av skipstrafikk i området Lofoten-Barentshavet. Directorate of Coastal Affairs, 157 pp (In Norwegian).

Anon. (2004c) Utredning av konsekvenser av ytre påvirkning. Klimaendring, forurensning og annen påvirkning fra kilder utenfor norsk del av Barentshavet. Norwegian Polar Institute, 144 pp. (In Norwegian).

Anon. (2005) Konsekvenser av samlet påvirkning på Lofoten-Barentshavet med dagens aktiviteter og i 2020 (Impacts of overall pressures on the Barents Sea–Lofoten area). Norwegian Polar Institute, 77 pp. (In Norwegian).

Anon. (2006) Report No. 8 to the Storting (2005–2006) Integrated management of the marine environment of the Barents Sea and the sea areas off the Lofoten Islands (Translation in English), Ministry of Environment, Oslo. Available at: http://www.regjeringen.no

Anon. (2007) Regjeringens miljøpolitikk og rikets miljøtilstand (The Government's environmental policy and state of the environment in the country). St.meld. nr. 26 (2006–2007). Norwegian Ministry of Environment (In Norwegian). Available at: http://www.regjeringen.no

CBD (2000) Decision V/6: Ecosystem approach. UN Convention on Biological Diversity. Available at: http://www.biodiv.org/decisions/default.asp

Dalpadado, P., Bogstad, B., Gjøsæter, H., Mehl, S. and Skjoldal, H.R. (2002) Zooplankton-fish interactions in the Barents Sea. In: Sherman, K. and Skjoldal, H. R. (eds) *Large Marine Ecosystems of the North Atlantic*. Elsevier, Amsterdam, pp. 269–291.

Fosså, J.H., Lindberg, B., Christensen, O., Lundälv, T., Svellingen, I., Mortensen, P.B. and Alvsvåg, J. (2005) Mapping of Lophelia reefs in Norway: experiences and survey methods. In: Freiwald, A. and Roberts, J.M. (eds) *Cold-Water Corals and Ecosystems*. Springer, Berlin, pp. 359–391.

Føyn, L., von Quillfeldt, C.H. and Olsen, E. (2002) Miljø- og ressursbeskrivelse av området Lofoten – Barentshavet ('The environment and natural resources in the Barents Sea–Lofoten area'). Fisken og havet, nr. 6-2002. Institute of Marine Research, Bergen, 84 pp. (In Norwegian).

Gjøsæter, H. (1998) The population biology and exploitation of capelin (*Mallotus villosus*) in the Barents Sea. *Sarsia* 83, 453–496.

Gjøsæter, H., Huse, G., Robberstad, Y. and Skogen, M. (2008) Havets ressurser og miljø ('Marine resources and environment'). Fisken og havet, særnummer 1-2008. Institute of Marine Research, Bergen, 189 pp. (In Norwegian, English summary).

ICES (2007) Report of the ICES Advisory Committee on Fishery Management, Advisory Committee on the Marine Environment and Advisory Committee on Ecosystems, 2007. ICES Advice 2007, Book 3. International Council for the Exploration of the Sea, Copenhagen.

IMM (1997) Statement of conclusions. Intermediate Ministerial Meeting on the Integration of Fisheries and Environmental Issues, 13–14 March 1997, Bergen, Norway. Ministry of Environment, Oslo. Available at: http://odin.dep.no/nsc

Kenny, A.J., Skjoldal, H.R., Engelhard, G.H. and Kershaw, P. (2008) An integrated approach for assessing the relative significance of human pressures and environmental forcing on the status of a Large Marine Ecosystem. Manuscript in review.

Kålås, J.A., Viken, Å. and Bakken, T. (eds) (2006) *Norsk Rødliste 2006 – 2006 Norwegian Red List*. Artsdatabanken, Norway, 416 pp.

NSC (2002) Bergen Declaration. Fifth International Conference on the Protection of the North Sea, 20–21 March 2002, Bergen, Norway. Ministry of Environment, Oslo. ISBN 82-457-0361-3.

Olsen, E. and von Quillfeldt, C.H. (2003) Identifisering av særlig verdifulle områder i Lofoten – Barentshavet (Identification of particularly valuable areas in the Barents Sea–Lofoten area). Report, Institute of Marine Research and Norwegian Polar Institute, 72 pp. (In Norwegian).

Olsen, E., Gjøsæter, H., Røttingen, I., Dommasnes, A., Fossum, P. and Sandberg, P. (2007) The Norwegian ecosystem-based management plan for the Barents Sea. ICES Journal of Marine Science 64, 599–602.

Skjoldal, H.R. and Rey, F. (1989) Pelagic production and variability of the Barents Sea ecosystem. In: Sherman, K. and Alexander, L. (eds) *Biomass Yields and Geography of Large Marine Ecosystems*. AAAs Selected Symposium 111. Westview Press, Boulder, pp. 173–186.

Sunnanå, K. and Fossheim, M. (2008) Forvaltningsplan Barentshavet – rapport fra overvåkingsgruppen 2008 (Management plan for the Barents Sea – report from the monitoring group 2008), 95 pp. Fisken og havet, særnummer 1 b-2008 (In Norwegian).

von Quillfeldt, C.H. and Dommasnes, A. (2005) Forslag til indikatorer og miljøkvalitetsmål for Barentshavet (Proposed indicators and environmental quality objectives for the Barents Sea). Institute of Marine Research and Norwegian Polar Institute, 170 pp. (In Norwegian).

15 Ecosystem-based Fisheries Management in Iceland: Implementation and Some Practical Considerations

JÓHANN SÍGURJÓNSSON

Marine Research Institute, Reykjavík, Iceland

Abstract

Managers and fisheries scientists providing advice have, for many years, discussed and argued definitions of an ecosystem-based approach to marine fisheries. The Reykjavík Conference on Responsible Fisheries in the Marine Ecosystem held in 2001 concluded that there was no reason to wait, since many of the measures that are being implemented under single-species management schemes are in the spirit of ecosystem-based fisheries management. We need simply to do it better.

In this chapter, some examples are given as to how such single-species approach has been exercised in Iceland. While it is important to study and define scientific criteria to be applied under the scope of an holistic view of the marine ecosystem in its greatest complexity, a more simple approach may provide some steps forward. For scientists involved in single-species assessment of fish stocks, a systematic mapping of various relevant aspects is suggested and discussed. The chapter reports on such pragmatic approaches, involving consideration of assessment methods and the basis for scientific advice. This includes the issue of discards of target and non-target species, the effects of fisheries on the physical environment and certain ecosystem components, multi-species considerations and the effects of environmental changes on target stocks. Such an approach is meant to help scientists to focus on aspects that are relevant in this context, to help identify gaps and research needs and to draw the attention of all stakeholders to these factors. Later, it may contribute to a more holistic ecosystem approach to the management of the fisheries and other ocean resources.

Introduction

Since the introduction of the precautionary approach concept in Rio de Janeiro in 1992 and the adoption of the FAO Code of Conduct for Responsible Fisheries in 1995, managers and fisheries scientists providing advice have for many years discussed and argued definitions of ecosystem-based fisheries management (EBFM) or ecosystem approach to fisheries (EAF). This concept has been on the agenda of international fora in recent years, and on several occasions

(eds G. Bianchi and H.R. Skjoldal)

dedicated international conferences and symposia have been held.The Reykjavík Conference on Responsible Fisheries in the Marine Ecosystem (FAO, 2002; Sinclair and Valdimarsson, 2003), held in 2001 addressed the scope of the concept and subsequently FAO produced basic guidelines for implementation (FAO, 2003). As a follow-up, immense efforts have been devoted to define indicators and scientific criteria to be applied, including the Bergen Conference on Implementing the Ecosystem Approach to Fisheries (Bianchi and Skjoldal, this volume). However, despite all these efforts, we still do not move very fast towards implementation, and there is even some misconception as to what this is about.

This chapter is written from the point of view of a marine research organization, which has a group of marine scientists onboard, as well as being principal advisory authority to the government and the fishing industry in Iceland on fisheries and ocean matters. The chapter is also written from a perspective of a modern fishing society with a modern industry and economy, which is heavily dependent on well-managed marine resources and environment. In this chapter some practical considerations on EAF, based on a single-species approach in Iceland that influence the daily work of assessment scientists, are suggested and discussed.

The Ecosystem Approach to Fisheries: the Essence

It is important to recognize that we need not find a single definition of the EAF, but rather accept the common sentiment, which involves managing human activities in such a way that we determine the course of action and have in advance predicted the consequences of our actions, with the aim of securing sustainability and optimum utilization of the resources and the marine environment. In principle, this implies taking note not only of the resources we are utilizing, but also of other resources that may be affected by our activity and potential effects on the physical environment that these resources inhabit. In reality, we are talking about a framework or mechanism for environmental risk assessment and management of activities that one has developed for resources on land for quite some time.

In essence, such a framework will consist of a mechanism that clarifies causative links between relevant components that are affected by the fisheries or other human activities. This involves measuring values of different interacting components or resources and weighing values of different resources one against another, in monetary, ecological or other terms, based on ruling, which the society determines at any given point in time. Then authorities make decisions on where to go and these evaluations are translated into management actions.

The 2001 Reykjavík Conference and the EAF

The 2001 Reykjavík Conference on Responsible Fisheries in the Marine Ecosystem (FAO, 2002; Sinclair and Valdimarsson, 2003) addressed the scope of the EAF concept, and subsequently the FAO produced basic guidelines for implementation (FAO, 2003). Here one identified EAF as integrated management; a holistic approach with a broad set of conservation objectives. But it was recognized that for a fully fledged system that would address this new approach, one would need far better understanding of the ecosystem components and dynamics of the ecosystem, one would need a much broader angle of view to fisheries management than was the current practice and one would need an institutional and legal framework that would suit the new situation, in addition to strong political commitment and stakeholder participation.

However, one of the main conclusions of the scientific symposium that was held in conjunction with the Reykjavík meeting was that by conducting our single-species management of fisheries with greater care and more discipline than up to now, we could have done much better with respect to ocean resource management than has been the case. Indeed, had fishing effort worldwide been less intense in the past, we would be well on our way towards EAF. Many of the problems that the new and broader concept is to address would not have been on the agenda under a more cautious fishing regime in recent decades.

Therefore, it was concluded that one should not wait until all conditions and equipment for EAF have been developed, since many of the measures that are being implemented under single-species management schemes are in the spirit of such an approach. We need simply to do it better. Also it was stressed that while a fully fledged EAF scheme of the ocean resources management is the ultimate goal, it needs to be understood that in order to achieve this we may have to undergo a lengthy incremental process. So it was advocated to start immediately but in a stepwise process as experience and circumstances would allow.

In recent years the scientific community, including the ICES scientists, has devoted much effort to define a more holistic framework for use in fisheries management. It should be recognized how much thoughtful work has been done in shaping the concept (e.g. Gislason *et al.*, 2000; FAO, 2003; Sainsbury and Sumaila, 2003; Sinclair and Valdimarsson, 2003; Cury and Christensen, 2005; Daan *et al.*, 2005; Garcia and Cochrane, 2005; ICES, 2005; O'Boyle *et al.*, 2005), developing the appropriate framework, and in taking steps towards implementation. So we can say good progress has been made, but it is complicated to determine objectives, criteria and appropriate/relevant indicators, and in general, the matter is still at a design or development stage. This certainly applies to Iceland, where the government, however, has published an official policy document on ocean matters (Anon., 2004) with EAF as part of the portfolio.

Some practical examples from Iceland

Traditional measures

So while this development has taken place, many countries have continued to elaborate on single-species approaches with standard ingredients such as those that have been developed and improved in Iceland and elsewhere in recent years. These comprise total allowable catches (TACs) to limit total fish removals, fishing gear spatial/temporal restrictions, restrictions on vessel size, selective mesh size and gear, season length and timing, multi-species interactions and short- or long-term area closures. We can say that all these elements are essential and in the spirit of EAF.

Geographical scope

The currents and topography around Iceland (Fig. 15.1) provide oceanographic conditions that give a relatively closed system on the continental shelf around the island. The geographical position of Iceland on the ocean ridges means that the country is in the vicinity of mixing areas of the warm and cold ocean currents. The warm North Atlantic current originating in the Gulf of Mexico meets the polar East Greenland current, flowing south along the East Greenland coast. Close to the coast there is a coastal current, which flows clockwise around Iceland and is formed by mixing of warm oceanic water with freshwater from the land.

So the area around Iceland has been defined as one of the large marine ecosystems (LMEs) due to these rather well-defined conditions (Ástthórsson and Vilhjálmsson, 2002). This encompasses the 200 nautical mile exclusive economic zone (EEZ) around Iceland, about 760,000 km^2, which is seven times the area of the island. The scope would, however, have to be adjusted, extended or reduced, with respect to migratory behaviour and biology of the species and stocks in question. For example, one would have to extend such an area beyond the Icelandic EEZ in order to address questions related to the Atlanto-Scandian herring (*Clupea harengus*) stock, a highly migratory species that occurs in the national waters of several countries, while less-mobile stocks such as Norway

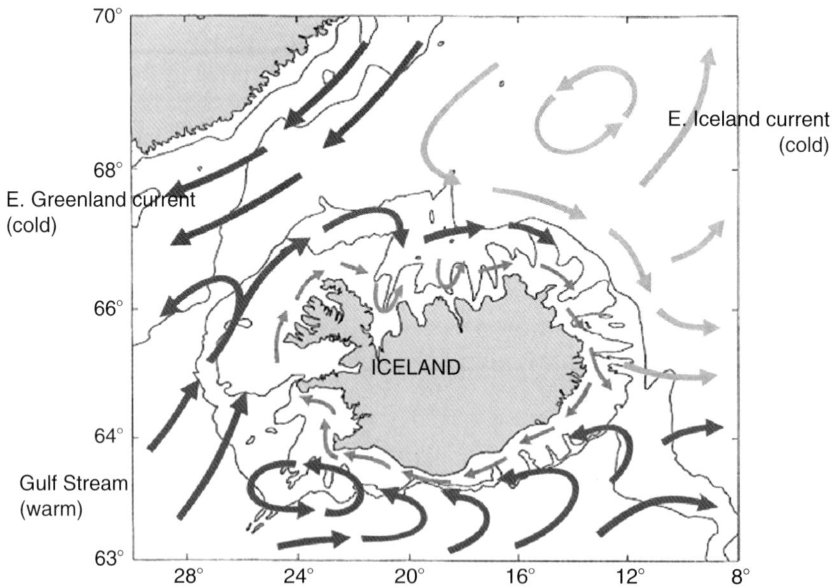

Fig. 15.1. Ocean currents and topography of the Iceland seas with the warm Gulf Stream coming from the southwest and cold currents reaching the northern part of the island with the East Greenland and East Iceland currents.

lobster (*Nephrops norvegicus*) and shrimp (*Pandalus borealis*) stocks would be managed on a smaller scale area basis within the Icelandic EEZ.

Area closures

Although marine-protected areas (MPAs) are a popular theme when discussing EAF, they are only one of many potential tools of a fully fledged EAF and are not likely alone to meet our goals in managing the stocks (Stefánsson, 2003). However temporary closures of areas have proved important tools in the Icelandic context and have been used extensively. For example, Fig. 15.2 shows a map of areas closed permanently or temporarily to bottom trawlers, either due to too high abundance of juvenile fish or unwanted by-catches of non-target species (Ragnarsson and Steingrímsson, 2003; Jaworski *et al.*, 2006). Also shown are areas with compulsory sorting grid provisions (shaded areas).

Likewise, extensive areas have been closed to long-lining fishing as shown in Fig. 15.3. In all cases we are dealing with measures imposed in order to conserve certain biological resources: spawning grounds, juveniles or unwanted by-catches. Here, criteria have been established to tell when and where areas need

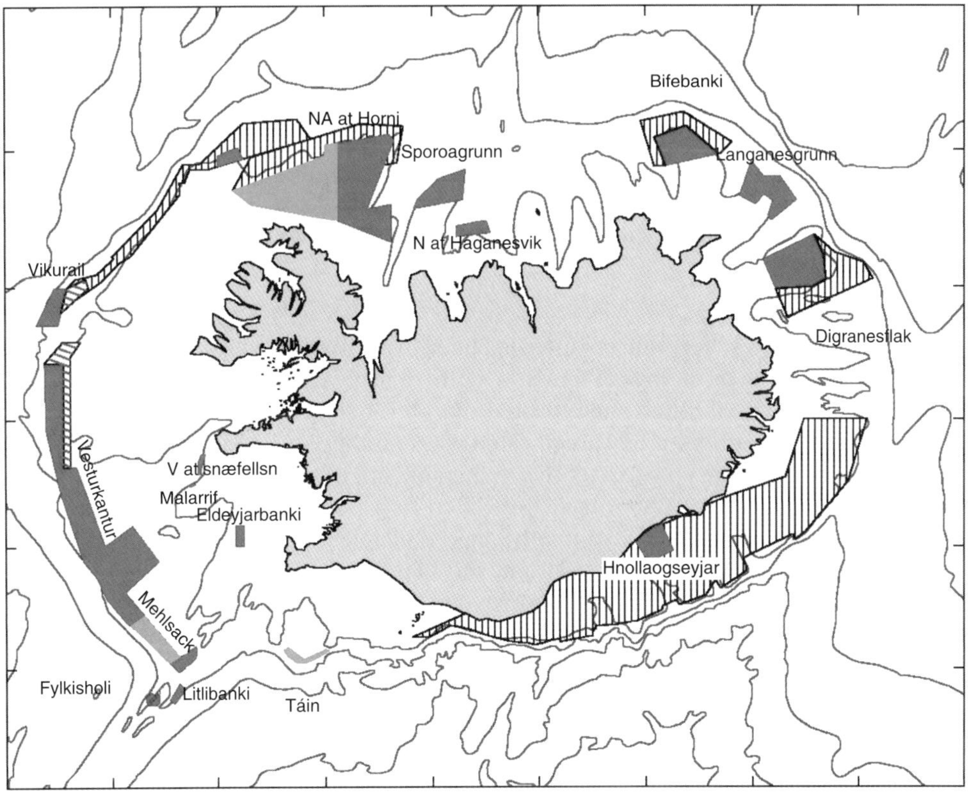

Fig. 15.2. Closure of fishing grounds around Iceland from bottom trawling. Year-round closures in dark grey (horizontal lines, part of the day only), closures part of the year in light grey, and areas with trawling permitted but sorting grids imposed shown with vertical lines.

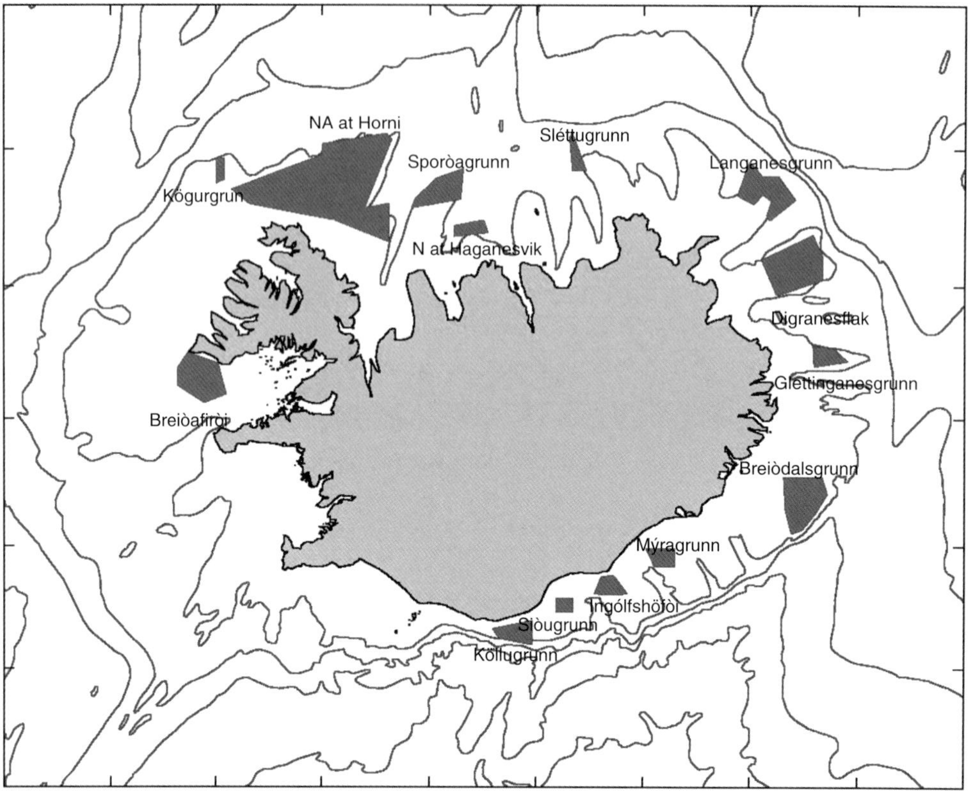

Fig. 15.3. Year-round closure of fishing grounds around Iceland from long-line fishing (dark grey areas).

to be closed, where one resource is valued on its own or weighed against another in economic or biological terms, or both. In many cases, long-term area closures have been established and repeated short-term closures have been set since too high a proportion of unwanted catch was found upon inspection by observers.

In the same spirit, extensive vulnerable coral reef areas on the south coast of Iceland (Anon., 2005) have now been fully protected from all fishing activities after ocean floor mapping became available and surveys were conducted. The gathering of information and the identification of areas to be closed or protected have always been done in close cooperation between all main stakeholders, i.e. the fishing industry, management authorities and scientific advisory institutions.

Multi-species management system

Weighing one resource against another in a multi-species context (see Fig. 15.4) is how authorities and stakeholders in Iceland have, with the help of harvest control rules, managed the economically valuable cod (*Gadus morhua*) stock that feeds on capelin (*Mallotus villosus*) and shrimp (Baldursson *et al.*, 1996; Daníelsson *et al.*, 1997; Jakobsson and Stefánsson, 1998). Every year sufficient

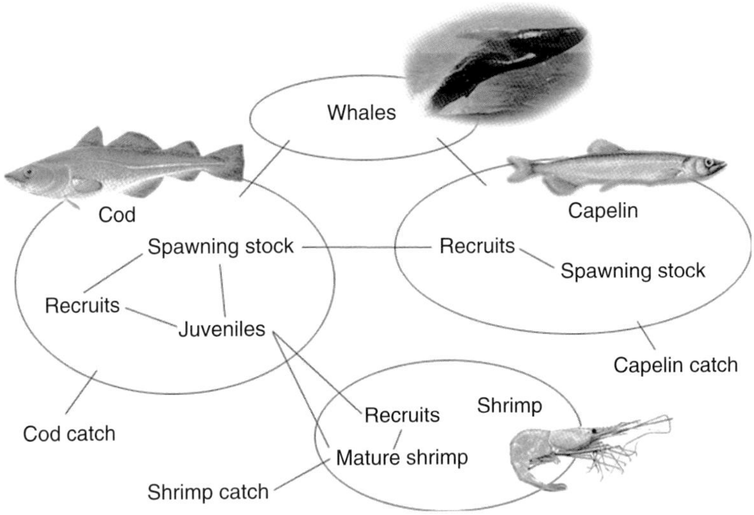

Fig. 15.4. Multi-species stock system and management in Icelandic waters.

quantities of capelin are left as fodder for the cod and sufficient quantities are left for capelin to spawn. Due to the close dependence of capelin as food for cod, short-term predictions for cod are significantly linked to predictions of the development of the capelin stock in the following year. Since cod is valuable in economic terms, the long-term strategy was to build up the cod stock at the cost of lower shrimp yields. Finally, while in terms of biomass, whales constitute a major component of the marine life in Icelandic waters (Sigurjónsson and Víkingsson, 1997) and may significantly influence the yield of the interacting fish stocks, different views arise as to how to value and manage the whale stocks (Stefánsson *et al.*, 1997). Here again weighing of components provides a basis for longer-term management strategies.

It is the task of society, so to speak, to put price tags on these resources, on whatever basis, and to take a well-balanced management action with predicted consequences. A well-founded EAF framework is an appropriate tool for society to deal with such questions.

Pragmatic Ecosystem Considerations

EAF as part of routine scientific fish stock assessments

The above cases are well worth referring to when discussing EAF. Here scientists and the fishing industry have been heavily involved. Many research activities in recent years have been directed towards a more holistic view such as bottom trawl surveys and other resource surveys that were initially targeted at certain important fish stocks, but are now also valuable sources of information on many related or non-targeted and often non-commercial species.

We need to note that for the scientists conducting their daily fish stock assessment work, the EAF is far from being on their routine agenda. While bits and pieces appear here and there in their work, a holistic framework is usually not in their minds; nor is it in fact available. It has been said that EAF is not a scientific undertaking to be run by scientists, but rather a management process involving authorities and other stakeholders, where the aim is to manage certain human activities rather than the ecosystem itself. However, for a successful EAF there is a need for well-founded scientific inputs, so unless scientists widen their scope and take this task as a part of their normal routine, they will not provide the necessary basis for authorities and stakeholders.

Scientists need to be prepared and get acquainted with this new world. At the Marine Research Institute in Reykjavík, this is being realized and a pragmatic approach to be applied in our current single-stock fish assessments is under development (Sigurjónsson, 2007a,b,c). Rather than waiting for a fully fledged EAF system to operate, the choice has been taken to address these questions in a wider context than in the past, within the present framework.

Routine EAF scientific considerations

For each species and stock that is assessed, the aim is to map relevant information both for research and management purposes, including the quality and nature of the assessment techniques used and the effects of the given fishery on the target stock. Further, the effects of the fishery in question with respect to discards of target and non-target species by gear and area will be mapped, as well as the potential effects of the fishery in question on the physical environment by area, and the potential effects of the fishery in question on different ecosystem components or species/stock complexes. Also, when relevant, multi-species considerations will be noted and special attention will be given to potential effects of environmental changes on the target stock in question. Finally, one would routinely allow for some special management considerations to be made, where they may seem needed.

All these additional ecosystem considerations would cast light on aspects that are relevant for EAF. This would be reflected in the assessment work itself, and in future plans of investigations. In addition to conventional advice to authorities on recommended TACs, a qualitative statement on important or relevant issues in ecosystem context would follow that puts the advice into a wider EAF context than conventional advice.

Nature and quality of assessment methods

In this approach it is of general interest and often of some importance to know the nature and quality of the assessment made for the stock in question, since it can have consequences for interpretation of the results and the level of risk. Thus, there is a major difference between assessments based on a wealth of high-quality data and advanced assessment techniques on the one hand, and assessments of stocks in data-poor or no-data situations on the other. Stocks may be assessed with the help of age-based techniques and managed on the basis of a well-defined long-term management strategy, or they may be assessed with age- or length-based techniques and catch data, and managed on an *ad hoc* basis. Finally, in data-poor situations, one would normally require special caution and notation of this would be relevant in this context.

Effects of fisheries on target stock, discards and indirect mortalities

Following this approach, the assessment scientist would first make note of the effect of the specific fishery on the target stock, i.e. a conventional single-species consideration, where the level of impact would be noted on a qualitative scale from 0 (no apparent impact) to 10 (serious impact). Figure 15.5 gives hypothetical examples of two fisheries under relatively high fishing pressure (A: cod trawl fishery) and relatively low fishing pressure (B: purse seine herring fishery), respectively.

Apart from population size, special attention would be given to population and genetic structure, reproductive capacity and geographic range. Is there a stock estimate available, and is the stock monitored regularly? If there is no estimate or limited/no data available, is this relevant or of no concern or no apparent importance? Finally, is there need for actions to be taken, e.g. conduct of a new or improved assessment, or are there management measures that have not been properly implemented, or advice that has not been followed? These qualitative aspects would be broadly classified into three categories (Fig. 15.5) with respect to level or status of knowledge and the need for actions to be taken:

- Information is available or the respective factor is not of any concern (+, Fig. 15.5).
- Information is not available but is needed, or the situation is of serious concern (–, Fig. 15.5).
- Intermediate situation (±, Fig. 15.5).

If the respective factor is not relevant for the given fishery, this would be so indicated (open circle in Fig. 15.5). Although well-defined scientific criteria would obviously be of great value for such an assessment scheme, gross qualitative assessment will suffice the purpose here to begin with. More precisely parameterized and developed criteria are, however, important for future use and can be developed as new data and methods become available.

When it comes to discards of target and non-target species that result from the fishery of the target species, the same general approach will be applied. As with the effects on the target stock, one would here need to examine the discards and other factors of ecosystem importance on area basis, and possibly for different fleets and fishing gear. Similarly, indirect mortalities of target and non-target species would need to be examined. This could involve fish escaping through mesh or off hooks, or fish escaping under the fishing gear, i.e. in trawl fishery. Again we would note assumed level of impact, availability of

A: Fishery – Icelandic cod
Sub-fishery: Trawl – North

+ Available or not of concern
– Of concern or needed/not available
± Intermediate
O Not relevant or not known

Effects of fishery on:	Level of impact	Estim. available	Monitored regularly	If no data– relevance	Action needed	
					Assessm.	Managem.
Target species						
Population size	8	+	+	O	+	±
Age/size/sex structure	8	+	+	O	+	±
Reproductive capacity	8	+	+	O	+	±
Geographic range	5	±	±	O	±	O
Genetic structure	5	±	–	O	±	O
Discards of target spp	5	+	+	O	+	±
Discards of non-target spp	5	±	–	–	–	O
Indirect mortality of target and non-target spp	5	±	–	–	–	O

B: Fishery – Icelandic herring
Sub-fishery: Purse seine

+ Available or not of concern
– Of concern or needed/not available
± Intermediate
O Not relevant or not known

Effects of fishery on:	Level of impact	Estim. available	Monitored regularly	If no data– Relevance	Action needed	
					Assessm.	Managem.
Target species						
Population size	6	+	+	O	±	+
Age/size/sex structure	6	+	+	O	±	+
Reproductive capacity	6	+	+	O	±	+
Geographic range	3	±	±	O	±	±
Genetic structure	3	–	–	±	±	O
Discards of target spp	2	±	–	±	±	+
Discards of non-target spp	1	±	–	±	±	+
Indirect mortality of target and non-target spp	1	±	–	±	±	+

Fig. 15.5. Hypothetical examples to show how the impact of (A) the bottom trawl cod fishery off Iceland, and (B) the purse seine herring fishery off Iceland could be classified with respect to the status of the stocks, discards and indirect mortalities during regular assessments of these fish stocks. The level of impact would be classified from 0 (no impact) to 10 (serious impact) while the level or status of knowledge and the need for actions to be taken would be classified into four categories (see text for further information).

estimates and monitoring series, and whether there is specific need for actions to be taken.

Figure 15.5A shows hypothetical example of how the cod trawl fishery off Iceland could turn out in such an exercise. Here it matters that the fishing intensity has been considerable for a long period of time, and it was important that authorities in 2007 reacted responsibly to secure sustainable long-term yield of the stock. The impact on the cod stock has been substantial (Anon., 2007), the level of knowledge or the quality of data and assessment is more or less sufficient, as are the monitoring activities, and the management actions that have just been taken seem sufficient (has moved from the – category to ± in the current season; becomes + if results prove successful!). When it comes to discards of cod and other species and indirect mortalities in this fishery, the impact is apparently relatively small, the level of knowledge is rather good although monitoring activities are at rather modest levels (apart from cod). It is important that authorities be aware of potential discards of cod and that the research focus is directed towards other species and indirect mortalities of all species.

Figure 15.5B demonstrates the corresponding mapping for the local summer-spawning herring fishery off Iceland. The fishing pressure on the herring stock for the last 35 years has been sustainable, and one would conclude that the direct impact of the fishery is rather modest (Anon., 2007). This stock is also very well studied, apart from the genetic structure, which has not been investigated. Since the herring stock has undergone major shifts in distribution during the years, this needs to be specifically addressed, but generally speaking one would not call upon any strong measures on behalf of the management authorities. The same applies to the possible impact of the herring fishing operation on herring discards, discards of other species, or indirect mortalities. Although these aspects are not regularly monitored, studies indicate that the impacts here are not of great importance and therefore no specific actions are required on behalf of the authorities.

Effects of fishery on ecosystem components and physical environment

It will also be relevant in this context to ask questions that traditional assessment scientists would not ask - is the fishery for the given species affecting specific ecosystem components, species/stock complexes or communities? Here one would examine benthic and zooplankton communities, seabirds, marine mammals and fish communities. Is the exploitation of the target species affecting the livelihood of other biological resources, e.g. due to lesser predation or competition, or is the target species an important food item for other important ecosystem components? We would note to what extent such effects can be assessed, whether studies are being conducted into this, and whether there are any indications as to whether the impact is low or high.

The effects of fishing activities (by fishing gear and area) on the physical environment also need to be on the checklist, e.g. fish and benthic habitats such as cold-water corals. Are there seabed maps available and have the potential effects been studied? Which measures are needed and which measures are in place? Here important fish habitats such as spawning grounds of cod, capelin, sand eel (*Ammodytes* sp.) and herring (*C. harengus*) are in focus, as well as nursery grounds of cod, haddock (*Melanogrammus aeglefinus*) and redfish (*Sebastes* sp). Cold-water corals and other three-dimensional habitats have potential importance as habitats for juvenile fish and other animal life and need special attention.

A: Fishery – Icelandic cod
Sub-fishery: Trawl–North

+ Available or not of concern
– Of concern or needed/not available
± Intermediate
O Not relevant or not known

Effects of fishery on:	Level of impact	Estim. available	Monitored regularly	If no data–relevance	Action needed	
					Assessm.	Managem.
Ecosystem components						
Benthos	5	±	–	–	–	O
Zooplankton	0	O	O	+	O	O
Birds	2	–	–	+	O	O
Marine mammals	3	–	–	±	±	O
Fish	3	–	–	–	–	O
Physical environment						
Fish habitats	5	±	–	±	–	O
Benthic habitats	5	±	–	±	–	O

B: Fishery – Icelandic herring
Sub-fishery: Purse seine

+ Available or not of concern
– Of concern or needed/not available
± Intermediate
O Not relevant or not known

Effects of fishery on:	Level of impact	Estim. available	Monitored regularly	If no data–relevance	Action needed	
					Assessm.	Managem.
Ecosystem components						
Benthos	0	O	O	+	O	O
Zooplankton	1	–	–	±	O	O
Birds	1	–	–	±	O	O
Marine mammals	2	–	–	±	±	O
Fish	3	–	–	–	+	O
Physical environment						
Fish habitats	0	O	O	+	O	O
Benthic habitats	0	O	O	+	O	O

Fig. 15.6. Hypothetical example to show how the impact of (A) the bottom trawl cod fishery and (B) the purse seine herring fishery off Iceland could be classified with respect to the effects on selected ecosystem components and physical environment, in regular evaluations of the respective fisheries. The level of impact would be classified from 0 (no impact) to 10 (serious impact) while the level or status of knowledge and the need for actions to be taken would be classified into four categories (see text for further details).

Figure 15.6 shows hypothetical examples where the possible impacts of the cod (A) and herring (B) fisheries on selected ecosystem components (species, stocks or communities, e.g. benthic animals, zooplankton, seabirds, sea mammals or fish) and physical environment have been mapped. Here the aim is to begin mapping what research has been conducted and what data are available with respect to these factors and the respective fisheries, and to attempt to tell whether the impact is potentially small or large. Also one would focus on the impact of the fishery on fish habitat, e.g. vulnerable spawning and nursery bottom habitats, such as hard bottom and coral areas. Here one would assess whether the habitats are mapped, whether potential impacts have been assessed and whether protective measures are in place or needed.

From Fig. 15.6A it is clear that the alleged impact of the cod fishery on the ecosystem components or habitats are generally less than the direct impact of the fishery on the cod stock itself, although it is evident that both research and monitoring of these factors are much lacking. There does not seem to be a strong need for immediate management actions based on this. The impacts of the herring fishery (Fig. 15.6B) on the ecosystem components and the physical environment seem even less, although it needs to be kept in mind that research and monitoring of these factors is very limited.

Other considerations

By this approach, which would be applied for each species caught in a single-species management scheme, we would ask several additional questions about availability of food-web data and modelling. Have models been developed and predictions been made? Whether there have been recent changes in environmental factors need special attention since our concern is not only effects of human activities on the ecosystem, but also what natural changes of the ecosystem are taking place that may affect the resources in question. Also some operational factors may be a cause of concern, such as sudden market incentives or technological shifts that may influence the basis and interpretation of our assessments. These need to be mapped as well in an ecosystem context.

Conclusions

EAF is an opportunity to manage human activities, particularly fishing activities, more successfully than has been achieved up to now. It aims at securing growth and sustainability of the fish stocks and their environment in the long term. It contributes to securing biodiversity and ecosystem health. But it requires knowledge and understanding of the nature of the ecosystem and its dynamic interactions, and much more commitment and research than at present. This may require some sacrifices in the short run, but beyond doubt, it will be beneficial in the long term. So it is important to continue developing EAF methodology, objectives, scientific criteria and indicators to be used in a fully fledged holistic and cross-sectoral approach.

Until such time that we have the perfect scheme in hand, each sector may choose to develop methods and means to implement this new broader approach. Broadened single-species considerations discussed in this chapter may provide a pragmatic approach to move stepwise forward in this respect. To start broadening the portfolio of items to be addressed by classical assessment scientists, will help integrate the concept into the institutional culture, and prepare and motivate the people, including the scientists, that need to be involved in the future work. An inventory of this kind will reveal that much of the information needed is not available, and it will take a prolonged period of time and great resources to collect these. However, the most important data deficiencies will be identified and appropriate sampling programmes can be initiated. Thereby one would gradually be moving from qualitative ecosystem considerations towards a more developed quantitative scheme of EAF.

Acknowledgements

My sincere thanks to my colleagues at the Marine Research Institute, Reykjavík for sharing their views and assisting while preparing this chapter, particularly Björn Ævarr Steinarsson, Höskuldur Björnsson, Ólafur S. Ástthórsson and Stefán Áki Ragnarsson. My thanks are also due to Michael Sinclair of the Department of Fisheries and Oceans, Canada for stimulating discussions and encouragement and to the editors for their very useful suggestions for improvements of the final manuscript.

References

Anon. (2004) The Ocean-Iceland's Policy. *Government of Iceland* 1-44.

Anon. (2005) Friðun viðkvæmra hafsvæða við Ísland (Protection of vulnerable ocean areas off Iceland). Niðurstöður og tillögur nefndar sem sjávarútvegsráðherra skipaði í október 2004. *Ministry of Fisheries* 1-60 (In Icelandic).

Anon. (2007) Nytjastofnar sjávar árið 2006/2007 og aflahorfur fiskveiðiárið 2007/2008 (State of marine stocks 2006/2007 and prospects 2007/2008). Hafrannsóknastofnunin Fjölrit 129, 1-180.

Ástthórsson, Ó.S. and Vilhjálmsson, H. (2002) Iceland shelf LME: decadal assessment and resource sustainability, 2001. In: Sherman, K. and Skjoldal, H.R. (eds) *Large Marine Ecosystems of the North Atlantic - Changing States and Sustainability*. Elsevier, Amsterdam, pp. 219-243.

Baldursson, F.M., Stefánsson, G. and Daníelsson, A. (1996) On the rational utilization of the Icelandic cod stock. *ICES Journal of Marine Science* 53, 643-658.

Cury, P.M. and Christensen, V. (2005) Quantitative ecosystem indicators for fisheries management. *ICES Journal of Marine Science* 62, 307-310.

Daan, N., Christensen, V. and Cury, P.M. (eds) (2005) Quantitative ecosystem indicators for fisheries management. *ICES Journal of Marine Science* 62, 479-560.

Daníelsson, A., Stefánsson, G., Baldursson, F. and Thórarinsson, K. (1997). Utilization of the Icelandic cod stock in a multi-species context. *Marine Research Economics* 12, 329-344.

FAO (2002) Report of the Reykjavík Conference on Responsible Fisheries in the Marine Ecosystem, Reykjavík, Iceland, 1-4 October 2001. *FAO Fisheries Report* No. 658, Rome, Italy, 128 pp.

FAO (2003) The ecosystem approach to fisheries. *FAO Technical Guidelines for Responsible Fisheries* No. 4, Suppl 2, FAO, Rome, Italy, 112 pp.

Garcia, S.M. and Cochrane, K. (2005) Ecosystem approach to fisheries: a review of implementation guidelines. *ICES Journal of Marine Science* 62, 311-318.

Gislason, H., Sinclair, M., Sainsbury, K. and O'Boyle, R. (2000) Symposium overview:

incorporating ecosystem objectives within fisheries management. *ICES Journal of Marine Science* 57, 468–475.

ICES (2005) Guidance on the application of the ecosystem approach to management of human activities in the European marine environment. *ICES Cooperative Research Report*, No. 273, 22 pp.

Jakobsson, J. and Stefánsson, G. (1998) Rational harvesting of the cod-capelin-shrimp complex in the Icelandic marine ecosystem. *Fisheries Research* 37, 7–21.

Jaworski, A., Sólmundsson, J. and Ragnarsson, S.Á. (2006) The effect of area closures on the demersal fish community off the east coast of Iceland. *ICES Journal of Marine Science* 63, 879–911.

O'Boyle, R., Sinclair, M. Keizer, P., Lee, K., Ricard, D. and Yeats, P. (2005) Indicators for ecosystem-based management of Scotian shelf: bridging the gap bewteen theory and practice. *ICES Journal of Marine Science* 62, 598–605.

Ragnarsson, S.Á. and Steingrímsson, S.A. (2003) Spatial distribution of otter trawl effort in Icelandic waters: comparison of measures of effort and implications for benthic community effects of trawling activities. *ICES Journal of Marine Science* 60, 1200–1215.

Sainsbury, K. and Sumaila, U.R. (2003) Incorporating ecosystem objectives into management of sustainable marine fisheries, including 'best practice' reference points and use of marine protected areas. In: Sinclair and Valdimarsson (eds) *Responsible Fisheries in the Marine Ecosystem*. FAO, Rome/CAB International, Wallingford, UK, pp. 343–361.

Sigurjónsson, J. (2007a) Ecosystem-based fisheries management in Iceland: implementation and practical considerations (abstr.). In: Nordquist, M.H., Long, R., Heidar og, T.H., Moore, J.N. (eds) *Law, Science and Ocean Management*. Martinus Nijhoff Publishers, Leiden, The Netherlands, pp. 169–170.

Sigurjónsson, J. (2007b) Frá fiskileit til vistkerfisnálgunar við stjórn fiskveiða: Breyttar áherslur í haf- og fiskirannsóknum (From fisheries exploration to ecosystem-approach to fisheries management: New emphasis in oceanographic and fisheries research). *Ægir* 100, 38–43 (in Icelandic).

Sigurjónsson, J. (2007c) Vistkerfisnálgun við stjórn fiskveiða: Breyttar áherslur í haf- og fiskirannsóknum (Ecosystem-based fisheries management: New emphasis in oceanographic and fisheries research). pp. 18–20 *In þættir úr vistfræði sjávar 2006-Environmental conditions in Icelandic waters 2006*. Hafrannsókastofnunin, Fjölrit 130: 1–39 (in Icelandic).

Sigurjónsson, J. and Víkingsson, G. (1997) Seasonal abundance of and estimation of food consumption by cetaceans in Icelandic and adjacent waters. *Journal of Northwest Atlantic Fishery Science* 22, 271–287.

Sinclair, M. and Valdimarsson, G. (eds) (2003) *Responsible Fisheries in the Marine Ecosystem*. FAO, Rome/CAB International, Wallingford, UK, 426 pp.

Stefánsson, G. (2003) Multi-species and ecosystem models in a management context. In: Sinclair and Valdimarsson (eds) *Responsible Fisheries in the Marine Ecosystem*. FAO, Rome/CAB International, Wallingford, UK, pp. 171–188.

Stefánsson, G., Sigurjónsson, J. and Víkingsson, G. (1997) On dynamic interactions between some fish resources and cetaceans off Iceland based on a simulation model. *Journal of Northwest Atlantic Fishery Science* 22, 357–370.

16 The Implementation of the Ecosystem Approach to Fisheries Management in the Benguela Region: Experiences, Advances and Problems

Kevern Laurence Cochrane,[a] Carel Johann Augustyn[b] and Michael John O'Toole[c]

[a]*FAO, Rome, Italy;* [b]*Marine and Coastal Management, Cape Town, South Africa;* [c]*BCLME Programme, Windhoek, Namibia*

Abstract

An ecosystem approach to fisheries (EAF) has been adopted by the nations of the world as being necessary for sustainable use of marine fisheries, and efforts are being made in most countries to make progress in its implementation. Angola, Namibia and South Africa, making up the coastal states of the Benguela Current large marine ecosystem (BCLME), are committed to implementation of EAF and are making use of the opportunities presented by the Global Environment Facility (GEF) BCLME programme to strengthen progress towards this end. This chapter describes a BCLME project which is at the core of these efforts and is examining the feasibility of implementing EAF in the Benguela region. The project, a cooperative effort by BCLME, the management agencies of the three countries and FAO, started in January 2004 and ended in December 2006. Focusing on several of the major fisheries in each country, it has pursued a structured and participatory approach to identify and prioritize the gaps in the existing, largely conventional, approaches to fisheries and potential management actions necessary to address those gaps. Again using a participatory approach that has attempted to engage the range of stakeholders in each case, preliminary estimates of the costs and benefits of those actions have been made. Costs and benefits are being measured in terms of the broad objectives applicable in each fishery. In addition to a large number of issues directly related to the target species and conventional management, gaps have been found, as examples, in relation to by-catch of retained and non-retained species, including impacts of fisheries on species of conservation concern, interactions between fisheries, potential impacts of some gear on habitat and the impacts of non-fishery sectors on fish habitats and species. The detailed results, including potential management actions and their costs and benefits, are still preliminary, but the issues and the broad management needs and possible actions that have been identified are highly informative. The process that has been developed provides a valuable framework for future refinement and implementation of EAF.

Introduction

The Benguela Current ecosystem stretches along the southwest Atlantic coast of Africa from central Angola through Namibia to the south coast of South Africa, bounded by the Angola-Benguela front in the north and the Agulhas Current in the South (from between roughly 14°S and 17°S to between 36°S and 37°S). As such, it covers the west coast of South Africa, the entire Namibian coast, and southern Angola to an extent depending on the position of the Angola-Benguela front (Fig. 16.1). The BCLME programme and this project address the Benguela Current region as a whole, which extends as far north as Cabinda.

The ecosystem is a highly productive one in terms of primary production and fisheries resources. It is also highly complex in relation to, for example,

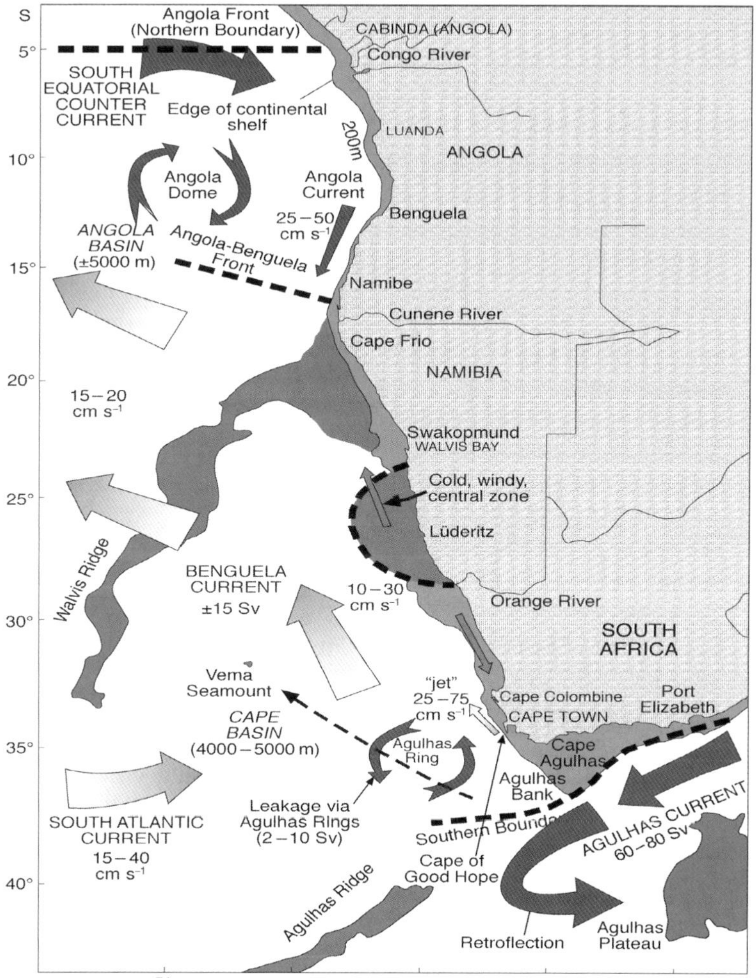

Fig. 16.1. The boundaries, major currents and physical features of the Benguela Current large marine ecosystem. (From BCLME, Windhoek, Namibia.)

its oceanographic features (Fig. 16.1), trophic structure and human activities such as mining, oil extraction and fishing, all of which impact upon its biodiversity and ecosystem health. These human activities have substantial social and economic significance, providing important job opportunities and incomes for the three developing countries. For the living marine resources to be managed sustainably and the social and economic benefits to be maintained, it is consequently critical that their dynamics should be adequately understood and that the countries should introduce management strategies that preserve ecosystem health and minimize the risk of overexploitation (Cochrane *et al.*, 2004; Roux and Shannon, 2004; Shannon *et al.*, 2004). In accordance with the Plan of Action of the World Summit on Sustainable Development, the BCLME countries aim to achieve this goal through implementation of an ecosystem approach to fisheries (FAO, 2003). This is also consistent with the conclusions of Wang (2004) that the complexity of large marine ecosystems requires an ecosystem approach to their management.

The region has long been at the forefront of ecosystem-based marine science. As long ago as 1981, South African institutions developed a multidisciplinary and multi-institutional research programme named the Benguela Ecology Programme (Moloney *et al.*, 2004). The programme integrated physical, chemical and biological oceanography, ecosystem modelling, fisheries biology and stock assessment approaches in a way that allowed the first steps to be taken away from single-species management approaches only to addressing management of the ecosystem in a more holistic way. This resulted from improvements in understanding of the processes involved in production, retention and enrichment, trophic structure and functioning and the impacts of fisheries on various components of the ecosystem. By 1986, major strides had been made and published in a seminal symposium volume (Payne *et al.*, 1987) followed up by subsequent work (Payne *et al.*, 1992; Pillar *et al.*, 1996). As government funding waned in the mid-1990s and the countries of the region began to work together in the post-apartheid period, it was realized that there was an enormous opportunity for improved understanding of the whole Benguela ecosystem by pooling resources across boundaries and tackling these issues on an ecosystem-wide basis. The answer was found in the establishment of a new marine science programme in 1996, the Benguela Environment Fisheries and Training programme (BENEFIT), initiated and funded by the three countries, but strongly supported by Germany (through GTZ) on the environmental side and Norway (through NORAD) on the resources side. The programme turned out to be highly successful, and further advances were made in the understanding of linkages between resources and the environment, as well as capacity building in these areas.

Aware of the complexity of sustainable management, the three countries with the assistance of the Global Environment Facility (GEF), subsequently jointly developed an integrated cross-sectoral programme to address transboundary human impacts on the ecosystem, namely the Benguela Current Large Marine Ecosystem Programme (BCLME). This initiative was developed over the period 1997–2001 and formally launched in 2002. Although the Programme considers the human impacts across all sectors, it particularly focuses on transboundary fisheries and management actions to derive sustainable economic benefits for the region. Further strides have been made in understanding environmental variability in the region and its impact on the productivity of resources. One of the key activities to be commissioned by the BCLME Programme was a project specifically designed to address the implementation of an ecosystem approach to fisheries management (EAFM) and this has allowed managers in the three countries to develop a philosophy and consider practical measures to deal with impacts of fisheries on the ecosystem and its components, both within their own areas of jurisdiction and in a regional manner where certain fish stocks are exploited across borders.This chapter outlines the approach used by the project and some of the results achieved.

During the developmental phase of the BCLME, there was an early realization that better understanding of the dynamics of the ecosystem and improved approaches in its management would eventually need to be backed, first, by political support at the highest level in order to achieve the economic benefits that seemed to be possible from these initiatives. Second, top managers would need to utilize the information being generated by the scientists to achieve the desired advances in management.The main output of the Strategic Action Programme of the BCLME initiative was to establish a formal Commission, which would allow managers to access information on the status of resources and the ecosystem as a whole, and to agree on sustainable levels of utilization and reduction of negative impacts.The Benguela Current Commission (BCC) was formally initiated in August 2006 with the signing of an Interim Agreement by the three countries and will allow managers to advise their governments on these matters. It is hoped that by the end of the second phase of the BCLME (2008–2012), a fully integrated BCC will have been developed and signed with a legally binding Convention that will set terms over which total allowable catches (TACs) for transboundary resources will be negotiated bilaterally within the Commission between the neighbouring countries, and outcomes will be enforced by the Commission.

The terms of reference of the BCLME programme on EAF

The main objective of the project described in this chapter has been to investigate the feasibility of EAF management in the BCLME region through examining the existing issues, problems and needs related to EAF, and considering different management options to achieve sustainable management of the resources at an ecosystem level. Its scope included the following tasks:

1. Review of all the major 'Target Resource Oriented Management' (TROM) fisheries from an ecosystem perspective.
2. Evaluate the consequences of continuing with TROM approaches to the fisheries.
3. Analyse the benefits and costs of implementing EAF and present them to managers and decision makers.
4. Propose operational goals and objectives to implement EAF.
5. Identify management measures and rules to achieve the best results within an EAF.

6. Liaise with managers and decision makers to formulate preliminary management plans for EAF at national and regional levels.
7. Develop improved techniques and approaches to strengthen the decision making process.
8. Identify useful ecosystem indicators and their application to characterize ecosystem states, changes and functioning.
9. Identify research needs for improved EAF.
10. Propose incentive measures to facilitate the implementation of EAF.
11. Recommend appropriate institutional arrangements for successful implementation of EAF.
12. Inform stakeholders of project results.

This chapter focuses on tasks 1–12 mentioned above.

The final report of the project has been published since its presentation at the Ecosystem Approach to Fisheries Symposium and presents a full description of the project results, conclusions and recommendations (Cochrane *et al.*, 2007).

Approach and Methods

EAF is still perceived by many to be essentially a scientific exercise and the debate is frequently dominated by scientific considerations. In reality, EAF is as much about people and policy as it is about ecosystems and it is essential that, from the outset, planning for EAF is conducted in a consultative and transparent manner, allowing for interaction between stakeholders, managers and those providing scientific and other information. The institutional structure of the project was therefore designed to ensure that societal goals and operational requirements of EAF were the guiding force, notwithstanding the essential role of scientific information and advice. National Task Groups (NTGs) were set up in each country to facilitate participation by and guidance from the range of stakeholders, including managers, decision makers, fishing industry members and conservation groups. The NTGs were supported by science and modelling groups to provide the crucial scientific advice and input to the process. Overseeing the project was a regional Steering Committee, made up of the Chairs of the NTGs and convenors of the science and modelling group in each country and the international project coordinator. Regional workshops were an essential mechanism to facilitate and maintain the regional perspective of the project and ensure good communication and coordination between the three countries. Three regional workshops were held during the course of the project.

The fisheries addressed

EAF can be considered from different entry points. For example, the ecosystem as a whole could be considered as the starting point or the analysis could start from an individual fishery or a particular community or other group of stakeholders. In the case of the Benguela, conventional fisheries management is well established in all countries, focusing on target species and fishing methods. In accordance

Table 16.1. Fisheries included in the EAF project.

Angola	Namibia	South Africa
Small pelagics	Sardine purse seine fishery	Small pelagics purse seine
Demersal trawl fishery (finfish and deep-water shrimp)	Hake trawl and long-line fisheries	Hake fishery
Small-scale fishery using gill nets and beach seine nets	Horse mackerel midwater trawl fisheries	West coast rock lobster

with an incremental approach, it was considered most effective to use those fisheries as the starting point. For a complete EAF, it would have been desirable to include all fisheries in the Benguela ecosystem in the project, which would have allowed all the ecological and technical interactions between the different fisheries to have been taken into account. However, this was not practical with the limited time and resources and it was therefore decided to focus on selected fisheries in each country. The fisheries that were addressed in the project include the most important fisheries in each country and collectively should cover most of the major impacts of fishing on the ecosystem (Table 16.1).

TROM reviews

The first important step in the study was to review the selected fisheries in order to identify the likely key ecosystem impacts if the existing conventional management approaches (also known as the TROM approach; FAO, 2003) continued to be implemented without change. More specifically, the objective was to identify any problems, issues and needs related to EAF in the existing management strategies.

The TROM reviews were undertaken largely by reviewing the available literature, both formal and informal. They included a description of the distribution and biology of the target species, the current status of the stocks, and the fishing methods and social and economic importance of the fishery. The effectiveness of management and the interactions of the fishery were evaluated including:

- Effectiveness of the current management measures in relation to the fishery itself, including their effectiveness in ensuring sustainable utilization.
- Associated impacts, including significance and risk of each impact on the ecosystem structure and/or function, on habitats or on the populations of associated species and on associated biological diversity and productivity.
- Problems being experienced in the fishery with respect to compliance and monitoring, and any complaints or dissatisfaction among fishers and rights holders.
- Details of direct interactions with other fisheries, e.g. competing for the same target species, target species taken as by-catch in another fishery, etc.
- Information on the nature and extent of by-catch (capture of non-target species) and extent of discards (the proportion of the catch not landed)

and unobserved fishing mortality (i.e. sources of mortality other than those mentioned above).

- The effects of supply and use of bait.
- Impacts on recognized protected, endangered or threatened species and management objectives in terms of impact identification and avoidance/reduction of these species.
- Details of direct interactions with the ecosystem (impact on sea bottom, pollution caused by fishery and effects of coastal zone development or land-based pollution).
- Physical impacts on habitat: gear and gear lost during fishing operations, e.g. ghost fishing.

The TROM reviews were used as a basic source of information for the EAF process that followed.

A process for evaluating the feasibility of EAF

Ecosystem management in all its guises, including EAF is still a confusing topic for many and there is still much debate on what it is and what it entails (Cochrane *et al.*, 2004; Wang, 2004). The approach used in this project to clarify the concept was to start by examining, fishery by fishery, the strategy currently being used to manage it and any problems or concerns, related to the ecosystem and the set of stakeholders for the ecosystem, that were not being satisfactorily addressed by the existing management strategy. Any factors beyond the mandate or control of the fishery managers that were impacting on the fishery were also considered. All of these were then prioritized and potential management actions to resolve the problems were identified. The overall goal of this process is to identify where the current management system may be failing to prevent or adequately mitigate impacts that are threatening the sustainability of the fishery itself, affecting other stakeholders, both within the wider fishery sector and outside it, or that may threaten the long-term sustainability and productivity of the ecosystem. The steps in this process can be summarized as follows (Fig. 16.2):

1. The background (TROM) reviews.
2. Identification of all issues of concern in the fisheries considered, within the scope of EAF, that were not being satisfactorily addressed under the existing management strategy and system.
3. Prioritization, through risk assessments, of the issues identified under point 2.
4. Preparation of performance reports, outlining an appropriate management response, for each issue of moderate or higher priority.
5. Aggregating issues into groups in which they could potentially be addressed by a common management measure or set of management measures.
6. Amalgamation and refinement of performance reports to produce a single performance report for each group of issues, including feasible management actions to address each group.
7. Benefit–cost analyses for the issues considered to arise and require action as a result of adoption of EAF, consisting:

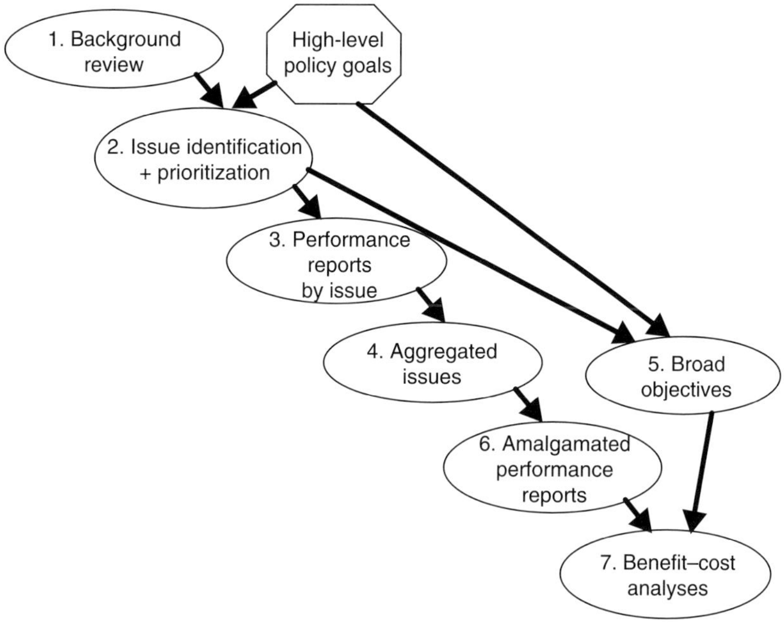

Fig. 16.2. The process followed in evaluation of the feasibility of ecosystem approach to fisheries (EAF) in the BCLME. The ovals represent outputs from activities undertaken within the project and the hexagon represents an underlying external input.

- identifying the broad objectives for the fishery against which costs and benefits needed to be evaluated; and
- performing preliminary evaluations, based on expert opinion, of the benefits and costs of alternative management responses for each group of issues.

The results from this process could form the basis of a policy report on the problems being experienced in a fishery, potential solutions to those problems and the benefits and costs of different management options. For this project, they represented an assessment of the feasibility of implementing EAF in the fisheries that were considered.

This is little different, in essence, from the standard process that should be followed in conventional fisheries management and which is currently applied, albeit in different guise, in the BCLME countries for at least some of the fisheries (e.g. management procedures for small pelagics in South Africa; De Oliveira *et al.*, 1998). There is, therefore, nothing fundamentally new in the process of moving from identification of priority issues to implementation of EAF. What will be new under EAF is that many more issues than are usually considered in conventional, target-resources management will have to be addressed, and these are likely to highlight more conflicts than are commonly recognized in conventional management.

Methods used in implementing the process

Identification and prioritization of issues and potential management responses for those issues

After the preliminary scoping exercise, it is necessary to identify and prioritize the problems, or issues, related to implementation of an ecosystem approach in each fishery. In order to ensure the best available information, to gain support for the process and its outcomes and thereby encourage compliance with any changes to management measures, the stakeholders should participate in this process. Any factor which is a cause for concern to any participant under the prevailing management regime is considered to be an 'issue'.

The approach used in the project to identify and prioritize issues followed that developed by the 'ecologically sustainable development' (ESD) initiative undertaken in a number of Australian Federal fisheries (Fletcher *et al.*, 2002). The ESD approach includes the following tools and steps:

1. A conceptual framework in the form of hierarchical trees (Fig. 16.3) and guidelines to facilitate the identification of issues in a particular fishery or ecosystem.
2. A versatile and informative means of prioritizing the identified issues on the basis of estimated or perceived risk. Risk is estimated as the product of the ordinal scores of the likelihood of the feared outcome from a particular issue occurring under the existing management strategy (e.g. the likelihood of current fishing leading to a serious decline in the abundance of a particular by-catch species); and the magnitude of the consequences of that outcome in terms of the goals of ecosystem management.
3. A template for a Performance Report which describes the best management response to reduce or eliminate the risk associated with a particular issue. The description of the response includes operational objectives, indicators and reference points and future management actions (Table 16.2).

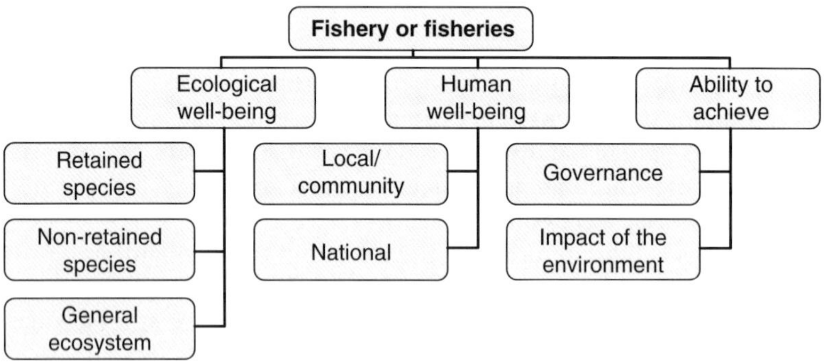

Fig. 16.3. The basic hierarchical tree (after Fletcher *et al.*, 2002) used to guide deliberations on the issues of concern in the fishery or ecosystem under consideration. Additional trees developed by those authors break down the boxes under each of the three second-level headings (Ecological well-being,etc.) into more and more detail to assist users to think broadly across all possible issues that could apply in the fishery or ecosystem under consideration.

Table 16.2. The structure of a Performance Report used in the BCLME programme to describe the potential management response to a particular EAF issue or group of issues where a number of issues could be addressed by a common management response. (After Fletcher *et al.*, 2002.)

	Report heading
1.	Issue or issues being addressed
2.	Objectives • Operational objectives • Subsidiary objectives (where appropriate)
3.	Indicators and robustness
4.	Reference points
5.	Data requirements/availability
6.	Fisheries management response • Current • Future
7.	Future research
8.	Comments and action
9.	External drivers

The approach developed by Fletcher *et al.* (2002) was considered by FAO to be a useful means of operationalizing EAF and was described in the FAO Technical Guidelines on the EAFM (FAO, 2003) as a valuable tool. It was implemented in this project in the form of a series of workshops, one for each fishery, that have been designated as Risk Assessment for Sustainable Fisheries (RASF) workshops (BCLME, 2006a).The workshops were intended to be participatory and to include representatives from the range of managers, science and information advisors and stakeholders, including representatives of fishery sub-sectors and conservation groups. Responses from stakeholders varied with good representation in a number of workshops, but disappointing in some other cases. Ensuring good stakeholder representation will be very important as the process is taken further across the region.

The RASF workshops generated three major results: (i) a list of issues of concern for each fishery; (ii) the estimated risk associated with each issue; and (iii) preliminary Performance Reports that proposed potential management responses to address the higher priority issues.

Trading-off detail and practicality

The purpose of EAF is to recognize and take into account in management of the fishery, within the context of the full range of human impacts, all the objectives being pursued within a given fishery or set of fisheries, without compromising the overarching objective of sustainable use. The large number of objectives across the different stakeholders and interest groups for each fishery quickly became apparent in this study and is likely to be a standard outcome in any attempt to implement EAF. Further, many of the objectives cannot be simultaneously met as they are inherently in conflict. Ideally, all of these objectives should be explicitly considered and the optimal trade-off between them identified and implemented through an optimal set of management measures. Computationally and psychologically dealing with such large numbers of issues would almost certainly be impossible and simplification, through aggregation, was found to be necessary. This was done at three primary points in the process.

- Rather than attempting to evaluate the benefits and costs of EAF against the objective associated with each issue that had been identified, broad objectives were described for each fishery based on the original policy goals and the issues identified in the RASF workshop. Approximately ten broad objectives were considered to be a practical number. If EAF is adequately represented in the high-level policy goals, these broad objectives should effectively reflect the policy goals as applied to the fishery under consideration.
- The ESD approach (Fletcher *et al.*, 2002) proposes preparation of performance reports for each medium- to high-priority issue that has been identified. This was found to be impractical in this study and an attempt was made to consolidate the issues into groups that could be addressed by a single set of management measures. For example, by-catch of commercially important and non-retained species of conservation concern would have been identified as separate issues in the RASF workshops but, in some cases, may require the same management response, for example, changes to gear or fishing practice, or the establishment of a closed area.
- In accordance with the aggregation of issues into groups that could be addressed jointly, a single aggregated performance report was prepared for each group.

Separating out the EAF issues

EAF encompasses but goes beyond conventional management. Any issue, even if arising purely from a single, target-species objective should fall within EAF and successful conventional management is necessary, although rarely sufficient, for successful EAF. However, this study was intended to investigate the feasibility of implementing EAF, which was interpreted, for the purposes of the project, as including only those issues that would not normally be addressed by effective conventional management. The RASF workshops intentionally did not make that distinction in order to allow for identification and prioritization of the full set of issues. For further consideration in the project, the EAF issues were subsequently separated from the others. These issues were defined as:

> any impact of the fishery on the wider ecosystem or any impact of the environment (human or ecological) on the fishery, apart from the direct interactions between a fishery and the species it targets.

It was recognized that some EAF issues are already being addressed in all three countries. In such cases, the evaluation of costs and benefits would be for any actions necessary to improve or strengthen the current approaches. If no additional action was required, the issues should either not arise or be given a low priority in the RASF workshops.

Benefit–cost analyses and aggregated performance reports

Effective implementation of EAF will result in benefits, which may be ecological, economic, social or some combinations of these three, but will frequently also invoke additional costs across the same dimensions. If EAF is to be accepted and sustained, it is essential that in the planning and implementation phases, decision makers and all stakeholders are well aware of the benefits and costs that will result from different options. The next step in the process was therefore to estimate the benefits and costs of any management actions being proposed.

In practice, the benefits and costs should be estimated using the best available information, which will often include both scientific information and stakeholder knowledge. In this feasibility study, with the wide range of issues and management actions being considered, it was not possible to investigate the benefits and costs thoroughly and rigorously. Instead, the project generated preliminary estimates based largely on expert opinion. This was done through a series of dedicated workshops, one for each fishery, referred to as the Benefit-Cost Workshops (BCWs). As with the RASF workshops, the BCWs were intended to include good stakeholder representation, but, again, this varied from case to case. The tasks of each workshop were as follows:

- As discussed above, to aggregate and develop:
  - a full set of detailed objectives for each fishery into broad objectives, against which the benefits and costs of each management action could be estimated;
  - issues according to their broad theme and on the basis of whether they could be addressed by similar management responses;
  - performance reports for each group of issues.
- Using the best information available, within the time and personnel constraints, to evaluate the expected benefits and costs of those management measures or rules in relation to the broad objectives. Benefits and costs were estimated for both the short term, which was defined as up to 3 years, and the long term, which was defined as 5–10 years.

In the study, benefit–cost analyses were undertaken in multi-stakeholder workshops. The benefits and costs were based almost entirely on the collective wisdom of the participants in each workshop, which would generally have included scientists with knowledge of the best available scientific information. The workshop was asked to provide the consensus estimate of benefits and costs for each action against each broad objective on a scale of 0–4 where: 0 indicates negligible cost or benefit; 1 is a small but noticeable impact; 2 is a moderate impact; 3 indicates a major improvement or will have major negative impact; and 4 indicates an immediate and long-term impact or will be unsustainable from the outset. The assumption was made that the difference in value between each score is constant across the range of scores (i.e. they are linearly related to actual impact). In addition, it was assumed that the sum of zero costs (i.e. negligible) across all broad objectives would generate a total cost for the measure of 1 (i.e. small). This was based on the assumption that no benefit would be achieved without some cost. With those assumptions, benefit and costs ratios could be used for comparative purposes. The explicit assumption in the results presented in this study is that all broad objectives have the same policy weighting. In practice, this is highly unlikely and it will be necessary in the future to consider the weighting of the different objectives and, again in a participatory manner, to try to reach agreement. This would best be done prior to and independently of decisions on specific management measures.

It is important to emphasize that, as with the performance reports, the benefit–cost analyses and the results that have been produced from them are preliminary only and that no focused scientific assessments (including the human sciences where appropriate) and validations were undertaken (Cochrane *et al.*, 2007). Such improvements and checks will still need to be done, where feasible, before this advice can be considered sufficiently reliable and accurate for use by decision makers in setting management regulations. Nevertheless, the results obtained are still considered to be informative, providing guidance on the possible options for and obstacles to implementation of EAF.

Use of the best available scientific information

One of the common arguments against the implementation of EAF is the need for increased information. The absence of good information will undoubtedly hinder progress towards EAF but, as with conventional fisheries management and natural resource use in general, the precautionary approach advises that lack of certainty should not be used as a justification for not taking appropriate action. The key principle is the use of the best scientific information available and appropriate use of precaution in the face of uncertainty (FAO, 1995).

This project was set out to make use of the best information available and was hoped that the science and modelling groups would be able to supplement existing information through undertaking new analyses to evaluate, for example, risks, the feasibility and impacts of changes to management measures and some of the costs and benefits. As a result of heavy commitments by all scientific staff participating in the project, this has, to a large extent, not been possible. As a result, most of the results generated by the project are based on existing scientific results, which commonly have had to be interpreted within a new context to provide the particular information required, and on expert opinion from all participants. This information unquestionably has at least indicative value and the results and conclusions are considered to be qualitatively valid and accurate, but not necessarily quantitatively so.

As the three countries move forward in implementation of EAF, it will be necessary to revisit results and conclusions that would benefit from precise quantitative information, such as estimated future TACs, or the risk of overfishing on retained and non-retained species. Where improved information can be provided in a timely and cost-effective manner it should be generated and used to improve the information obtained in this feasibility study.

Results to Date: EAF Issues and Potential Management Responses in the BCLME

Issues and priorities

The RASF workshops held in the three countries identified a large number of issues relevant to an ecosystem approach that were considered to be inadequately addressed under the prevailing management in the various fisheries (BCLME, 2006a). Many of the issues were applicable within a conventional approach alone and were not only a consequence of consideration of EAF. The full lists of issues are provided in the reports of the workshops (BCLME, 2006a). The list of EAF issues and risk values from the South African hake fishery is shown in Table 16.3 as an example of the results obtained in all the workshops. A total of 96 issues were identified for this fishery, of which the 58 listed in Table 16.3 were considered to be EAF issues. Of the 58 EAF issues, 10 were categorized as 'retained species' issues, 14 as non-retained species, 10 as general ecosystem issues, 1 as a community issue and 23 as governance issues. Three were considered to be of extreme priority, 14 of high priority, 22 of medium priority and 19 were of low priority. The extreme issues included the lack of suitable baseline information on the social and economic aspects of the fishery as well as inadequate management and research capacity in the management agency.

High-priority ecological issues identified in the hake fishery included the implications of fishing on the size structure of the *Merluccius capensis* stock and the possible effects of trawling on benthic habitat and biota. For the small pelagic fishery, the potential impacts of removing forage fish eaten by predators and uncertainty around decadal-scale fluctuations were considered to be issues of high priority. In the west coast rock lobster fishery, issues relating to human well-being, especially those relating to small operators, were generally considered to be of higher priority than human and ecosystem well-being issues.

In the case of Angola, examples of high-priority issues included the impacts of the banda-banda fishery (small-scale fishery utilizing a fine-meshed beach seine of 10–12 mm mesh size; Tchikulupiti, 2005) on the sustainability of exploited pelagic species. This fishery exploits juveniles of many species, including those of key pelagic species such as horse mackerel and sardinellas. Challenges faced by the demersal fishery were considered to be mainly related to the multispecific nature of that fishery and the need to develop suitable indicators. A high risk was also perceived for the possible impacts of bottom trawls on epibenthic organisms. In the small-scale fisheries, which use gill nets, ghost fishing and incidental capture of vulnerable species (e.g. sharks and sea turtles) were considered as high-risk environmental issues, while lack of infrastructure and of organization of the sector (e.g. through cooperatives) were recognized as key issues affecting the development of the sector.

Allocation of fishing rights, collection of reliable fishery data, inadequate monitoring and control systems and the lack of effective management plans for all the species exploited were considered as major governance issues in the fisheries of Angola examined in the workshop. Management plans are in place, but their effectiveness in rebuilding the stocks is still unclear. In all cases, oil exploration and exploitation activities including the resulting oil spills and pollution were considered as important threats to the resources and the environment and the communities depending on them. The fishery sector also seems to be threatened by a number of social issues such as the increased use of alcohol and drugs by fishermen, with both health consequences and negative impacts on safety at sea. Lack of infrastructure and high oil prices were seen as major threats to sector development.

Table 16.3. An example of EAF issues for a fishery: the South African hake fishery.

Objective	Type	Category	Subcategory	ID	Issue	Cons.	Like.	Risk	Cat.
1. Fund further EAF research and model dynamics									
1	EAF	Eco well-being	Retained spp	3	Both hake sp: Uncertainty about the estimation of natural mortality (predation and cannibalism)	3	6	18	H
4	EAF	Ability to achieve	Governance	78	Currently biodiversity audits for marine species are not being done	2	6	12	M
4	EAF	Gen ecosystem		34	Removal of predators may have an effect on the abundance of smaller pelagic species and mesopelagics	4	3	12	M
4	EAF	Gen ecosystem		33	Trophic effects of removing a proportion of a high-level predator, with no obvious replacement species	4	3	12	M
4	EAF	Eco well-being	Non-retained	24	Mortality of *Galeorhinus and Mustelus* in the inshore trawl fishery (these species are commercially harvested)	2	6	12	M
4	EAF	Gen ecosystem		35	Change in size structure of hake leads to a switch in prey preference	3	3	9	M
1	EAF	Gen ecosystem		37	Hake are a component of the diet of marine mammals and other top predators (seals, swordfish – possible, snoek)	1	6	6	L
4	EAF	Eco well-being	Non-retained	23	Mortality of *Galeorhinus and Mustelus* in the long-line fishery (these species are commercially harvested)	1	6	6	L
4	EAF	Eco well-being	Retained spp	17	Lack of understanding and quantification of the impact on linefish (kob, white stumpnose, etc.)	1	6	6	L
4	EAF	Gen ecosystem		42	Disturbance of sediments may change water chemistry (oxygen,etc.)	0	5	0	N
2. Enforce responsible fishing practices									
2.1 Enforce appropriate permit conditions to minimize seabird mortality									
2	EAF	Eco well-being	Non-retained	21	Threatened species of seabirds (also protected) caught/injured/killed by trawling	3	6	18	H
2	EAF	Gen ecosystem		39	Distribution patterns and behaviour of seabirds are being affected by the availability of offal	2	6	12	M

2	EAF	Eco well-being	Non-retained	20	Threatened species of seabirds (also protected) caught/injured/killed in long-line operations	2	6	12	M
2	EAF	Eco well-being	Non-retained	22	There is directed catch of seabirds in the hand-line fishery for the pot	1	5	5	L
2	EAF	Eco well-being	Non-retained	32	Potential soaking of gannets from fish meal factory vessels	1	2	2	L
2.2 Enforce appropriate permit conditions to manage by-catch utilization									
2	EAF	Eco well-being	Non-retained	27	By-catch of 'protected' linefish (in MLRA) on soft ground available to the inshore trawling – silver kob, dusky kob etc.	3	6	18	H
2	EAF	Eco well-being	Retained spp	15	Monk, kingklip stocks are overexploited	4	4	16	H
2	EAF	Eco well-being	Retained spp	18	Impact on other commercial species (skates, rays, gurnards, sharks, jacopever, john dory, angel fish, bellman, chokka, etc.)	2	6	12	M
2	EAF	Eco well-being	Non-retained	26	By-catch of wreckfish	3	3	9	M
2	EAF	Eco well-being	Retained spp	16	Snoek stock is being impacted	2	4	8	M
2	EAF	Eco well-being	Non-retained	29	By-catch of other benthic species that have been recorded in the trawl catch (see 60–65 species in S. Walmsley, Ph.D. Thesis)	1	5	6	L
2	EAF	Eco well-being	Non-retained	28	By-catch of 'protected' linefish (in MLRA) on hard ground	2	1	2	L
2.3 Enforce appropriate permit conditions to minimize shark mortality									
2	EAF	Eco well-being	Retained spp	18	Impact on other commercial species (skates, rays, gurnards, sharks, jacopever, john dory, angel fish, bellman, chokka, etc.)	2	6	12	M
2	EAF	Eco well-being	Non-retained	30	By-catch of other sharks, rays and skates (not threatened but not assessed) are caught	2	6	12	M
2	EAF	Eco well-being	Non-retained	25	Mortality of all other threatened sharks in long-line and trawl (see Petersen report)	2	6	12	M

Continued

Table 16.3. Continued

Objective	Type	Category	Subcategory	ID	Issue	Cons.	Like.	Risk	Cat.
2.4 Enforce appropriate permit conditions to minimize impact on seal populations									
2	EAF	Gen ecosystem		40	Seals benefit from offal discards	1	6	6	L
2	EAF	Eco well-being	Non-retained	19	Seals (protected sp) are killed in trawling operations	1	6	6	L
2	EAF	Eco well-being	Non-retained	31	Shooting of seals interacting with gear	0	6	0	N
2.5 Enforce appropriate permit conditions to minimize impact on benthic substrate									
2	EAF	Gen ecosystem		38	Impact of trawls on the benthic biota habitat and biota	3	6	18	H
2	EAF	Gen ecosystem		36	Ghost fishing by net fragments	1	2	2	L
2	EAF	Gen ecosystem		41	General pollution associated with fishing vessels and harbour activity is considered across all fisheries			0	N
3. Maintain socio-economic well-being through management measures									
3.1 Rebuild hake stock									
3	EAF	Eco well-being	Retained spp	2	Both hake sp: fishing mortality is underestimated due to discarding and survival after escapement	3	6	18	H
3	EAF	Eco well-being	Retained spp	5	Both hake sp: uncertainty about variability in recruitment	3	6	18	H
3.2 Develop economic parameters									
3	EAF	Human well-being	Community	44	There is a lack of baseline socio-economic information	5	6	30	E
3.3 Mitigate negative social impacts									
4. Maximize sustainable yield by monitoring biological trends									

5. Ensure MCM institutional structures (Chief Directorates, RMWG, SWG, National Task Groups) are in place and effective									
5.1 MCM institutional structures									
5.1.1 Improve enforcement of compliance									
5	EAF	Ability to achieve	Governance	59	Inspector coverage is inadequate and possibly biased geographically biased and per sector	2	6	12	M
5	EAF	Ability to achieve	Governance	58	Compliance is inadequately enforced – occasional examples are made but the coverage is low	3	3	9	M
5.1.2 Improve Resource Management Capacity									
5	EAF	Ability to achieve	Governance	68	Lack of management capacity (no-one appointed to manage demersal fishery at present) and institutional knowledge	4	6	24	E
5	EAF	Ability to achieve	Governance	60	There is no Resource Management Working Group	3	6	18	H
5	EAF	Ability to achieve	Governance	85	The requirements of the MSC are possibly beyond the abilities of management's resources (for those conditions that require MCM to play a role)	2	6	12	M
5	EAF	Ability to achieve	Governance	79	No institutional reviews of research and management	1	6	6	L
5.1.3 Improve research efficacy									
5	EAF	Ability to achieve	Governance	67	Inadequate research capacity and institutional knowledge	4	6	24	E
5	EAF	Ability to achieve	Governance	62	Catch data is not available for real time response	3	6	18	H
5	EAF	Ability to achieve	Governance	63	Observer data has not been properly analysed or reconciled with catch records	3	6	18	H
5	EAF	Ability to achieve	Governance	69	Inadequate coordination of research (nationally, regionally and internationally)	3	6	18	H
5	EAF	Ability to achieve	Governance	64	Problems with the validity of scientific observer data in portraying the real picture	2	6	12	M
5	EAF	Ability to achieve	Governance	79	No institutional reviews of research and management	1	6	6	L

Continued

Table 16.3. Continued

Objective	Type	Category	Subcategory	ID	Issue	Cons.	Like.	Risk	Cat.
5.2 Improve consultative mechanisms									
5.2.1 Access and coordinate formation of RIBS									
5	EAF	Ability to achieve	Governance	61	There are no formal or informal lines of communication with industry bodies and other stakeholders	3	6	18	H
5	EAF	Ability to achieve	Governance	57	Conflict between sector users	2	6	12	M
5	EAF	Ability to achieve	Governance	83	Industry is not particularly interested in some broader management issues, focusing on direct issues	1	6	6	L
5.2.2 Formalize and improve communication with other stakeholders									
5	EAF	Ability to achieve	Governance	86	NGOs not involved in management and scientific working groups	3	6	18	H
5	EAF	Ability to achieve	Governance	70	Inadequate communication with other government departments – specifically with Mineral and Energy Affairs or Petroleum Agency	2	6	12	M
5	EAF	Ability to achieve	Governance	84	The fact that the long-line and hand-line industries are not MSC certified hampers the certification of the Trawl fishery	2	6	12	M
5	EAF	Ability to achieve	Governance	71	Inadequate coordination with National Ports Authority with regard to facilities and services for fishing vessels	2	5	10	M
5	EAF	Ability to achieve	Governance	77	There is no formal peer-review of management plans	1	6	6	L
5	EAF	Ability to achieve	Governance	82	Criteria for representation on SWGs should be reviewed; difficulties in weighting representation	1	6	6	L
5.3 Improved legal and policy framework									
5.3.1 Develop and implement legal and policy actions									
5	EAF	Eco well-being	Retained spp	8	*M. paradoxus*: stocks are shared between Namibia and South Africa	3	6	18	H
5	EAF	Eco well-being	Retained spp	10	*M. capensis*: stocks are shared with Namibia	3	3	9	M
5	EAF	Ability to achieve	Governance	75	The MLRA needs to be revised; CAF, consultation	1	4	4	L

For the Namibian fisheries examined, a central issue was considered to be the lack of reconciled and approved management plans. Thirteen issues, mainly within the 'Governance' component, received an 'Extreme' risk rating for the hake fishery. Also within the hake fishery, two issues from the 'Ecological well-being' category were considered to be of 'Extreme' risk: the by-catch (or incidental mortality) of threatened seabirds in both long-line and trawl operations; and the potential impact of the by-catch of monkfish by the hake fishery on the sustainability of the monkfish fishery. Under the heading of 'Human well-being', the close link between the living standards of the fishing community and fishery service providers and the state of the fishery and of the stock were identified as concerns that required careful management consideration. High levels of unskilled labour and lack of training and development opportunities within industry structures were also considered to need attention. International economic factors such as fuel prices and exchange rates, as well as local health issues such as HIV and AIDS were among the extreme risk 'External impacts' that could hinder the fishery attaining its objectives.

In the midwater trawl fishery it was concluded that little was known about the trophic position of horse mackerel in the ecosystem and the dependence of certain predators on it. Furthermore, the impact of this fishery on several by-catch species had not been quantified; not least, several species of sharks and seabirds that have a threatened conservation status. A very specific management issue that was considered to need attention was the justification for the regulation prohibiting trawling within the 200 m depth contour.

The most striking feature of the purse seine fishery was the apparent low abundance and variability in the biomass of the target species, sardine and the implications of this for the TAC of the species. There is also an urgent need to rebuild the stock. Another significant feature of this fishery was considered to be the keystone trophic position occupied by the target species, which has led to changes in conservation status of dependent species as well as possible long-term changes in the trophic structure of the ecosystem. This fishery was previously the largest employer in the fisheries sector, and the current depressed state of the stock is leading to significant social and economic hardships.

Performance reports

The risk values (likelihood multiplied by consequence) estimated for all the issues were ranked and classified on a scale from negligible to extreme according to the value in each case (see Fletcher *et al.*, 2002 for details). Preliminary performance reports were developed at the RASF workshops for issues with an estimated risk value of moderate or higher (BCLME, 2006a).

The large number of issues identified in most of the fisheries resulted in a considerable number of performance reports and therefore also a large number of independent management measures. This is likely to be a common outcome for fisheries and ecosystems wherever management is still dominated by conventional approaches. In theory, each management response could and often should be developed independently as the optimal means of addressing a given issue and then all the management measures for all the issues reconciled in order to arrive at an optimum strategy for the fishery or ecosystem as a whole. In practice, this will frequently be logistically and scientifically very difficult.

As discussed in the description of methods above, for the purposes of this study, therefore, the scope of the problem of reconciliation was simplified. The first step was to identify and separate the 'EAF issues' from the conventional issues. Considering only the EAF issues reduced the number that had to be considered substantially. For example, the total number of issues in the Namibian midwater trawl fishery was 54 of which 13 were considered to be EAF issues (Cochrane *et al.*, 2007). Of course, the other issues are also important, as indicated by their risk values, and would still need to be addressed by the relevant management agencies. The full list of issues and risk values will be supplied to the management agencies for their consideration.

The second step was to aggregate the EAF issues on the basis of whether they could be addressed by similar management responses. Each group was then treated collectively. This step, while potentially losing detail, reduced the task to a manageable scale. For example, in the case of the South African hake fishery, a total of 96 issues were identified in the RASF workshop (BCLME, 2006a). Of these, 58 were considered to be EAF issues and these were grouped into the following eight categories at the BCW on the South African hake fishery (Table 16.3; BCLME, 2006b; Cochrane *et al.,* 2007):

- EAF research & model dynamics.
- Responsible fishing (including impacts on non-retained species).
- By-catch of commercial species.
- Socio-economic considerations.
- Hake management issues (going beyond those currently considered in the existing management strategy).
- Database maintenance.
- Research capacity issues.
- Policy issues.

While this list will probably be found to be too aggregated to be translated directly into a practical management strategy and measures, in most cases, where implementation of EAF is being planned, there is likely to be a need to reduce the complexity of the problem, reflected in the number of issues, to a manageable scale and it will be important to find an appropriate balance in each case between practicality and ensuring that important detail and considerations are not lost by the amalgamation.

A performance report was then produced for each group of issues. The performance reports at this stage are still advisory documents, potentially providing information to assist decision makers to identify and set the management measures necessary for an EAF management strategy. The reports therefore did not necessarily specify a particular measure or set of measures, but considered different options to address the issues, each of which would have unique advantages and disadvantages or benefits and costs. The management measures proposed within the project are still broad and generalized because of time constraints and the magnitude of the task. For example, in the South African hake fishery, the management measures proposed to address the group of issues under the heading 'by-catch of commercial species' were: (i) where feasible, assess status of and develop management plans for targeted by-catch species; (ii) manage fishing effort; (iii) manage and monitor by-catches (includes coordinating with linefish management); and (iv) investigate (and implement) zoning of sector-specific fishing areas. The management measures considered for 'compliance and management issues' were: (i) develop capacity for Resource Management, to include both training and appointment of new staff (Resource Management is one branch within Marine and Coastal Management, the national management agency); (ii) establish effective communication between stakeholders (e.g. through a Resource Management Working Group); and (iii) enhance compliance by improving and increasing the capacity of fishery control officers.

In the future, if this work is taken further by any of the national management agencies or the BCC, it will be necessary to translate those broad measures into clearly and precisely specified measures, for example, specifying the exact size and location of suitable closed areas or the precise reduction in effort required and in which fishing sectors. This will require more detailed scientific analysis than has been possible during this study and, as was attempted in this study, should also be carried out in consultation with the stakeholders.

Benefit–cost analyses

Establishing the broad objectives

Broad objectives were described for each fishery. The benefits and costs of different management options were measured in terms of their implications for satisfying these broad objectives. Examples are provided in Table 16.4.

Considering the benefits and costs of management measures

The performance reports contain different options for management measures to address each group of issues. In order to establish an effective and acceptable management strategy, it is necessary to provide the decision makers with the best available information on the impacts of each option, positive and negative, for the range of objectives underpinning the fishery. This will enable them, ideally in a fully transparent and participatory manner, to consider the trade-offs and arrive at a strategy that, in implementation, will come closest to achieving those objectives.

An example of the output from a benefit–cost analysis for a single management measure in the South African hake fishery is shown in Table 16.5 and a summary of the costs and benefits of different management measures proposed to address the group of issues related to 'by-catch of commercial species' is shown in Fig. 16.4. Figure 16.4 provides a comparative view of the average implications of the different options. Management measure 13 – controlling existing effort in the fishery in order to manage impacts on the retained by-catch species, instead of focusing only on the hake as at present – stands out from the others as having the higher benefits in the short and long term. However, the short-term

Table 16.4. The broad objectives identified for example fisheries in each country.

Angola demersal trawl fishery	Namibia hake fishery	South Africa demersal trawl fishery
Restore biomass of commercially important demersal species to optimal levels of productivity	Ensure sustainable exploitation of the hake stocks (rebuilding, optimize yield, maintain size structure, etc.)	Maximize long-term economic sustainability of the fishery (e.g. improve catch rates and size structure)
Maintain demersal fish community structure in terms of size structure and species composition	Ensure sustainable exploitation of the other stocks (e.g. monkfish, sole, kingklip, etc.)	Manage hake stocks to ecologically sustainable levels (trophic interactions)
Reduce impacts of bottom trawl fishery on vulnerable species (sea turtles, sharks)	Maintain biodiversity	Rebuild hake stocks to minimize risk to the resource (recruitment, etc.)
Reduce impacts of bottom trawling on bottom substrate	Maintain ecosystem functioning	Minimize loss of biodiversity due to seabed damage
Promote development of the artisanal fishery	Avoid environmental damage (habitats and substrate)	Minimize incidental mortality of seabirds, sharks, marine mammals, etc.
Promote the development and Angolization of the industrial sector	Ensure optimum economic return to industry/country processing value added, etc.	Minimize discard and loss of target species and manage by-catch
Promote reliable supply of fish products to the population	Optimize social returns, employment, food security, empowerment and social upliftment	Develop appropriate management measures for multiple and or shared stocks
Contribution of the fishery to improvement of the local economic infrastructure and social base	Namibianization of the sector	Optimize socio-economic benefits across sectors
Ensure economic stability of the Angolan demersal fishing industry		Maintain adequate research and management capacity
Increase the contribution of the fishery to the national economy		

Table 16.5. An example, taken from the South African hake fishery, of the output from a benefit–cost analysis for higher priority EAF issues related to commercial by-catch. The proposed management actions are different options, which could be applied independently or in various combinations. The results, which are preliminary and based primarily on expert opinion, are shown here for only one of the possible management actions (Action 1: the development of species-oriented by-catch plans). See text for explanation of the scores.

	Management action			
	Where feasible, assess status of, and develop management plans for, 'commercial' by-catch species			
	Short term		Long term	
Broad objectives for the fishery	Cost	Benefit	Cost	Benefit
Maximize long-term economic sustainability of the fishery (e.g. improve catch rates and size structure)	2	2	1	3
Manage hake stocks to ecologically sustainable levels (trophic interactions)	0	1	0	1
Rebuild hake stocks to minimize risk to the resource (recruitment etc.)	0	0	0	0
Minimize loss of biodiversity due to seabed damage	0	0	0	0
Minimize incidental mortality of seabirds, sharks, marine mammals, etc.	0	0	0	0
Minimize discard and loss of target species and manage by-catch	0	2	0	2
Develop appropriate management measures for multiple and or shared stocks	0	0	0	0
Optimize socio-economic benefits across sectors	0	1	0	1
Maintain adequate research and management capacity	0	1	0	1

costs were considered to be higher than for measure 12 - developing individual management plans for each impacted species - and measure 14 - implementing supplementary by-catch limits. The long-term benefits for all four of the measures were considered to be greater than those in the short term, while the costs were estimated to diminish in the long term. Before making a choice in practice, the decision makers would need to look at the details of the costs and benefits for each objective (e.g. Table 16.5) to identify where the costs and benefits would be felt most strongly. A final decision would be a policy choice and should take into account the weightings applied to each broad objective.

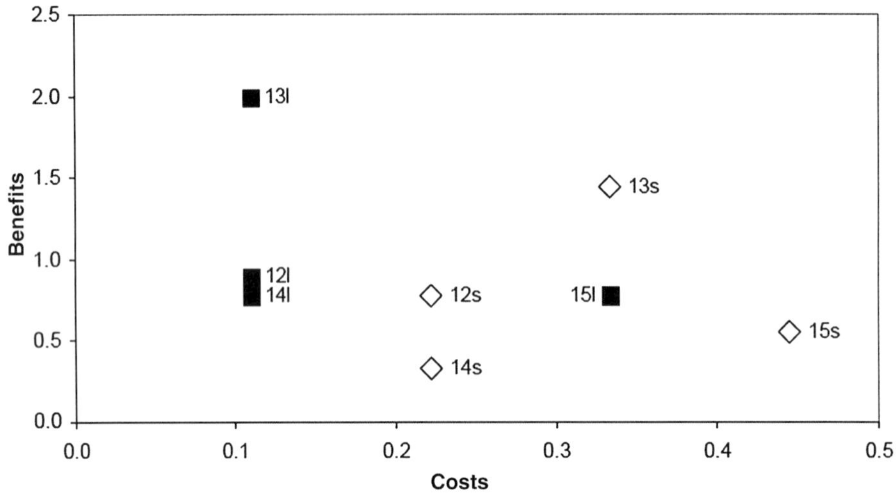

Fig. 16.4. A plot of the average benefits and costs across the nine broad objectives for the South African hake fishery for each of four management measures identified as potentially suitable to address the issues grouped under the heading 'by-catch of commercial species'. The label numbers refer to: 12 – where feasible, assess status of, and develop management plans for, 'targeted' by-catch species; 13 – manage fishing effort; 14 – manage and monitor by-catches (includes coordinating with linefish management); and 15 – investigate and implement zoning of sector-specific fishing areas. The letters following each number indicate whether the benefits and costs are evaluated in the short term (s) or long term (l).

Towards Implementation

Integrating the results across issue groupings and fisheries

As the field of view is expanded under EAF, inevitably the number of interactions increases. As a result, the disciplines of provision of scientific advice and of decision making will need to iteratively narrow and broaden their fields of view in order to take into account the objectives at the level of individual priority issues and still reconcile them at the ecosystem level (Fig. 16.5). This will require comparing the management measures for each issue group and their costs and benefits across the different issue groups within each fishery and identifying conflicts and redundancies between the groups. The same will then need to be done between fisheries and, recognizing the need to integrate EAF within the broader field of integrated coastal and ocean management, across other users of the ecosystem as well. At each expansive step, it may be necessary to go back to the original management measures and consider adjustments, or total change, to address other issue groups and other users. For example, the proposed management of fishing effort in the South African hake fishery to reduce negative impacts on other commercially important retained species could be optimal for that group of issues but,

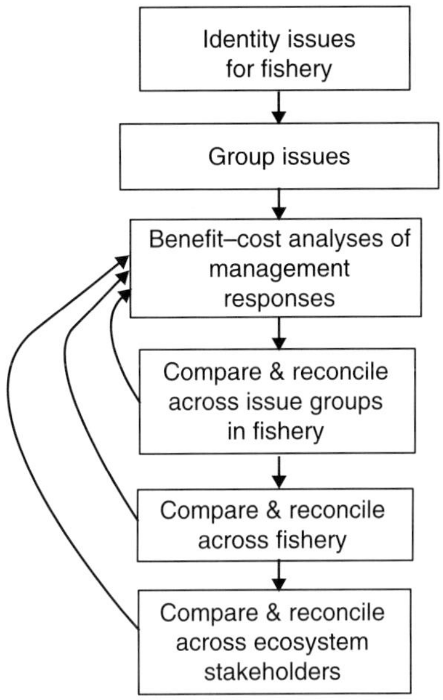

Fig. 16.5. Scaling up and down the ecosystem and user groups.

hypothetically, may not address sufficiently the impacts on, for example, some low production shark species of conservation concern. It may also not take into account the ecological benefits of closed areas resulting from oil and gas exploration. When these considerations are taken into account, an alternative management measure, or more likely a combination of several management measures, may be found to be preferable with an improved benefit/cost ratio and distribution.

There are no simple or recipe-book approaches to resolving the multi-criteria and multidimensional features of EAF, integrated coastal management (ICM) and integrated ocean management (IOM). The iterative approach here involves a breaking down of the problem into its smallest components and then rebuilding to a level of aggregation that enables the scientists, stakeholders and decision makers to wrestle with the complexity. There will be alternative methods and some of these may be found to be better, but the experience gained in the project to date indicates that the approach followed in here is undoubtedly informative, feasible and practical.

Implementation and review

The project is intended to evaluate the feasibility of implementation of EAF and does not include implementation itself. The implementation process that may follow should include three consolidated tasks: reconciling the objectives and measures, followed by implementation of the agreed measures and periodic reviews of progress.

The intention of EAF is to ensure that all fishery uses of and impacts on an ecosystem are collectively sustainable and, under ideal circumstances within a wider system of ICM or IOM, result in optimal use of the products and services from that ecosystem. Direct and indirect interactions within an ecosystem mean that individual issues cannot be addressed independently, as attempts to manage any one issue are likely to have impacts, positive or negative, on other issues. The management plan and strategy for any fishery and ecosystem must therefore simultaneously address all objectives, as far as practical. The different objectives will usually not be fully compatible and there will be conflicts between them, which will be reflected in the costs and benefits of management actions already in place and those being considered for implementation of EAF. These conflicts need to be reconciled, which will frequently require trade-offs between the priority given to different objectives and, in some cases, entire objectives may be found to be unattainable when combined with other higher priority objectives and will have to be abandoned. The benefit–cost analyses are intended to provide decision makers with information on how the different objectives will be impacted by given management actions. This is essential information for the proactive and objective-driven fisheries management that is required for implementation of EAF. Failure to reconcile objectives in the planning will lead to conflicts in implementation with attendant management problems and costs that may affect a number of different goals and objectives.

The provision of information does not solve the problem and ultimately the decision makers and stakeholders will have to decide on the exact form of the final set of objectives and the management actions required to achieve those objectives. This should be done with ongoing, iterative provision of scientific advice and the process can be facilitated by using suitable tools and aids for decision making. The benefits and costs of EAF will frequently be substantive and will also frequently be inequitable, with costs being required from some users in order to provide sustainable benefits for others, including the societal benefit of halting unsustainable use. Agreeing on objectives and management actions is therefore unlikely to be a quick and easy process in EAF and a pragmatic, step-by-step approach, implemented over years rather than months, may often be required until the full goal of EAF is achieved.

Once the objectives and management actions have been agreed upon, they need to be implemented with suitable enforcement to complement voluntary compliance, and the performance of the fishery monitored and periodically reviewed. Adjustments should be made to the management actions where found to be necessary in the review. The indicators and reference points identified in the performance reports will be central to this process. It needs to be recognized that uncertainties in all the information can lead to suboptimal decisions being made and that the ecosystem is likely to change with time. Failure to make progress towards objectives therefore needs to be identified early and appropriate management responses made to avoid serious and long-lasting damage to the ecosystem and the goods and services it can provide.

Discussion

Progress and constraints

In the opinion of the authors the project has made very good progress in several key areas. Most important of these has been the increased comprehension of what EAF implies among those who have participated in the project. The first training session on issue identification and prioritization, which took place during the first Regional Workshop in Namibia in 2004, was a turning point for many of the participants. The practical exercise of identifying issues for local fisheries undertaken there clarified the concept of EAF and gave a clearer indication of what its implementation would entail. The subsequent RASF and BCWs provided more detailed analysis and insights and have served to highlight the key issues and shortcomings in the existing management strategies and where greatest attention needs to be given in the future. The process of analysis, awareness-building and priority setting is therefore considered to have been very successful for those who have participated in all or most of the project activities.

Within this process, the participation of stakeholders has been seen to be very important. The process, from the initial identification and prioritization of issues through to the formulation of an EAF management strategy, must include a mixture of science, natural and human, and policy. What is considered an issue, the severity of consequences of different issues, the objectives and the estimated costs and benefits are heavily influenced by human choice and values. Involvement of all relevant stakeholders from the outset is therefore essential if the final EAF management plan is to be realistic, include the best available information and be likely to enjoy widespread support and credibility. Science can help to inform and advise EAF: the better the science, the lower the level of precaution required (Cochrane, 1999) and the less the probability of being surprised by unexpected outcomes; but the common misperception that EAF is mainly about and dependent on good science is turning the problem upside down. Fundamentally, EAF is a product of and a requirement for sustainable human use.

The BCLME project has been partially successful in engaging stakeholders. It must be acknowledged that the process has been dominated by natural scientists in all countries, probably because of their professional interests and responsibilities. As a result, the priorities, objectives and costs and benefits may be biased towards the perspectives and priorities of natural science. Nevertheless, there has been participation by members of the fishing industry, conservation groups and managers in the various workshops, albeit it less representative than had been hoped for. Representatives of the South African fishing industry and two conservation NGOs were particularly active in the project.

The project has highlighted, as expected, substantial unknowns and uncertainty in knowledge and information, including biological, ecological, social and economic knowledge. The process described in this chapter is designed to take uncertainties into account and, for example, the method of risk analysis is sufficiently flexible to apply to circumstances ranging from expert opinion to data-intensive analyses, as are the benefit–cost analyses. Nevertheless, improved information leads to improved results and decisions and, with the amount and types of information that were used throughout the project, many of the outcomes must be considered to be preliminary. The very broad scope of the project, the fact that it was, in many cases addressing new questions and problems, and the limited time and resources available to provide the necessary information meant that best guesses had to be used frequently but, with even moderate additional time and resources, a number of these could be improved upon. As discussed in the results, this is particularly applicable when it comes to the more complex and multidimensional tasks of formulating detailed management responses to address issues and in estimating the benefits and costs of the different management options. Nevertheless, the groundwork done is considered to be valuable and to establish a platform for subsequent improvement and refinement as necessary. Cochrane *et al.* (2007) is the final report of the project.

EAF in the BCLME: the future

As described in the Introduction, the BCLME countries are familiar with the interactions between fisheries and the ecosystem and have a long history of research into these interactions. Management of fisheries in the region is also well established and generally of a high quality although a number of problems are being experienced at present as a result of different combinations of environmental influences and overfishing. This project has built on that local expertise and knowledge. It has done so by bringing a formalized and structured process to considering the goals and objectives of an ecosystem approach and evaluating the weaknesses in the current management strategies and systems. In some cases, for example, in the pelagic and demersal fishing sectors in South Africa and others, some issues that were identified in the TROM reviews are already being addressed by means of practical management measures. However, much remains to be done if sustainable use of the BCLME ecosystem is to be assured for the long term. The project has explored, in a preliminary manner, broad approaches to address those weaknesses. In any follow-up to the project, which it is hoped will further progress towards implementation of EAF, it will be essential to build and improve upon the involvement of the stakeholders. This would almost certainly follow automatically from a clear intention by governments or the BCC to begin serious actions towards implementation of EAF.

The results of this project will be brought to the attention of the managers and decision makers in the national management agencies and to the BCC. The prioritized issues should give them good insight into the problems that need to be addressed and those that require greatest urgency. Follow-up action should include:

- Reviewing any uncertain or contested risk values and priorities using the best available information, including new investigation and analysis where necessary and attainable within acceptable time frames.
- For the moderate, high and extreme priority issues, re-examining the issue groupings and performance reports and, again with improved and updated

information where relevant, refining those to ensure that they reflect the current state of knowledge and uncertainty and identify optimal and precisely specified management responses.

- Revising in the same manner the estimates of cost and benefits.
- Reconciling the proposed management responses across the different issue groups, fisheries and ecosystem as outlined in Fig. 16.5.
- Using this information in a participatory and transparent manner to decide on the management responses to be implemented.
- Proceeding with implementation, followed by review in due course.

It is hoped that these steps will be followed as rapidly as the priorities require and that the second phase of the BCLME Programme will provide both impetus and resources to facilitate this. It is also hoped that that it will be recognized by the countries that the long-term benefits achieved through the implementation of management measures will outweigh the costs. As pointed out by Wang (2004) this will require the political will and cooperation of the three states that share the resources of the Benguela Current ecosystem and that bear the responsibility for their sustainable use.

Acknowledgements

The project is a regional, team effort and the progress is a result of substantive and invaluable contributions from many people. Particular thanks go to Filomena vaz Velho and Kumbi Kilongo (INIP, Angola), Jean-Paul Roux, Angie Kanandjembo and Ben van Zyl (MFMR, Nambia), Lynne Shannon and Tracey Fairweather (MCM, South Africa), Gabriella Bianchi (FAO), Dave Japp (CapFish, South Africa), Deon Nel (WWF, South Africa) and Pedro de Barros (Universidade do Algarve, Portugal).

References

BCLME (2006a) Annual Report January–December 2005. Project LMR/EAF/03/01 Ecosystem Approaches for Fisheries (EAF) Management in the BCLME. FAO, Rome, Italy, 238 pp.

BCLME (2006b) Report of the Second Regional Workshop. Project LMR/EAF/03/01 Ecosystem Approaches for Fisheries (EAF) Management in the BCLME. Luanda, Angola, 20–24 March 2006. FAO, Rome, Italy.

Cochrane, K.L. (1999) Complexity in fisheries and limitations in the increasing complexity of fisheries management. *ICES Journal of Marine Science* 56, 917–926.

Cochrane, K.L., Augustyn, C.J., Cockcroft, A.C., David, J.H.M., Griffiths, M.H., Groeneveld, J.C., Lipinski, M.R., Smale, M.J., Smith, C.D. and Tarr, R.J.Q. (2004) An ecosystem approach to fisheries in the Southern Benguela context. *African Journal of Marine Science* 26, 9–35.

Cochrane, K.L., Augustyn, C.J., Cockcroft, A.C., Lipinski, M.R., Smale, M.J., Smith, C.D. and Tarr, R.J.Q. (2007) Results and conclusions of the project Ecosystem Approaches for Fisheries Management in the Benguela Current Large Marine Ecosystem. *FAO Fisheries Circular No 1026*, 167 pp.

De Oliveira, J.A.A., Butterworth, D.S., Roel, B.A., Cochrane, K.L. and Brown, J.P. (1998) The application of a management procedure to regulate the directed and bycatch fishery of South African sardine *Sardinops sagax*.

In: Pillar, S.C., Moloney, C.L., Payne, A.I.L. and F.A. Shillington (eds) *Benguela Dynamics: Impacts of Variability on Shelf-Sea Environments and Their Living Resources. South African Journal of Marine Science* 19, Sea Fisheries Research Institute, Cape Town, South Africa, pp. 449–469.

FAO (2003) The ecosystem approach to fisheries. FAO Technical Guidelines for responsible Fisheries, No. 4 (Suppl. 2). FAO, Rome, Italy, 112 pp.

FAO (1995) Code of Conduct for Responsible Fisheries. FAO, Rome, Italy, 41 pp.

Fletcher, R., Chesson, J., Fisher, M., Sainsbury, K., Hundloe, T., Smith, T. and Whitworth, B. (2002) National ESD reporting framework for Australian fisheries: the 'how to' guide for wild capture fisheries. FRDC Project 2000/145, Canberra, Australia, 120 pp.

Moloney, C.L., Van Der Lingen, C.D., Hutchings, L. and Field, J.G. (2004) Contributions of the Benguela Ecology Programme to pelagic fisheries management in South Africa. In: Shannon, L.J., Cochrane, K.L. and Pillar, S.C. (eds) Ecosystem Approaches to Fisheries in the Southern Benguela. *African Journal of Marine Science* 26, 37–51.

Payne, A.I.L., Gulland, J.A., and Brink, K.H. (1987) The Benguela and Comparable Ecosystems. *South African Journal of Marine Science* 5, 957 pp.

Payne, A.I.L., Brink, K.H., Mann, K.H. and Hilborn, R. (1992) Benguela Trophic Functioning. *South African Journal of Marine Science* 12, 1108 pp.

Roux, J.-P. and Shannon, L.J. (2004) Ecosystem approach to fisheries management in the northern Benguela: the Namibian experience. In: Shannon, L.J., Cochrane, K.L. and S.C. Pillar (eds) Ecosystem Approaches to Fisheries in the Southern Benguela. *African Journal of Marine Science* 26, 79–93.

Shannon, L. J., Cochrane, K.L., Moloney, C.L. and Fréon, P. (2004) Ecosystem approach to fisheries management in the southern Benguela: a workshop overview. In: Shannon, L.J., Cochrane, K.L. and Pillar, S.C. (eds) Ecosystem Approaches to Fisheries in the Southern Benguela. *African Journal of Marine Science* 26, 1–8.

Tchikulupiti, M. (2005) Patterns in the catches of banda-banda from Benguela Province. Paper presented at the 12th Scientific Meeting of the National Institute for Fishery Research: the Decade of Fishery Research in Angola in an Ecosystem Approach: Perspectives and Challenges. Luanda, Angola, 17–18 November 2005.

Wang, H. (2004) Ecosystem management and its application to large marine ecosystems: science, law and politics. *Ocean Development and International Law* 35, 41–74.

17 Implementing the Ecosystem Approach in Australian Commonwealth (Federally)-managed Fisheries

Richard McLoughlin,[c] Nick Rayns,[a] Anthony D.M. Smith,[b] David Johnson,[a] Alistair Hobday[b] and Tim Smith[a]

[a]Australian Fisheries Management Authority, Canberra, Australia; [b]CSIRO Division of Marine and Atmospheric Research, Hobart, Australia; [c]Australian Government Department of the Environment, Water, Heritage and the Arts, Australia

Abstract

In the 5 years since the 2001 Iceland conference on 'Responsible Fisheries in the Marine Ecosystem', Australia developed, researched and made good progress on implementing many of the elements of an ecosystem approach to fisheries management. However, it was in early 2007 that the integration of all the relevant elements of the ecosystem approach would take place for Australia's 21 Commonwealth-managed fisheries. It is noteworthy that the decision to do this was effectively a major policy decision of the Australian Government, with significant inputs from science, fishery management and industry.

The main elements of the approach include: implementing formal harvest strategies for target and by-product stocks in every fishery; undertaking fishery-level ecological risk assessments (ERAs) and developing an ecological risk management response; implementing large-scale spatial management (via both fisheries regulation and conservation-oriented MPAs); enhancement of fisheries data collection and management; and enhancing liaison and communication capacity for the EBFM approach.

In particular, the ecological risk assessments will guide fishery-level priorities for research, data collection and management. These assessments identify the level of risk of causing undesirable impacts to all components of the marine ecosystem in each fishery, including species, habitats and ecological communities. The ERA methodology we have developed can be applied in both data-rich and data-poor fisheries, and uses a hierarchical approach that moves from qualitative through to a fully quantitative assessment.

This chapter will provide an overview of the policy development and implementation approach being taken, as well as examples of how the Australian Fisheries Management Authority (AFMA), with scientific support from the Commonwealth Scientific and Industrial Research Organization (CSIRO), is implementing ecosystem-based fisheries management (EBFM) in Australian fisheries.

Introduction

Australia has actively been pursuing the principles of ecologically sustainable development (ESD) in the way fisheries have been managed for well over a decade. While there have been advances in many areas, the focus has continued to broaden in scope as we respond to changing policy drivers. In particular, the policy drivers have shifted from sustainability of target (and by-product) stocks to the integration of target stock harvest strategies with ecological risk assessments (ERAs) and consequent risk management, in the context of targeted commercial fishing.

This chapter provides an overview of the progress that is being made in Australian Commonwealth-managed fisheries, and the tools that are being developed and implemented to support ecosystem-based fisheries management (EBFM).

Australia's Commonwealth-managed Fisheries

The Australian Fisheries Management Authority (AFMA) is responsible for managing those Australian fisheries that fall within the jurisdiction of the Australian Government, that is, fisheries generally beyond 3 nautical miles (nm) of the coast. Fisheries within 3 nm of the coast are generally managed by the relevant State and Territory governments.

The Australian Fishing Zone is an area of ocean that extends 200 nm from the coastline, a coastline that is >25,000 km long. Commonwealth-managed fisheries span a vast area across approximately 100 degrees of longitude and 50 degrees of latitude from the tropics to the Antarctic (Fig. 17.1). Despite the large area of the Australian Fishing Zone – the world's third largest – Australia ranks only around 50th in world fisheries in terms of tonnes of fish landed (Larcombe and McLoughlin, 2007). The total value of all Australian fisheries in 2005/2006 was about AUS$2.13 billion, from a harvest of around 241,000 t.

EBFM Policy Drivers in Australia

During the early 1990s, Australia became actively involved in the pursuit of ESD across all areas of government. This led to the introduction of new national (and state) fisheries legislation, which had ESD embedded as a key objective. The Australian Fisheries Management Act 1991 (FMA 1991) included powers and functions to address the clear recognition that fishing could have a negative impact on the environment.

In the late 1990s the pursuit of ESD also led to the commencement of new national environmental legislation, which explicitly included additional provisions to direct that Australia's fisheries are managed sustainably. The Environment Protection and Biodiversity Conservation Act 1999 (EPBC Act 1999) requires that every Commonwealth-managed fishery (and those State-managed fisheries, which export product) must be 'strategically assessed' for its ecological sustainability by the Minister for the Environment and Heritage, as an external check on the performance of AFMA or the relevant State fisheries management agency. This 'strategic assessment' is effectively an ecosystem-level environmental

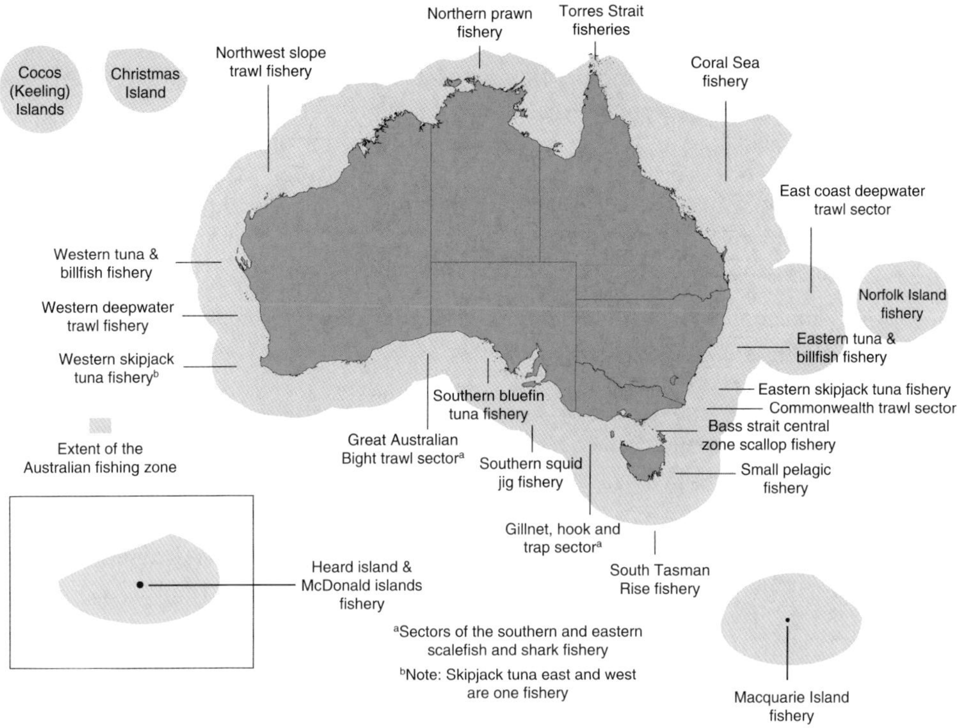

Fig. 17.1. The Australian fishing zone.

impact assessment of the fishery undertaken independent of the fishery agencies. The assessment also potentially has real impact if it finds that the fishery is unsustainable – exports of products from the fishery are banned until the fishery meets the criteria both for acceptable catch levels and environmental impacts.

However, while Australia had the legislation in place, a major deficiency was the lack of relevant scientific tools to support the necessary decision making. The ability of AFMA and the State- and Territory-based fisheries agencies to pursue the legislation was also constrained by the need to bring stakeholders along with these new and evolving concepts. In short, the policy drivers for EBFM had raced ahead of the scientific and operational tools to support it, resulting in policy managers and fishery management stakeholders increasingly becoming frustrated with a perceived lack of progress on EBFM, while fishery managers and industry were equally frustrated by a lack of understanding of the technical (and socio-economic) difficulties and risks involved in decision making at the ecosystem level.

AFMA's Approach to EBFM

While our pursuit of ESD continued throughout the 1990s, the evolving requirements of the *EPBC Act 1999* and a diverse range of other policies and legislative requirements resulted in a fragmented and inconsistent approach to managing the environmental effects of fishing. We have continued to refine our approach and now describe EBFM as the way we integrate the minimization of the impacts of fishing on the marine environment, while ensuring that fish stocks are sustainable for future generations.

Thus, managing fishery ecosystems means not only managing the impact of fishing on target species, but also on the broader marine ecosystem. This includes by-catch species (including threatened, endangered and protected species), habitats and communities.

There are four main threads to AFMA's EBFM approach. These include:

1. Management actions (including standardized sets of decision rules) to reduce ecosystem impacts to an acceptable level, both for target stocks and the ecosystem that supports them.
2. Ecological and stock assessments to inform management.
3. Information and data collection to support the assessments.
4. Education and capacity building to bring the fishing industry and other key stakeholders along in the process.

This chapter covers the relatively recent developments AFMA and CSIRO have made in the development of key assessment tools to support EBFM.

AFMA's Key EBFM Assessments

In pursuing EBFM, AFMA undertakes two broad types of assessments.

For whole ecosystems, we undertake ERAs on a fishery-by-fishery basis. These risk assessments provide a comprehensive basis for prioritizing the relative ecological risks from a range of fishing activities and developing appropriate management responses for the high-risk components.

For target and by-product fish stocks, we aim to prevent or end overfishing and promote rebuilding of depleted stocks. Detailed stock assessments are carried out for our key target species and we have implemented formal harvest strategies for every target (and by-product) fish stock in all fisheries from 2008.

While stock assessment methods have been around for many years, we needed to develop new tools to prioritize the ecological risks and develop harvest strategies for our key species. In both cases, it has been a close collaboration between fisheries managers and fishery scientists to develop these new tools, with frequent reference to stakeholder groups (mainly the commercial fishing industry and environment NGOs) to test ideas and methods.

Developing the Tools: 1. Ecological Risk Assessment

ERA is a key tool, which we have developed with CSIRO to prioritize ecological risks within each fishery; they are a central tool in support of EBFM. An ERA is an assessment of the impacts of fishing on all ecosystem components, including species (target, by-catch and protected species), habitats and marine communities.

ERAs provide a way to prioritize ecological risks, at least in a relative sense. This priority list can then be used to ensure appropriate investment in research, data collection and management, focusing on the key (ecological) issues facing the fishery. It also ensures that time, effort and resources are not wasted on pursuing ecological issues, which may appear important, but which are generally insignificant in the context of the broader fishery.

Development of the ERA methods and their application has been a major investment from both AFMA and CSIRO, over a period of about 5 years.

The unique features of the ERA methodology, which make it an effective and efficient tool for fisheries managers are that it is:

- Comprehensive, covering all aspects and components of each fishery.
- Eigorous and scientifically defensible, by making use of stakeholder/expert input in a transparent manner.
- It uses a hierarchical approach to risk assessment, involving three assessment levels.
- Cost- and time-efficient, screening out lower risks so that more detailed analysis only follows where necessary.
- Cost-efficient through making use of existing data and information.
- Precautionary in approach and recognizes the inherent uncertainty in many aspects of fisheries management.
- Flexible because it can apply to all types of fisheries, regardless of the fishing methods involved or the scale or value of the fishery.
- Transparent, with all steps in the process being openly documented and discussed.
- Understandable to stakeholders.
- Informs management responses to assist better decision making.

The ERA process is hierarchical and provides scope for screening out minor ecological impacts at the earliest stage in the process. This ensures that only those aspects, which need to be assessed in detail are afforded the highest levels of assessment. In moving through the hierarchical process, the fishery is faced with increased data and information needs, and increasing time and costs to undertake the assessment, with the benefit being decreased uncertainty. At any point in the process, once an ecological impact has been identified, the fishery has the choice to further assess the impact (to reduce uncertainty) or to develop an appropriate management response.

The hierarchical approach of the ERA process involves a scoping level and three levels of assessment, each with increasing complexity and use of existing data (Fig. 17.2).

The scoping phase involves preparing a description of the fishery, the management objectives and identifying and listing all of the species, habitats and communities that are impacted by the fishery (Hobday *et al.*, 2006; Smith *et al.*, 2007).

The level 1 ERA involves a comprehensive assessment of all activities in a fishery. This is an expert judgement-based process involving a 'plausible worst-case scenario' approach. If an activity is found to be high or medium risk, the fishery can choose to either mitigate the risk directly, or to further assess the risk at level 2.

The level 2 ERA involves a semi-quantitative assessment of each species, habitat or marine community. At this level, the productivity and susceptibility

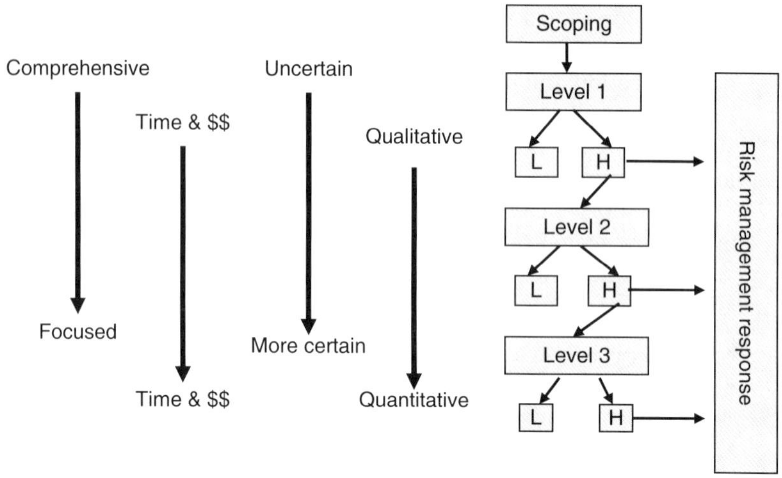

Fig. 17.2. Schematic diagram of ecological risk assessment (ERA) hierarchical structure.

characteristics of each species (or other component) are determined. If a species or component is assessed as high or medium risk, you can choose to either mitigate the risk directly, or to further assess the risk at level 3.

The most detailed level of assessment, level 3, involves a full quantitative assessment and is currently mainly undertaken for our key target species where we have the necessary data to undertake the assessment, and occasionally for some species where environmental legislation classifies them as Threatened, Endangered or Protected (TEP).

Over the past 5 years AFMA and CSIRO have jointly developed the methodology (Hobday *et al.*, 2006) and undertaken ERAs, culminating in the finalization of ERAs for 31 Commonwealth-managed Australian fisheries. For most fisheries, the assessments have been completed to level 2 for all species. Assessments at level 2 have been carried out for about one-third of fishery habitats with a focus on bottom-contact gear, because there is a paucity of habitat data in most of our fisheries. The community level 2 analysis is also a continuing work in progress, because the supporting science to characterize marine communities and carry out assessments at this level has not yet caught up with the policy needs. These two areas will be priorities for attention in coming years.

In addition to the assessments themselves, we have developed an extensive database with relevant information on over 2000 species. This database will continue to be enhanced and will serve as a solid basis for further ERAs in Australia and elsewhere.

Draft ERA reports were developed and discussed in the industry during 2007 and were expected to be publicly released in 2008. The key part of the process then commenced in determining appropriate management responses to address each of the high (and some medium) priority ecological risks. This ongoing work will be carried out in consultation with AFMA's key stakeholders, including our Management Advisory Committees for each fishery. This partnership approach is a key feature of the way AFMA implements fisheries management on behalf of the Australian Government.

Developing the Tools: 2. Harvest Strategy Policy

AFMA has implemented formal harvest strategies across all Commonwealth fisheries. Harvest strategies involve a catch-setting process that incorporates strategies for ongoing monitoring of agreed characteristics of target fish stocks and other aspects of the fishery, a process of periodic assessment and review, and follow-up action consistent with pre-agreed decision rules. They provide a transparent, consistent and predictable process for decision making (Smith *et al.*, 2008).

For example, in the multi-species (and multi-method) fisheries in south-eastern Australia, AFMA and CSIRO have adopted a four-tier system for dealing with uncertainty about target stocks. This approach allows for decision rules to be based on robust quantitative assessments at tier 1, preliminary quantitative assessments at tier 2, estimates of fishing mortality rate at tier 3 and trends in catch per unit effort (CPUE) at tier 4.

The system produces recommended biological catches (RBCs) for each of 32 species and stocks, which are then used to determine total allowable catches (TACs), which can be shared among fishermen according to their individual transferable quota (ITQ) holdings for each species.

This approach is precautionary by ensuring that more conservative RBCs and TACs are set when there is increasing uncertainty and less information available.

Work on refining the harvest strategy policy continued into late 2006, and the default settings were applied in late 2007 and early 2008, involving a '20/40' strategy. That is, the exploitation rate and catch will be reduced when the biomass drops below 40% of pre-fished levels. Target fishing will cease when estimated biomass levels drop below 20% of the pre-fished level.

The aim is to limit fishing mortality to achieve target biomass levels at maximum sustainable yield (MSY), while avoiding overfishing (or the fishery becoming overfished) with a probability of at least 80%. For some species, such as slow-growing and long-lived species, more conservative strategies may be implemented; this was particularly the case during 2007 for more vulnerable species (such as deep-water stocks) and highly migratory fish stocks where RFMOs are involved. The harvest strategies developed in the southern and eastern scalefish and shark fishery were being used as the starting point for developing a harvest strategy policy that applies to all Commonwealth-managed fisheries, as of January 2008. The current policy settings and explanatory documents can be obtained from http://afma.gov.av

Conclusion

Australia's experience in implementing EBFM in Commonwealth-managed fisheries was prompted by the changing international policy focus during the late 1990s and the need to address a range of diverse national initiatives in a more holistic and strategic approach.

AFMA, together with the CSIRO, have made a major investment over the past 5 years in developing two key tools to support EBFM: ERA and harvest strategies, both supported by a sound and widely consulted policy framework approved by government.

Now that we have developed the tools, AFMA will be shifting focus to comprehensively applying these tools across all Commonwealth-managed Australian fisheries over the next few years.The next steps will be working with stakeholders to prioritize research, monitoring and management actions to demonstrate compliance not only with the policy framework mandated by government, but also that we are demonstrably achieving the goal of ecologically sustainable fisheries.

Acknowledgements

While there are multiple authors for this chapter, we wish to acknowledge the efforts of many other participants and stakeholders in the processes and actions described above. The following is in no particular order and is not exhaustive: The Board of AFMA (since 2001 and Dr Wendy Craik in particular), Mr Geoff Richardson of AFMA, the Department of Agriculture, Fisheries and Forestry, Australia, the CSIRO team involved in developing and applying the ERA methods, the CSIRO team involved in developing and implementing the harvest strategies in the SESSF, the Commonwealth Fisheries Association and the South East Trawl Fishing Industry Association.

References

Hobday, A.J., Smith, A., Webb, H., Daley, R., Wayte, S., Bulman, C., Dowdney, J., Williams, A., Sporcic, M., Dambacher, J., Fuller, M. and Walker, T. (2006) Ecological Risk Assessment for the Effects of Fishing: Methodology. Report R04/1072 for the Australian Fisheries Management Authority, Canberra, Australia.

Larcombe, J. and McLoughlin, K. (eds) (2007) Fishery Status Reports 2006: Status of Fish Stocks Managed by the Australian Government. Bureau of Rural Sciences, Canberra, Australia.

Smith, A.D.M., Fulton, E.J., Hobday, A.J., Smith, D.C. and Shoulder, P. (2007) Scientific tools to support practical implementation of ecosystem-based fisheries management. *ICES Journal of Marine Science* 64, 633–639.

Smith, A.D.M., Smith, D.C., Tuck, G.N., Klaer, N., Punt, A.E., Knuckey, I., Prince, J., Morison, A., Kloser, R., Haddon, M., Wayte, S., Day, J., Fay, G., Pribac, F., Fuller, M., Taylor, B. and Little, L.R. (2008) Experience in implementing harvest strategies in Australia's south-eastern fisheries. *Fisheries Research*, doi: 10.1016/j.fishres.2008.06.006

18 The Ecosystem Approach to Fisheries Management in the USA

Galen R. Tromble

National Oceanographic and Atmospheric Administration, USA

Abstract

United States marine fisheries are highly diverse, exploiting resources in ecosystems ranging from the arctic to the tropics, and in both the Atlantic and the Pacific Oceans. Protecting, restoring and managing the use of coastal and ocean resources is one of the strategic goals of National Oceanographic and Atmospheric Administration (NOAA). NOAA Fisheries is the agency with primary responsibility to achieve this goal. A variety of ecosystem approaches have been used in fisheries management, but improvements are needed in the understanding of ecosystems and their dynamics. The general approach is to establish strategic fishery ecosystem plans (FEP) that describe goals, the current state of information and knowledge and priorities for research; establish a management framework to utilize existing ecosystem knowledge; improve ecosystem science and models on an ongoing basis; and utilize the best available scientific information in management decisions. Important challenges still remain and include providing management with decision support tools to deal with increased complexity of objectives and information, the need for better communication and outreach to the public and to policy makers and the need to strengthen the statutory basis for the ecosystem approach to fisheries management (EAFM). Finally, ocean governance issues are challenging, involving multiple levels of government with overlapping, but differing, geographic scope and legal authority.

United States Marine Fisheries Management Policy Background

The National Marine Fisheries Service (also called NOAA Fisheries), an agency of the National Oceanographic and Atmospheric Administration (NOAA) within the United States Department of Commerce, has responsibility for marine fisheries management and science in federal waters (generally from 3 to 200 nautical miles (nm) off the coast). Coastal states and territories manage resources in

 The Ecosystem Approach to Fisheries (eds G. Bianchi and H.R. Skjoldal)

nearshore waters. A number of governance arrangements exist to coordinate management of fish stocks that cross jurisdictional boundaries.

The principal legal authority for marine fisheries management in the United States is the Magnuson-Stevens Fishery Conservation and Management Act (MSFCMA). The MSFCMA was first enacted in 1976, extending jurisdiction over marine fisheries to 200 nm and establishing a management system with eight regional fishery management councils (Councils). The MSFCMA has been amended several times in the ensuing years, including a major amendment in 1996 known as the Sustainable Fisheries Act, and the most recent major amendment, passed by Congress in December 2006.

The eight Councils, funded by the federal government, are mandated to develop fishery management plans and recommend management measures and regulations for adoption by the federal government. The jurisdictions of the eight councils correspond well to a set of large marine ecosystems that have been proposed for implementing EAFM. In most cases, each managed fish stock is contained within a single region. In some cases, where stocks straddle regions, management is done cooperatively by two Councils, with one designated as the lead Council.

Ecosystem considerations have been a part of marine fisheries management since the 1970s. The National Environmental Policy Act (NEPA), enacted in 1969, applies to all Federal actions, including fishery management measures. The NEPA requires consideration of a range of reasonable alternatives and analysis of the effects on the environment, including cumulative impacts from this and other actions. The Marine Mammal Protection Act (MMPA) of 1972 affords protections to all marine mammal species. The Endangered Species Act of 1973 protects endangered and threatened species and their habitats. All fisheries management actions have to meet the requirements of the NEPA, the MMPA and the ESA, affording ample opportunity for analysis and consideration of ecosystem impacts.

The MSFCMA governs marine fishery management in the United States and contains numerous provisions related to the ecosystem approach to management. It requires that overfishing be prevented, and that the optimum yield for a fishery not exceed the maximum sustainable yield (MSY). The optimum yield may be reduced from MSY for issues that include protection of marine ecosystems. The MSFCMA requires that by-catch be minimized, essential fish habitat (EFH) be designated, and that management actions consider the needs of fishing communities. While these provisions incorporate some elements of EAFM, they are not complete, and a more comprehensive framework for EAFM is needed.

In 1996, Congress directed NOAA to establish an Ecosystem Principles Advisory Panel to inform the Secretary of Commerce and Congress on ways to incorporate ecosystem principles into fisheries conservation and management. The panel developed a set of principles, goals and policies to evaluate current applications of EAFM and made recommendations for future expansion of ecosystem approaches. The panel recommended interim measures for the Secretary of Commerce to develop demonstration fishery ecosystem plans (FEPs) and voluntary adoption by Councils and NMFS of the principles, goals and policies included in the report (Ecosystem Principles Advisory Panel, 1999). Several Councils have developed FEPs, and others are under development. For example, the North Pacific Council recently released its FEP for the Aleutian Islands region (http://www.fakr.noaa.gov/npfmc/current_issues/ecosystem/AIFEP12_07.pdf).

In 2001, NOAA's Marine Fisheries Advisory Committee formed an Ecosystem Approach Task Force. This group identified key issues that need to be addressed in order to implement EAFM. The five key issues were: (i) enhancing intra- and inter-agency cooperation and communication; (ii) delineating geographic areas of the ecosystem; (iii) preparing quantified natural resource goals and objectives; (iv) identifying and applying specific indicators; and (v) considering socio-economic issues to evaluate management trade-offs. The Task Force recommended implementation of several pilot projects to illustrate the benefits and challenges to ecosystem-based fishery management (Busch *et al.*, 2003). As part of the 2004 NOAA budget, Congress included US$2 million to advance ecosystem approaches for the fishery management councils in the Atlantic Seaboard and Gulf of Mexico areas. Each of four Councils (New England, Middle Atlantic, South Atlantic and Gulf of Mexico) were provided funding to survey and understand ecosystem issues relevant to their activities. These reports are currently being compiled by the various Councils, and identify the particular issues in their respective areas requiring ecosystem approaches to management.

Subsequently, two influential reports were issued to address ecosystem issues in the broader marine environment: *America's Living Oceans* (Pew Ocean Commission Report, 2003) and *An Ocean Blueprint for the 21st Century: Final Report of the U.S. Commission on Ocean Policy to the President and Congress* (US Commission on Ocean Policy, 2004). With regard to fisheries management issues, both these reports call for increased use of ecosystem approaches. In support of the recommendations of these reports, over 200 scientists and policy experts issued a consensus statement in 2005 calling for the conservation and management of marine systems through a more integrated ecosystem approach (McLeod *et al.*, 2005).

In 2005, NOAA's Science Advisory Board formed the External Ecosystem Task Team (EETT) of eight members to provide advice on how to improve NOAA's ecosystem science programmes. The July 2006 EETT report concluded that:

> The incorporation of more general ecosystem principles into traditional management approaches for coastal and marine issues has progressed substantially in recent years. In particular, ecosystem approaches to fisheries management have progressed from the theoretical to the implementation stage. This has occurred because of the growing realization that fisheries management is imbedded (sic) in a larger set of ocean policy decision making involving living marine resources and attributes of their supporting ecosystems.
>
> (EETT report, Appendix 5, p. 80)

Status of EAFM in United States Marine Fisheries

As the primary federal agency responsible for marine fisheries management in the USA, NOAA is committed to implementing EAFM. One of NOAA's strategic goals is to: 'Protect, Restore, and Manage the Use of Coastal and Ocean Resources Through an Ecosystem Approach to Management'.

NOAA's Ecosystem Goal Team has identified seven characteristics of an ecosystem approach to management (Murawski, 2005):

1. Geographically specified.
2. Adaptive.
3. Takes account of ecosystem knowledge and uncertainty.
4. Considers multiple external influences.
5. Strives to balance diverse societal objectives.
6. Incremental.
7. Collaborative.

A defining characteristic of EAFM is that it considers a broader range of ecological, social and economic information about the ecosystem than traditional management approaches for individual species or activities. It considers interactions between target and non-target species and the impacts of all sectors within a geographic area, not just fishing activity. It emphasizes the protection of ecosystem structure, functioning and key processes, and recognizes multiple governance structures that operate within the geographic area of the ecosystem (Crowder *et al.*, 2006).

United States marine fisheries are highly diverse, occur in ecosystems that range from the arctic to the tropics and in both the Atlantic and Pacific Oceans, and are managed under 46 different fishery management plans. The various commercial fisheries include large vessels in corporate fleets as well as small family owner-operated vessels. In addition to commercial fisheries, recreational and subsistence fisheries are significant. Fisheries plan development is regionalized and conducted primarily by the eight regional fishery management councils, which include representatives of the coastal states and territories in the region. A wide range of management programmes is currently in use, including open-access, limited entry permits, cooperatives, community quotas and individual fishing quotas. The availability of biological, social and economic data from the nation's diverse fisheries varies widely, and few fisheries have completely adequate data.

The EAFM recognizes that fishing - through direct removal of target fish species, by-catch of other fish, birds, mammals and reptiles, and impacts on habitat - impacts not just the targeted stock, but also the ecosystem in which the stock lives and the fishery operates. It also recognizes that factors in the ecosystem can affect productivity of a stock in a variety of ways, and these effects are important in understanding the amount of a species that can be sustainably harvested. Ecosystem approaches have not been more extensively implemented, not because the effects were not recognized, but rather because the science, data and models to effectively incorporate ecosystem effects into decision making have not been adequate.

None the less, a number of specific EAFM measures have been implemented in United States marine fisheries, including measures to quantify and minimize by-catch, definition of EFH, designation of numerous marine-protected areas, including bottom trawl closures in areas off the Atlantic and Pacific coasts and in Alaska. While a comprehensive EAM approach is still being developed, the following outline measures within existing fishery planning represent progress towards EAM in US fisheries.

By-catch

By-catch of marine mammals, reptiles, birds and unwanted fish and invertebrate species is one of the primary impacts of fishing on the ecosystem. Minimizing by-catch is a management goal in US fisheries. The MSFCMA requires that 'conservation and management measures shall, to the extent practicable, (A) minimize bycatch and (B) to the extent bycatch cannot be avoided, minimize the mortality of such bycatch.' The MSFCMA also requires every fishery management plan to include a methodology for determining the quantity of by-catch. However, the availability of reliable data on by-catch varies widely among fisheries. Improved standards for by-catch information are currently being developed by NOAA Fisheries based on a comprehensive review of federally managed fisheries.

Essential Fish Habitat

Amendments to the MSFCMA in 1996 required definition of EFH for all marine fishery management plans. Criteria for EFH designation were developed and all fishery management plans now have approved EFH definitions. EFH can consist of both the water column and the sea floor of a particular area. Factors considered in defining EFH may include properties of the water column (e.g. temperature, nutrients and salinity), bottom types (e.g. sandy or rocky bottoms), vegetation (e.g. seagrasses or kelp) or structurally complex coral or oyster reefs. EFH includes those habitats that support the different life stages of each managed species. A single species may use many different habitats throughout its life to support breeding, spawning, nursery, feeding and protection functions. Impacts on EFH must be identified and analysed in fishery management planning.

Fishery Ecosystem Plans

The Ecosystem Principles Advisory Panel, in its 1999 report, addressed the question of how to move from single-species management to ecosystem-based management and recommended development of an FEP for every ecosystem. The FEP would describe the known components and interactions of the ecosystem and increase managers' awareness of how their decisions affect the ecosystem. The FEP would not have specific regulatory effect, but would be the basis for specific management actions in a fishery management plan.

Fishery ecosystem plans have been implemented or are under development for the Chesapeake Bay, the south Atlantic area, the Aleutian Islands (Alaska) and for five western Pacific archipelagos. In addition, in 2004, Congress funded ecosystem pilot projects in the New England, mid-Atlantic and Gulf of Mexico areas to engage regional stakeholders in defining goals and objectives for EAFM. An overview of some of these plans is provided below.

Chesapeake Bay FEP

Chesapeake Bay is the world's largest estuary. It is also a heavily used recreational area subject to the stresses of growing regional population and development. Fisheries for species such as oysters, blue crab, striped bass, menhaden and other species are some of the oldest in the nation. Predator–prey relationships among the species are critical characteristics of the ecosystem and management of the Bay's resources depends on understanding the trophic dynamics of the ecosystem. The bay watershed drains large agricultural and urban areas, and nutrient pollution is a significant issue. NOAA's Chesapeake Bay office developed an FEP in 2004 that describes the structure and function of the Chesapeake Bay ecosystem, including key habitats and species interactions. It is intended to be an umbrella document to support ecosystem-based approaches in individual fishery management plans, and recommends specific research to enhance knowledge of the ecosystem and its fisheries to support long-term management objectives.

Western Pacific Archipelago FEPs

The Western Pacific Regional Fishery Management Council is responsible for federal fishery management in the Hawaiian Islands and a number of island archipelagos in the Pacific. The Council is in the process of developing archipelagic FEPs that would eventually replace its separate Pacific-wide fishery management plans for coral reef fish, precious corals, bottomfish, seamount groundfish and crustaceans. The FEPs outline how bottomfish, coral conservation and socioeconomic considerations can be integrated in a geographically explicit series of plans (e.g. for the Mariana Archipelago, the Hawaiian Island Archipelago, Samoa Islands, Guam and the Pacific Remote Islands). Large pelagics would continue to be managed on a Pacific-wide basis given the scale of their migrations.

North Pacific FEPs

The North Pacific Fishery Management Council has, for a number of years, incorporated ecosystem considerations into its fishery management plans. Ecosystem-based measures including reduction of by-catch, accounting for trophic relationships among species and conservative long-term management approaches have been part of its ecosystem approach. In 2006, the Council added conservation of cold-water coral habitats by implementing fishery closed areas explicitly to protect these fragile habitats. Other habitats of particular concern have been reserved from fishing activities for various purposes including integrated management of protected species including sea lions and other mammals. The Council routinely utilizes a wide variety of ecosystem data and indicators as part of its annual groundfish management planning. The Council, along with the State of Alaska and stakeholder groups, has been considering how to implement area-based ecosystem plans. This process resulted in the Council developing an FEP for the Aleutian Islands area, recognizing that the Aleutian Islands contain unique ecological values. Recent scientific evidence indicates a clear ecological difference between the eastern Bering Sea shelf ecosystem and the western Aleutian Islands archipelago.

The Aleutian Islands FEP describes the ecosystem, including spatial boundaries, predator–prey interactions, habitat needs of the significant food web components and current and historic states of the ecosystem. Indices of ecosystem health will be used to assess all impacts, natural and anthropogenic, on the ecosystem. The FEP considers aggregate, cumulative impacts on the ecosystem – from fishing and non-fishing sources.

The FEP will help the Council consider each ecological component of the region (e.g. seabirds, marine mammals, communities and industries) in the sustainability of the whole, when making decisions on fishery management actions.

Atlantic Seaboard and Gulf of Mexico FEPs

The South Atlantic Fishery Management Council is developing an FEP for its fisheries. The South Atlantic FEP is being developed from the Council's current Habitat Plan. The transition from single-species management to an ecosystem approach will involve incremental steps to better characterize the ecosystem and understand the complex relationships among humans, harvested fish and prey, all other marine life and essential habitat and environmental characteristics. The FEP will provide the Council with a more comprehensive understanding of habitat and biology of species, fishery information, ecological consequences of conservation and management, and economic and social impacts.

Conclusion

The USA is committed to increasing application of EAFM to its diverse marine fisheries. Ecosystem principles have been applied in US fisheries for many years, and their use continues to increase as scientific understanding of the ecosystems and their dynamics improves. EAFM has been adopted by NOAA as a key policy emphasis in both science and management, with the goal of sustainable use of marine ecosystems, so their structure and function are preserved for future generations.

Several Councils have developed the first generation of FEPs and are beginning to use them to inform specific fishery management actions. NOAA Fisheries will continue to work with the Councils to support EAFM. Because the Councils are regionally based and involve stakeholders in the decision-making process, they are well positioned to implement EAFM and to take advantage of new scientific information and ecosystem modeling capabilities that are being developed.

References

Busch, W.-D.N., Brown, B.L. and Mayer, G.F. (eds) (2003) Strategic Guidance for Implementing an Ecosystem-based Approach to Fisheries Management. United States Department of Commerce, National Oceanic and Atmospheric Administration, NMFS, Silver Spring, Maryland 62 pp.

Crowder, L.B., Osherenko, G., Young, O.R., Airamé, S., Norse, E.A., Baron, N., Day, J.C., Douvere, F., Ehler, C.N., Halpern, B.S., Langdon, S.J., McLeod, K.L., Ogden, J.C., Peach, R.E., Rosenberg, A.A. and Wilson, J.A. (2006) Resolving Mismatches in US Ocean Governance. *Science* 313, pp. 617–618.

Ecosystem Principles Advisory Panel (1999) Ecosystems Based Fishery Management: A Report to Congress by the Ecosystem Principles Advisory Panel. US Department of Commerce, National Oceanic and Atmospheric Administration, National Marine Fisheries Service.

External Ecosystem Task Team (2006) Evolving an Ecosystem Approach to Science and Management Throughout NOAA and its Partners. US Department of Commerce, National Oceanic and Atmospheric Administration, Science Advisory Board, 85pp. Available at: http://www.sab.noaa.gov/Reports/eETT_Final_1006.pdf

McLeod, K.L., Lubchenco, J., Palumbi, S.R. and Rosenberg, A.A. (2005) Scientific Consensus Statement on Marine Ecosystem-Based Management. Published by the Communication Partnership for Science and the Sea. Available at: http://compassonline.org/?q = EBM

Murawski, S. (2005) Strategies for Incorporating Ecosystems Considerations into Fisheries Management. A white paper by the Director, Office of Science and Technology, NOAA Fisheries, 1315 East West Hwy, Silver Spring, Maryland 20910.

Pew Oceans Commission (2003) America's Living Oceans: Charting a Course for Sea Change. Pew Oceans Commission, Arlington, Virginia.

US Commission on Ocean Policy (2004) An Ocean Blueprint for the 21st Century. Final Report of the US Commission on Ocean Policy to the President and Congress, Washington, DC.

19 Are the Lake Victoria Fisheries Threatened by Exploitation or Eutrophication? Towards an Ecosystem-based Approach to Management

Jeppe Kolding,[a] Paul van Zwieten,[b] Oliva Mkumbo,[c] Greg Silsbe[d] and Robert Hecky[e]

[a]*University of Bergen, Bergen, Norway;* [b]*University of Wageningen, Wageningen, The Netherlands;* [c]*Lake Victoria Fisheries Organisation, Jinja, Uganda;* [d]*University of Waterloo, Waterloo, Canada;* [e]*University of Minnesota, Duluth, USA*

Abstract

Lake Victoria's ecosystem has shown fundamental changes over its past recorded history in terms of nutrient loadings, productivity, faunal composition and fisheries. As yet, however, no attempt has been made to link the driving processes of eutrophication and fisheries to understand the feedback observed in fish stocks, food webs, exploitation patterns and trade. Single- and multi-species stock assessments, based on steady-state models with effort (and/or predation) as the only driver – still used in the region to advise on management – uniformly indicate overfished stocks of Nile perch that are in danger of collapse. These current views of overfishing are not validated by empirical observations. This chapter presents a holistic integrated ecosystem approach which combines a phenomenological analysis of key processes with a comprehensive set of simple indicators, covering physical, biological and human development, where directionality in time is made explicit to understand ongoing processes in the changing ecosystem. This new approach results in: (i) no signs of overfishing in any of the verifiable indicators; and (ii) biological production increasing over time together with effort and yield as a function of increased eutrophication. The results indicate that continued eutrophication presents a much graver risk to the resource base and thus livelihoods of Lake Victoria's coastal populations than fishing pressure. Lake Victoria can serve as an interesting case study for the inherent risk of using traditional fish stock assessment in changing ecosystems, and for the development of holistic monitoring systems for ecosystem-based management.

Introduction

Lake Victoria is perhaps best known as the location of 'the greatest vertebrate mass extinction in modern era' (Baskin, 1992) and the associated expansion of the introduced 'voracious top-predator' (Ogutu-Ohwayo, 1990; Kudhongania *et al.*, 1992; Kitchell *et al.*, 1997), the Nile perch (*Lates niloticus*). Accordingly, the vast literature on its fish and fisheries is strongly focused on the changes in the lake's biodiversity and their causes. The changes in fish, zooplankton and phytoplankton communities as well as changes in the physico-chemical environment have primarily been attributed to the large food web changes instigated top-down, both by the Nile perch introduction and by (local) overfishing (Barel *et al.*, 1985; Kaufman, 1992; Goudswaard *et al.*, 2008). Food webs, however, and with that the fishery resources of Lake Victoria, are driven both by changes in top-down processes (Nile perch predation and increased fishing pressure) and bottom-up processes (nutrients and eutrophication; Fig. 19.1). It is perhaps less well known that nutrient levels in Lake Victoria are rising rapidly (Hecky, 1993; Hecky *et al.*, 1994), but recent increasing awareness on the ongoing eutrophication has led to the recognition that both eutrophication and fisheries may be important drivers of biodiversity change (Balirwa *et al.*, 2003; Witte *et al.*, 2007a). The top-down perspective, however, still dominates within fisheries assessments and their associated management implications (van der Knaap

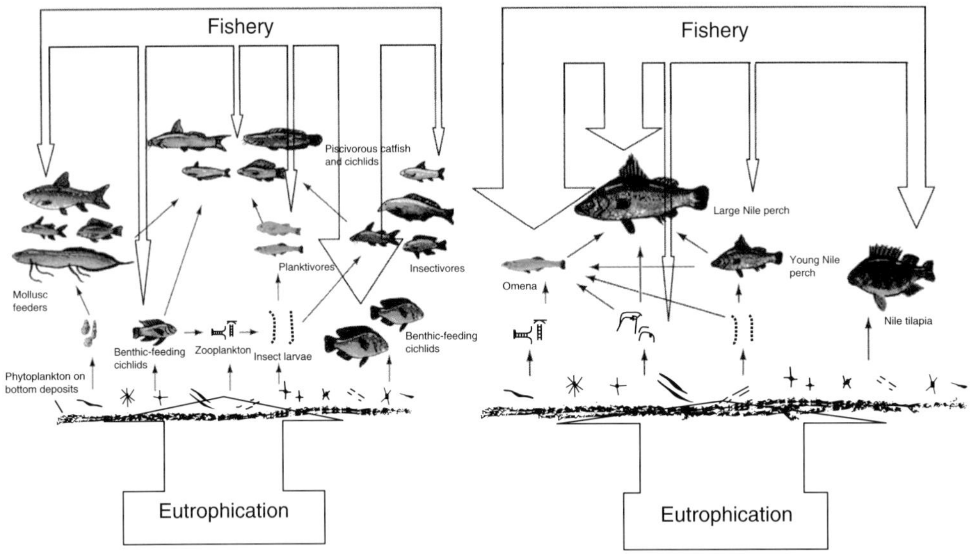

Fig. 19.1. Top-down and bottom-up processes in the Lake Victoria ecosystem before and after the loss of the indigenous haplochromine cichlids and the explosion of the introduced Nile perch. Fish communities are represented by the major trophic interactions between fish species. Fishing pressure and eutrophication, drivers of changes in the fish community, are represented by the open arrows. The fishery in its turn is driven by trade demands while resource-use opportunities are a main driver for choices made by the fishery community. (Adapted from Ligtvoet and Witte, 1991.)

2002; Matsuishi *et al.*, 2006), where the likely effects of eutrophication on the fish stocks are largely ignored. Assessments of the fishery are still done under the tacit assumption that top-down processes are the only drivers of change in Lake Victoria's ecosystem and they all – standard and advanced – have been done under the (untested) precept of an otherwise steady-state that most fisheries models require (Table 19.1). This lack of integration with contemporary results from limnological studies has therefore lead to different, and often highly conflicting, views on the direction and needs of fisheries and ecosystem management. Lake Victoria, however, is not in steady-state, and both types of drivers can interact significantly. But to what extent they reinforce or compensate each other, and what this means for the type and direction of fisheries management has not been investigated. To do so requires a more holistic approach at a higher analytical scale than observations at the level of the stocks (Kolding, 1994). Such an approach, based on carefully chosen indicators of ecosystem drivers, stock states and fishing pressures, is more likely to give guidance to the development of monitoring systems that can form the basis in ecosystem-based assessment and management (Jul-Larsen *et al.*, 2003; Choi *et al.*, 2005). We believe that a procedure that combines an analysis of key processes in Lake Victoria, based on time series of observable indicators of important phenomena, will aid in comprehension of system changes and in assessing unbiased causality of drivers, something that the application of predetermined models cannot achieve. Long time series also prevent what Tufte (2006) calls 'recency bias', or the amplification of recent events in short-term analyses, that otherwise would disappear as natural variations in the long-term perspective.

The main aim of this chapter is to assess the relative importance of the main drivers of fish production in Lake Victoria, with a focus on the important fishery on the Nile perch. To do so, a series of indicators were initially developed based on the understanding of processes by Lake Victoria researchers and a number of other stakeholders from the three riparian countries Kenya, Tanzania and Uganda (Kolding *et al.*, 2005). These indicators, traced as far back in time as possible, also served as a compilation and consolidation of available data on the physical environment, the fishery and the fish stocks. Together, they give a comprehensive picture of states, trends and interactions of the Lake Victoria ecosystem. Based on basic theoretical considerations on the behaviour of stocks under impact from both eutrophication and fishing, the indicators can be linked and interpreted. The following analysis will show that the dynamics of fish production in Lake Victoria, exemplified by Nile perch for which most data exist, to a large extent are environmentally driven; that the perception that fishing is the sole or main driver of stock dynamics is wrong; and that fish stock assessments based on steady-state assumptions, both in the interpretation of trends and in the use of fishery assessment models, therefore can be highly misleading. Only when viewing indicators of the environment, the stocks and the fishery in combination can an empirically based understanding of the relative impact of the processes that drive the Lake Victoria fishery and ecosystem be derived. In order to separate between top-down or bottom-up processes, we will formulate expectations of change in relevant indicators based on the assumption that only one driver (fishing) is the cause of change in the Nile perch stocks, and then

Table 19.1. A non-exhaustive list of published material (main literature, grey literature, PhD and MSc theses, that have the fundamental underlying assumption that the Lake Victoria ecosystem is in a steady-state with regard to trophic status. Analyses in which non-linear changes in species composition and stock sizes are predicted to occur through trophic interactions (e.g. ECOPATH and ECOSIM) extend the current conditions of the lake to their future maximum, but do not take into account changes in carrying capacity.

Study	Objectives	Model/observations	Recommendation
Pitcher and Bundy, 1995	Assessment of the Nile perch fishery	Surplus production, yield-per-recruit and a range of approximate models (Pauly, 1982; Beddington and Cooke, 1983)	All assessments indicate that the current Nile perch fishery is overexploited. Projections indicate stock collapse within a few years if present expansion of effort continues and immature fish continue to be harvested. Effort should be reduced to around 14,000 boats. A minimum size limit of at least 50 cm should be introduced
Kitchell *et al.*, 1997	Estimate predation rates by Nile perch and fishery yields to evaluate the consequences of previous, current, and future fishery exploitation patterns and their ecological implications	Bioenergetics model that estimated fishery harvests based on size susceptibility of Nile perch to different harvest equipment and estimated haplochromine predation rates based on the daily energy requirements of Nile perch	A combination of beach seine and gill net fisheries resulted in the lowest predation rates on haplochromines, but potentially could lead to an unsustainable Nile perch fishery. Development of fisheries based on large-mesh gill nets reduced total predation by Nile perch to ~40% of that estimated during the late 1970s. Large-meshed gill nets provided greatest yields to the fishery. Expansion of recent intensive beach seine and small-mesh gill net fisheries for juvenile Nile perch could reduce total predation to ~25% but would lead to recruitment overfishing. The combination of fishing methods could reduce total predation to ~10% of previous levels. The reported doubling in primary production rates is insufficient to account for the disparities in fishery yield estimates. Fishing is among the most important regulators of trophic dynamics in Lake Victoria
Schindler *et al.*, 1998	Fishery sustainability and reduced haplochromine predation rates	Same as Kitchell *et al.* (1997)	Nile perch harvests were maximized with minimum gill net mesh sizes between 6 and 10 in. Universal enforcement of a 5 in mesh size would reduce both Nile perch cannibalism and predation on other important fishes with little (10%) decrease in harvests. A restriction to 5 and 6 in would lead to ~35% decrease in Nile perch harvests

Kaufman and Schwartz, 2002	Fishery sustainability and reduced haplochromine predation rates	Used an ecosystem-based trophic-mass balance model (ECOSIM; Walters *et al.*, 1997)	An intermediate level of fishery effort was beneficial both to the fishery harvest size and to haplochromine population size. The increased population size of haplochromines contributed to a faster growth rate of Nile perch, and thus increased Nile perch harvest
Getabu *et al.*, 2003	Acoustic abundance estimates of fish in Lake Victoria between 1999 and 2001	Acoustic surveys and partitioning of echo-integrals between four target groups	Over the survey series (1999, 2001), Nile perch biomass showed a consistent decline, while the stocks of small pelagic species increased. Fishing pressure is a primary reason for the decline but environmental changes could also be important
Mkumbo, 2002 Getabu, 2003 Okaronon, 2004	Assessment of Nile perch stocks	Yield-per-recruit and virtual population analysis based on trawl surveys in Tanzania, Uganda and Kenya	Fishing effort for Nile perch should be reduced by approximately 50% to attain E_{max}, the exploitation level which attains the maximum yield-per-recruit
Matsuishi *et al.*, 2006	Review of trends in catch and effort of the Nile perch fishery and model scenarios at 90% and 120% of current fishing effort	Use an ecosystem-based trophic-mass balance model Ecopath to do scenario studies (ECOSIM)	The fishery exhibits classic indicators of intensive fishing erring towards overexploitation: (i) decline in catch since 1990; (ii) fishing down; (iii) increased fishing effort; and (iv) decline in catch per boat from 80 to 45 kg/day. Predictive modelling under a scenario of increased fishing effort suggests an unsustainable fishery and decline in the long term. Solutions: access restriction, mesh size and gear restrictions to protect younger life stages, reduction of postharvest losses
Page, 2006	Predict future population size and extinction risk in order to study fishery management options for fishery sustainability and biodiversity conservation)	Population viability analyses (PVAs) based on Nile perch life history parameters (age-structured grouping of individuals with similar survivorship and fecundity rates)	Prevent harvest of Nile perch below 50 cm total length, or ban beach seine fisheries to improve the sustainability of the Nile perch population. An open-access fishery with increasing effort likely leads to fishery collapse. Management other than fishery gear type restrictions are needed to aid the recovery of haplochromine populations as the proportion of Nile perch in piscivorous-age classes does not respond to gear management

examine these against available evidence. If the expectations are not met, then this would indicate an influence from another driver. Lastly, we will argue that whereas nutrient enrichment up till now has sustained the increased Nile perch fishery, it is highly likely that the ceiling has been reached and that continued eutrophication will bring deterioration to the fishery in the future. Before that, however, we will give a short historic background by briefly describing the changes in Lake Victoria's ecosystem over the past decades.

Lake Victoria's profound ecosystem changes

The equatorial Lake Victoria in East Africa is the second largest freshwater lake in the world with a surface area around 68,800 km^2, a maximum depth of 84 m and a mean depth of 40 m. Its drainage basin is 236,000 km^2 and includes large portions of its three riparian countries Tanzania, Uganda and Kenya and the neighbouring states of Rwanda and Burundi. The Lake in its current form is relatively young: since its formation less than 1 million years ago it has dried up completely three times possibly related to ice-age periods; the last period was around 17,000 years ago and the lake filled up again around 14,000 years ago (Johnson *et al.*, 1996; Stager and Johnson, 2007). Despite its young age and variable history, a large and highly diverse native fauna consisting of hundreds of cichlid species as well as around 46 cyprinids, catfish and other species developed (Greenwood, 1974; Witte *et al.*, 2007a), which supported several small-scale fisheries. Today, however, the lake supports one of the largest freshwater fisheries in the world and over the past four decades, the pace in ecological changes has increased exponentially, with far-reaching consequences for fishery-based livelihoods and fish trade. Though climatic influences may also play a role, the changes are mainly human-induced: water-level changes as a result of damming the lake for electricity production; deliberate and inadvertent exotic introductions of fish and plant species; intensive fishing; and not least eutrophication as a result of increased population densities and changes in land use.

The potential to regulate water levels arose when the Owens Falls Dam in the lake's only northern outlet, the Nile River, was completed in 1954. The addition of increased channel outflow capacity in the second phase of Owen Falls development early in the present century has further enhanced the potential for hydroelectric demands to affect lake levels. Despite this, relative lake-level fluctuations can be still be considered a proxy for changes in rainfall in the catchment area and thereby an important indicator for climatic changes (Nicholson, 1998; Yin and Nicholson, 1998). Recordings started in 1950 and relative water levels of the lake ranged almost 3 m largely due to a sudden upward jump during a period of exceptional rainfall between 1962 and 1964.

In the mid-1980s, Lake Victoria's ecosystem suddenly changed profoundly (Fig. 19.2) when the complex fish fauna with a high biodiversity of haplochromine cichlids was reduced concomitantly with the explosive increase of Nile perch that was introduced in the 1950s (Pringle, 2005). Since then, the lake's ecosystem has been simplified to consist mainly of four fish and a shrimp species (Witte *et al.*, 1992), though important remnants of the haplochromine

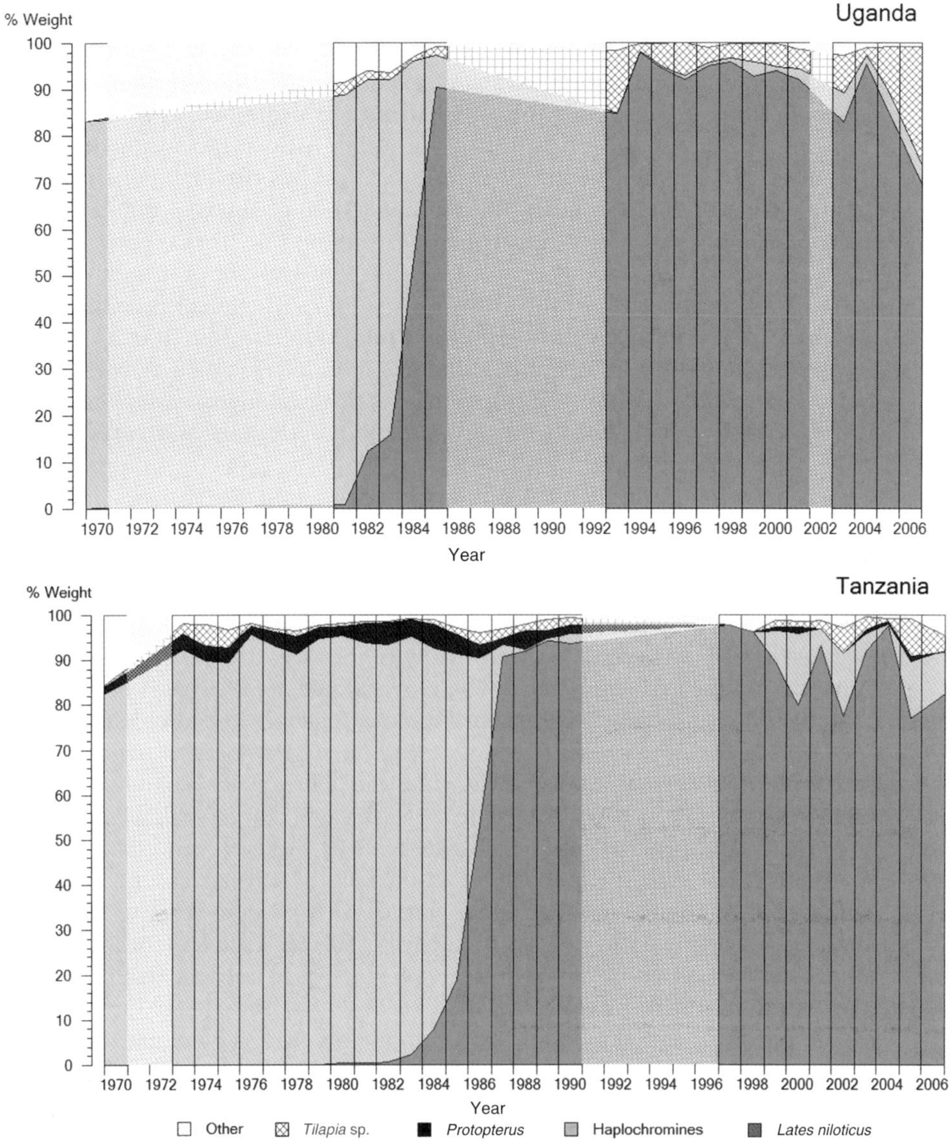

Fig. 19.2. Relative fish species composition based on experimental trawl survey data in Uganda (top) and Tanzania (bottom). Light-coloured areas lacking bars: no data available. *Rastrineobola* or *Caridina* do not appear in this diagram as they are not caught by experimental trawlers.

species flock remain. The sudden shift drastically changed the fishery, both in species composition and quantities. For instance, in the 1970s a bottom-trawl fishery catering for an emerging fishmeal industry in the southern part of the lake caught 1000–1500 kg/h of haplochromines (Witte and Goudswaard, 1985). From around the mid-1970s, however, the catch rates declined and from the mid-1980s a new fishery on Nile perch and existing fisheries on the small native cyprinid *Rastrineobola argentea* rapidly expanded. The new fisheries generated higher values and attracted many migrants to take up fishing, processing and trading – confirming the fast dynamics in fishing effort as a result of changing economic opportunities seen in other African freshwater fisheries (Jul-Larsen *et al.*, 2003; Zwieten *et al.*, 2003). Effort has grown exponentially in the past 40 years, and catches, now reaching a total of almost 1 million t/year, are dominated by Nile perch, *Rastrineobola* and the Nile tilapia (*Oreochromis niloticus*), the latter also introduced in the 1950s. Recently a new fishery started on a freshwater shrimp (*Caridina nilotica*) for meal reduction and poultry feed. The changes in the fishery of Lake Victoria resulted in a flourishing export industry of Nile perch, a domestic industry aimed at Nile tilapia and *Rastrineobola*, and the development of a fishmeal industry dependent on *Rastrineobola* and freshwater shrimp (Mkumbo *et al.*, 2002; Njiru *et al.*, 2005; Budeba and Cowx, 2007).

The invasion of the non-indigenous water hyacinth, *Eichhornia crassipes*, caused changes that perhaps were less drastic ecologically but disrupted local economies. Water hyacinth first appeared in Lake Victoria in 1989 (Twongo *et al.*, 1995). By 1995 in the Ugandan waters of Lake Victoria, stationary mats were estimated to cover 2200 ha along 80% of the shoreline; much of the Kenyan Winam (Nyanza) Gulf was covered, as were many other areas around the lake. Water hyacinth disappeared almost completely by the late 1990s. This is believed to be the result of mechanical and manual removal, changes in hydrological conditions during the 1997–1998 El Niño, ecological succession and the introduction of the weevils *Neochetina eichhorniae* and *N. bruchi* for biological control (Williams *et al.*, 2005). However, nutrient enrichment may cause resurgence and it is expected that periodic outbreaks may reoccur (Balirwa *et al.*, 2003).

Since the 1960s, the population in an area of 100 km around the lake has increased rapidly with an annual growth of 3.1% compared to the 2.5% African average (Fig. 19.3). One consequence of this increase, and the associated changes in land use in the drainage area, is the increased nutrient loadings to the lake. Increased phytoplankton production is observed from the 1930s onwards, which parallels the demographic developments and agricultural activities (Hecky, 1993; Verschuren *et al.*, 2002). Loss of deep-water oxygen from eutrophication started in the early 1960s (Hecky *et al.*, 1994; Verschuren *et al.*, 2001). Since then, the seasonally fluctuating area and volume of hypoxic and anoxic water layers below 20 m depth has increased significantly, leading to variable habitat availability and increasing fish kills of large Nile perch (Ochumba, 1990; Schofield and Chapman, 2000). This and other effects of eutrophication, such as decreased light penetration, may have contributed to the collapse of endemic fish stocks by eliminating suitable habitats (Seehausen *et al.*, 1997), as well as changes in the lower food web from changed phytoplankton and zooplankton communities.

The profound changes in Lake Victoria's ecosystem provided new opportunities for 1.3 million people (1999 estimate) to make a livelihood in the fishing industry (Witte *et al.*, 1999). As usual, however, threats to the fishery are perceived as imminent, and overfishing in particular is identified as a major

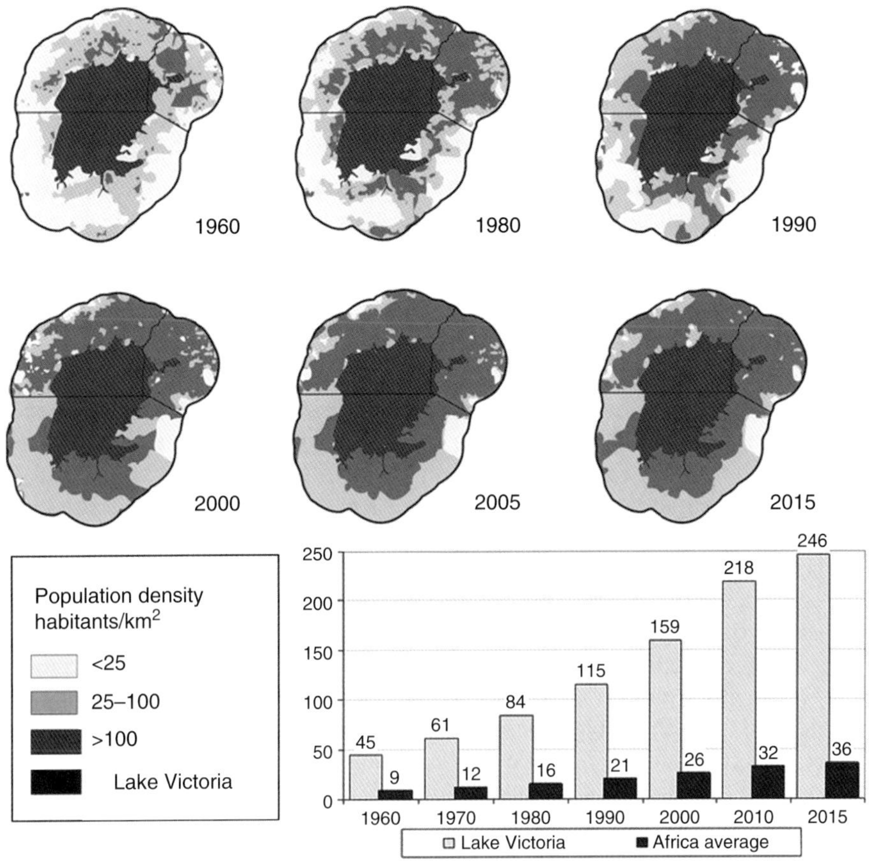

Fig. 19.3. Observed and predicted population density in an area of 100 km from the shore of Lake Victoria. The inset shows the increase in total population in habitants per square kilometre in the same area around the lake compared to the African average. (Redrawn from UNEP, 2006.)

threat among fisheries scientists. A collapse has been predicted if the increase in exploitation pressure is not halted (Pitcher and Bundy, 1995; Mkumbo *et al.*, 2002; Balirwa *et al.*, 2003; Cowx, 2005; Njiru *et al.*, 2005; Matsuishi *et al.*, 2006; Njiru *et al.*, 2007). In fact, overfishing on Lake Victoria from failures to control fishing effort have been reported since the early 1970s (Jackson, 1971; Fryer, 1973), and even the first scientific assessment of the Lake Victoria fishery (Graham, 1929) was a response to concern about declining catch rates. For others, however, eutrophication is considered as an even more risky driver of change because continued nutrient loading could lead to hyper-eutrophication if the loading is not addressed. The associated deoxygenation of large portions of the lake will have far-reaching consequences for the availability of suitable habitats of many fish species including Nile perch (Verschuren *et al.*, 2002; Silsbe, 2004; Silsbe *et al.*, 2006).

States, Trends and Processes: Indicator Selection

States and trends: indicators

The selection of relevant indicators that can support management decision making in Lake Victoria requires a process of exchange of knowledge between researchers and users of these indicators about the ecosystem, the characteristics of the fisheries and the social environment (Degnbol and Jarre, 2004; Rice and Rochet, 2005). In case of Lake Victoria, the main objective of the work that formed the basis of this chapter, was to synthesize information and knowledge gained during the Lake Victoria Environmental Management Program (LVEMP), a large World Bank-funded research project to describe current states, trends and processes affecting the lake. A series of indicators (Kolding *et al.*, 2005) were developed during the second half of 2005 through a number of workshops held in Uganda, Tanzania and Kenya with around 100 researchers and other stakeholders involved in the Fisheries Research and Management components of the LVEMP (Appendix 1). A similar synthesizing process had been initiated by the Water Quality components of the LVEMP (Mwanuzi *et al.*, 2005) that gave the required information to assess the impact of environmental changes on stocks and possible causes of species change.

To guide the process of indicator selection, a hierarchy of indicators was derived from an analytical framework, starting with the basic relation that catch (C) is a fraction ($F = f{\cdot}q$) of mean stock abundance or biomass (B) or biological production (P) over a period of time as:

$$C = F \cdot S(\bar{B}) = f \cdot q \cdot S(\bar{B}) = X \cdot E(P) \qquad (19.1)$$

with f = numerical fishing effort, q = catchability coefficient, and E = exploitation rate. Stock abundance and the amount of biological production are furthermore expressed as a function of the environment (S). In the formulas, the *state* of a stock, expressed as biomass (B), is a function of the environment or (eco)system *drivers* (S) and is affected as well by human *pressures* through fishing mortality ($F = f{\cdot}q$). Information then can be grouped as follows (Jul-Larsen *et al.*, 2003):

- System: indicators of ecosystem *drivers*, including habitat change, meteorological variables (wind and precipitation), physical variables (lake levels, light penetration and temperature) and chemical variables (nutrients and eutrophication).
- Stocks: indicators of *states* of fish stocks based on biomass levels, generalized life history parameters and production characteristics of fish species and communities.
- Exploitation: indicators of *pressures* through human activities in ecosystems including effort, selectivity and fishing patterns; drivers of fishing effort as investment in the fishery and effort allocation; trade, consumption and export of fish; as well as monitoring, control and surveillance indicators and other social and management indicators are included here.

Expectations in states and trends of fish stocks

Effects of fishing on single stocks are now well described and can be generalized from experience around the world. Thus, starting from the assumption that fishing pressure is the main driver of change in the Lake Victoria stocks (with focus on Nile perch), and that the lake has a constant carrying capacity, a number of expectations regarding the direction of state indicators - such as size-based indicators, life history parameters and fishery production indicators - can be expressed (Welcomme, 1999; Shin *et al.*, 2005, Table 19.2; Fig. 19.4). The expectations are that with increased effort in the fishery: (i) the stock of large spawners will decrease; (ii) the size of the exploited fish will decrease; (iii) the mean size in the population will decrease, both because the relative abundance of small fish caught increases while the relative abundance of larger individuals decreases; (iv) the slope of the abundance-size distribution will decrease - i.e. become more negative - while its intercept remains the same. The life history parameter length-at-maturity is also expected to decrease. Fishing mortality (F) will increase as well as the

Table 19.2. Expected, assuming fishing is the only driver, and observed long-term trend directions in length-based indicators of Nile perch caught in the experimental trawl fishery in Lake Victoria. $L_{max,\,10}$, $L_{5\%}$ and L_{mean} are respectively the mean length of ten largest specimens, the length of the exploited population defined as the 5% largest fish in the catch in numbers by length and the overall mean length of the catch. S_{asd} and I_{asd} are the slope and intercept of the abundance-size distribution; $CPUE_{\geq 40\,cm}$ and $CPUE_{<40\,cm}$ are catch rates of specimens larger and smaller than 40 cm (see text for extended definitions and Fig. 19.7 for observations). All indicators are in weights or frequencies per year. A grey area indicates an observed trend that does not conform to expectation.

Indicator	Expected trend with increased fishing effort	Observed trend (by 10 m depth categories)							
		Tanzania (1997–2006)				Uganda (1993–2005)			
		<10	<11–20	<21–30	>30	<10	<11–20	<21–30	>30
$L_{max,\,10}$	↘	↗	—	—	—	—	—	—	—
$L_{5\%}$	↘	↘	↘	↘	↘	—	—	↘	↘
L_{mean}	↘	↘	↘	↘	↘	↗	—	—	↘
S_{asd}	↘	—	—	—	—	—	—	—	—
I_{asd}	—	↗	↗	↗	↗	—	—	—	—
$CpUE \geq 40$ cm	↘	—	—	—	—	—	—	—	—
$CpUE < 40$ cm	—	↗	↗	↗	↗	↗	—	↗	—

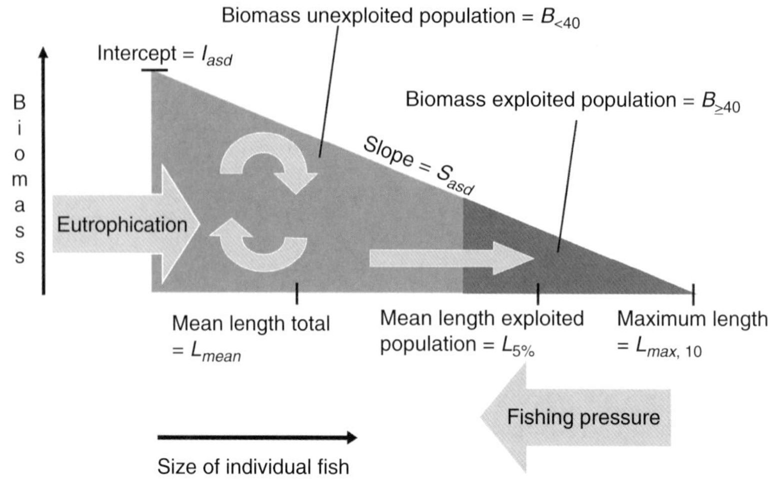

Fig. 19.4. A conceptual representation of the biomass-size distribution of a fish community or a fish population (e.g. of Nile perch) indicating the position of the various length-based indicators used in the analysis as well as the major external pressures and drivers (fishing and eutrophication) and internal processes (competition and predation) acting on, and occurring in, the fish community of Lake Victoria. (Adapted from Jul-Larsen *et al.*, 2003.)

exploitation rate (*E*) being the ratio between fishing mortality and total mortality (*F*/*Z*). Lastly, catch rates (*CPUE*) - as an index of biomass - of targeted specimen and size classes will decrease; while catch rates of non-targeted size classes remain the same, or decrease when the reproductive capacity of the stocks is affected. These expected trends under increased fishing pressure can then be examined and compared against observed trends and developments. Discrepancies will mean that the assumption is not met and that other processes play a role: e.g. the carrying capacity of the Lake Victoria ecosystem may not be in a steady-state. The assumption can then be tested by examining relevant indicators of ecosystem drivers related to eutrophication and climate change, and hypotheses on impacts of these drivers on stocks can be formulated.

Data: Origin and Analysis

System indicators

Meteorological data over Lake Victoria were obtained from the National Centres for Environmental Prediction-Department of Energy Reanalysis 2 (NCEP-DOE R2) that compiles and interpolates meteorological data from global data sets (Kanamitsu *et al.*, 2002). Water levels are measured near the dam near Jinja. Data and information on limnological indicators as nutrient concentrations, chlorophyll concentrations and spatial distribution of the oxycline were obtained from literature and are referenced in the text where appropriate.

Stock indicators

Particular attention was given to the construction and consolidation of data sets from trawl surveys. Experimental trawl surveys on the lake began in 1969 (Kudhongania and Cordone, 1974) and have since then been conducted on a fairly frequent basis in the three countries. Survey data from Uganda and Tanzania, representing independent replicates from the same lake, were used to construct time series of species composition, relative stock sizes and - for Nile perch - length-based indicators. Although they exist, no data were made available from Kenya. Potential sources of bias are: (i) in the two countries different research vessels have been used over time. Although a vessel effect may exist, this error is assumed to be systematic and relatively small. (ii) Changes in the trawl mesh sizes have occurred that are not accounted for very well. However, an unpublished experiment with the M/S Kiboko in Mwanza Gulf conducted by P.C. Goudswaard showed no difference in the size structure of Nile perch >30 cm between different cod ends of 25, 40, 60 and 100 mm and no difference in size structure of specimens >10 cm between cod ends of 20 and 40 mm (P.C. Goudswaard, personal communication, Dordrecht, 2006; data available with the authors). As all surveys in Uganda have used 25 mm cod ends, and most surveys in Tanzania have used 20–25 mm cod ends (with some additional hauls of 50 mm cod ends in the years 2000–2005), the relative changes in size structure can safely be compared over time from lengths >10 cm. (iii) Nile perch larger than 40 cm are thought to be fast enough swimmers to be able to escape the trawl[1] which means that the probability of catching large Nile perch decreases with its size. However, this probability can be assumed constant between surveys and therefore insignificant for indicators that describe relative changes over time (such as size structure and catch rates).

Nile perch in Lake Victoria has a wide depth range and has been caught down to 50–60 m depth. Though all sizes appear at all depths the relative abundance of smaller specimens of Nile perch is highest in shallower waters (Tumwebaze *et al.*, 2002). The species generally stays near the bottom or just above the oxycline (Goudswaard, 2006). As fishing on Nile perch mainly takes place in inshore waters of less then 20–30 m, large parts of offshore lake have no or low fishing pressure. In these offshore areas, the size structure of the Nile perch stock is therefore not directly influenced by fishing and changes in size of fish and biomass at different depths therefore will be indicative of different processes. Thus, apart from selective fishing, biophysical processes related to depth may play a role. Initially in the 1990s, experimental trawl hauls were stratified by depth categories of 10 m (an extensive description of the data is available with the authors) and we have therefore differentiated our analysis of length and catch rate indicators by 10 m depth categories up to 30 m and one category >30 m.

[1] If the sustained swimming speed of Nile perch is assumed to be around 2.5 body lengths/s (comparable to cod and pollack (Videler and Wardle 1991)) then large specimens have a good chance of escaping a trawl with a speed of 2 nautical mile/h – the speed of the RV Kiboko (P.C. Goudswaard, personal communication, Dordrecht, 2006).

Annual time series of length-based indicators (Shin *et al.*, 2005) were constructed from experimental trawl catches in Tanzania in the periods 1984–1990 (by RV Kiboko), where information on sampling and sites can be found in Goudswaard (2006), and 1997–2006 (except 2002), by RV Explorer. In Uganda the available data covered the period from 1993–2005 (except 2002) by RV Ibis. The indicators by 10 m depth categories (<11, 11–20, 21–30, >30) include:

1. Mean length of Nile perch in the experimental catches by year (L_{mean}) representing relative abundance of small and large individuals and indicating a change in number of large specimens and/or in recruitment.
2. Maximum observed length in the population defined as the mean of the ten largest specimens in the experimental catches per year ($L_{max,10}$)[2] – indicating the abundance of large spawners.
3. Mean length of the exploited part of the population, which for Nile perch is 40 cm and upwards. In the experimental trawls over all years and both countries, this size range largely coincides with the 5% largest-sized fish of the total number caught per year ($L_{exploited}$).
4. The slope (S_{asd}) of the relative abundance-size distribution of Nile perch, quantifying the relative overall size structure.
5. The intercept (I_{asd}) of the relative abundance-size distribution of Nile perch, quantifying the relative productivity level of a population (Gislason and Rice, 1998; Shin and Cury, 2004).
Furthermore, we examined:
6, 7. Changes in annual experimental trawl catch rates as indicators of the relative biomass of the unexploited ($CPUE_{<40\,cm}$) and exploited ($CPUE_{\geq 40\,cm}$) part of the Nile perch population.

The length-frequency and catch rates from the experimental fishery were adjusted to a haul of 30 min based on the parameters from a regression analysis between ^{10}log-transformed catch rates and trawl duration.

The time series of indicators were examined with the following separate slopes model:

$$Y_{ij} = \alpha + \beta \cdot year_{ij} + \alpha_i \cdot depth_{ij} + \beta_i\,(year \cdot depth_{ij}) + \varepsilon_{ij} iid \sim N(0, \sigma^2) \quad (19.2)$$

where the independent parameter Y_{ij} = indicators 1–7 above is regressed over year by $depth_{ij}$, $\alpha + \alpha_i$ is the intercept and $\beta + \beta_i$ is the slope over year at $depth_i$. Only significant variables and interactions were retained. Parameters for the models for catch rates (6, 7) and for slopes and intercepts of the abundance-size distribution (4, 5) were estimated through ordinary least squares

[2] The definition of 'largest' specimens is not self-evident, and the present definition suffers from dependency on the number of samples taken (i.e. hauls by depth) per year. We examined this by bootstrap re-sampling of the data set which did not give rise to changing the present analysis. Furthermore, we examined additional options: largest specimen caught per quarter and largest specimen caught per haul. The first option gave no different results than what is presented; the latter definition also included numerous hauls in which only small-sized fish (<20–30 cm) were caught and therefore did not serve our purpose.

estimation. Catch rates were transformed by a box-cox transformation with power = 0.25 to conform to the assumptions of least squares estimation in linear models. Abundance-size distributions (4, 5) were constructed for each year as $\log_2$ frequencies by $\log_2$-transformed length classes (cm). To reduce correlation between intercept and slope both axes were orthogonalized by subtracting the mean. Length frequencies are counts and for all length indicators (1–3) the model was performed as a general linear model with a Poisson error term and a log-link function. Over-dispersion was handled by estimating a scale parameter from the residual deviance. All dependent and independent variables were orthogonalized by subtracting the overall mean of the variable.

Catch and effort indicators

Total catch of Lake Victoria and catch by country were obtained from the respective fisheries research institutes and fisheries departments of the three riparian countries through mediation of the Lake Victoria Fisheries Organization (LVFO). Total effort in terms of numbers of boats and fishermen was obtained through frame surveys and the data were provided by the LVFO. Both catch and effort indicators have been published elsewhere (Reynolds *et al.*, 1995; Cowx *et al.*, 2003; Matsuishi *et al.*, 2006; Mkumbo *et al.*, 2007). Since 1998, changes in numbers of gear – gill nets, long-lines, other gear – are all based on frame surveys and catch assessment surveys in the three countries. Catch rates from the Nile perch fishery are calculated by dividing the total catch with the total number of boats or fishermen. This is based on the assumption that the increase in total number of fishermen is proportional to the increase in number of fishermen in the Nile perch fishery. Total numbers of boats and fishermen are interpolated where no data are available.

Examination of annual production and fishing mortality

Total annual biological production and fishing mortality rates of Nile perch from Uganda and Tanzania were estimated for the years where both *CPUE* (experimental trawling) and *Yield* (from the Catch and Effort Data Recording System) were available as follows: the catch rate (C/f) or catch per unit of effort (*CPUE*) is proportional to the standing biomass (B) with a constant factor of q (the catchability coefficient):

$$CPUE = \frac{C}{f} = q \cdot B \qquad (19.3)$$

An estimate of q can therefore be obtained from independent observations of *CPUE* and biomass (B). As average annual q can be assumed constant for experimental fishing, a series of mean annual biomass estimates can then be calculated from the mean annual *CPUE*. Next, the annual fishing mortality (F, which is an invariant measure of effort) can be calculated as it is simply defined as the fraction of the average biomass taken by fishing in a year:

$$F = q \cdot f = \frac{C}{\bar{B}} \tag{19.4}$$

Lastly, the total annual biological production can be estimated as the total annual mortality multiplied with the average standing biomass ($P = Z{\cdot}B$; Allen, 1971), where Z is the sum of the fishing mortality (F) and the natural mortality (M). Thus, in summary:

q (catchability coefficient) = $CPUE/B$ = constant

B_y (biomass) = $q{\cdot}CPUE_y$

F_y (fishing mortality) = $Yield_y/B_y$

M (natural mortality) = 0.35/year (Rabour *et al.*, 2003) = constant

P_y (total production) = $Z_y{\cdot}B_y = (F_y + M){\cdot}B_y$

where the subscript y indicates the annual estimate of the variable. Mean densities (t/km^2) were obtained in year 2000 from acoustic and swept area surveys (Cowx, 2005; Mkumbo *et al.*, 2007) for Uganda (9.44 t/km^2) and Tanzania (10.18 t/km^2). Multiplied by the areas covered by the surveys this resulted in a biomass estimate of 276,026 t (Uganda) and 353,042 t (Tanzania). Mean experimental trawl catches for the year 2000 were 228 kg/h (Uganda) and 253 kg/h (Tanzania). From these the experimental trawl catchability coefficients can be calculated.

Key Processes: Trends and Developments in States, Pressures and Drivers

State changes 1: introduction of Nile perch and developments in the fish community

Nile perch was introduced on several occasions between 1954 and the early 1960s with the aim to develop a (sport) fishing industry (Welcomme, 1988; Pringle, 2005). Since the sudden shift in fish communities in the 1980s (Fig. 19.2), this species now dominates the fishery in addition to *R. argentea*.[3] While the Nile perch spread relatively early all around the lake (Kudhongania and Cordone, 1974; Pringle, 2005), the dramatic population upsurge started in the northeastern part of the lake from around 1979 when the experimental trawlers in the Winam Gulf caught 46 kg/h, with Nile perch of all stages from juveniles to adults. Catches of ≥50 kg/h were observed in Uganda by 1983 and in Tanzania in 1984 (Fig. 19.5), while juveniles of <10 cm length appeared for the first time in Tanzania only in 1985 (Goudswaard, 2006). Catches above 100 kg/h were reached in Uganda by 1984 and in Tanzania by 1987. Goudswaard *et al.*

[3] *R. argentea* is not caught by experimental trawls and hence does not appear in the graphs based on the data collected through this method. The increasing abundance is monitored by hydroacoustic surveys.

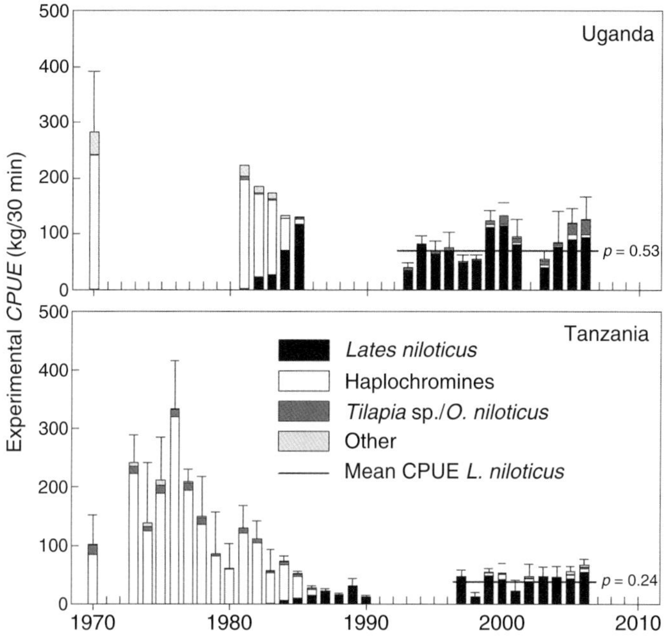

Fig. 19.5. Standardized catch per unit effort (kg/30 min) in experimental trawls in Uganda and Tanzania for all stations less than 40 m depth. Experimental trawlers do not catch *Rastrineobola* or *Caridina*, hence the apparently low relative abundance of the total stocks following the collapse of the haplochromines; *p*-values indicate the significance of the trend line for Nile perch: as trends are non-significant the mean over the time series is shown.

(2008) hypothesizes that Nile perch started to migrate from the Winam Gulf after local depletion of the main prey, haplochromines, due to fishing. Thus, while the Nile perch upsurge started in the Kenyan part of the lake as a result of either reduced predation on eggs and larvae of Nile perch, and/or an increased availability of zooplankton, shrimp and insect larvae for small (<10 cm) Nile perch – both perceived to be the result of reduced haplochromine stocks from fishing – the rest of the lake was then assumed to be occupied largely through migration and competitive advantage of the larger predator. Alternatively, ECOSIM models (Kitchell *et al.*, 1997; Walters and Kitchell, 2001) have suggested that the initial slow invasion of Nile perch was possibly inhibited by competition/predation by the original fish community of the lake. Accordingly, population growth rate increased only when Nile perch became abundant enough to depress the other community components, which then allowed increases in *Caridina* and *Rastrineobola* that subsequently became the Nile perch's dominant food. These authors hypothesize that by depressing the haplochromine community, the Nile perch also maintains the conditions to protect its own juveniles from predation, a process termed 'cultivation effect'. What caused the condition for Nile perch to overcome competition/predation and initiate the process of depressing haplochromines is, however, still unexplained.

The original fish community in Lake Victoria, prior to the sudden shift to a sub-littoral and pelagic ecosystem dominated by Nile perch, consisted mainly of a large number of endemic haplochromines occupying all of trophic positions (Greenwood, 1974; Witte, 1984). After the faunal shift, the detrivorous and phytoplanktivorous haplochromines, that constituted more than 40% of the ichthyomass, have largely been replaced by the small freshwater shrimp *C. nilotica*, while *R. argentea* replaced the zooplanktivorous haplochromines,[4] and Nile perch replaced the piscivorous catfishes and haplochromines. Prior to the Nile perch explosion, the introduced *O. niloticus* had already replaced the indigenous tilapiine species *O. esculentus* and *O. variabilis* that disappeared earlier (Fryer and Iles, 1972; Witte *et al.*, 1992; Kudhongania and Chitambwebwa, 1995). Decreases in overall abundance of haplochromines prior to the Nile perch boom have been reported from the Winam Gulf (Kenya), from Uganda and from Mwanza Gulf (Tanzania; references in Goudswaard, 2006), though experimental catch rates from Mwanza Gulf showed considerable fluctuations around the long-term (15 year) decline (Fig. 19.5). In recent years, Nile perch has become slightly less dominant and a number of species have increased in abundance (Fig. 19.2). This shift has been attributed to the selective harvesting of larger Nile perch which may have lead to reduced predation and hence the recovery of some haplochromine species (Balirwa *et al.*, 2003; Witte *et al.*, 2007b). Since the faunal shift the populations of zooplanktivorous *R. argentea* and *C. nilotica* have become the dominant prey of Nile perch (Wanink, 1998; Goudswaard *et al.*, 2006).

Recently, a resurgence in zooplanktivorous haplochromines has been observed as by-catch in the *R. argentea* fishery (Budeba and Cowx, 2007; P.A.M van Zwieten, personal observation, Mwanza, 2006; P.C. Goudswaard, Dordrecht, 2007 and F. Witte, Moutreal, 2007, personal comments), and Nile perch from 25cm is now preying on haplochromines again (F. Witte, personal communication, Moutreal, 2007). While the return of (some of) the haplochromines has been explained by the putative decrease in numbers of large Nile perch as a result of harvesting (Balirwa *et al.*, 2003; Witte *et al.*, 2000, 2007b; but see Table 19.2 and Fig 19.7 which show no such indication), the alternative hypothesis that increased eutrophication has resulted in shifts to species better adapted to eutrophic conditions has not yet been considered. This hypothesis is particularly interesting as stocks of some *pelagic* zooplanktivorous haplochromine species (e.g. *Haplochromis (Yssichromis) pyrrocephalus*) have increased. *H. pyrrocephalus* may adapt more easily to low ambient oxygen as a result of eutrophication as it presently has a gill surface area that is 70% larger than those from specimens sampled in the 1970s (Witte *et al.*, 2007a). These observations raise the question whether recruitment of Nile perch stocks again could become suppressed by competition/predation by haplochromines (*sensu* Walters and Kitchell, 2001).

[4] It should be noted though that the pelagic haplochromines (*Yssichromis* sp.) may not have been decimated to the same extent as the demersal haplochromines (Goldschmidt *et al.*, 1993), and they were by far the most abundant species in offshore pelagic beam trawl hauls in 1995–1996 (Tumwebaze, 1997) and 2000 (Tumwebaze *et al.*, 2002).

Human pressures: fishery catch and effort, resource opportunities and trade

In the early 20th century, the fishery was entirely on the inshore tilapiine species (Graham, 1929). Later, from the mid-1960s, a small trawl fishery developed on the haplochromines, which lasted to the faunal shift in the mid-1980s. The faunal shift represented new resource opportunities for the local fishermen and all four dominant species - Nile perch, Nile tilapia, *Rastrineobola* and *Caridina* - have been exploited since. By the early 1990s, total yields had increased by a factor of 6 since the start of the Nile perch upsurge in the 1980s and recently it has reached a level of nearly 1 million tonnes due to the rapidly expanding Dagaa/Omena (*Rastrineobola*) fishery (Fig. 19.6B). The Nile perch fishery reached a maximum around 300,000 t in 1990 and has since fluctuated around 230,000 t forming the basis of a large export industry. Trade patterns of Nile perch in Kenya shifted around 1984/1985 from domestic to international supply and at present around 75% of the reported total landings of Nile perch are exported.

The fishery on *R. argentea* first catered for local consumption and a few years later for a developing fishmeal industry (Abila and Jansen, 1997). It now forms the dominant part of the catch and reached more than 500,000 t in 2006. The fishmeal industry also started using the freshwater shrimp *C. nilotica*, which resulted in a new separate fishery for which no catch data are yet available. The fishery is driven not only by the new resource opportunities, but also by specific demands of the industry. For example, fish filleting factories in all countries are legally required to take only Nile perch within the 'slot size' of fish larger than 50 cm and smaller than 85 cm. These sizes are therefore selectively targeted by the export fishery.

The upsurge of new resources is reflected in the fishing effort: in 2004, the number of boats was approximately 51,500 (six times higher than in the 1970s), while number of fishermen increased 4.5 times to 166,000 (Fig. 19.6A; Matsuishi *et al.*, 2006). A clear change in trend can be seen from 1988 onwards when the

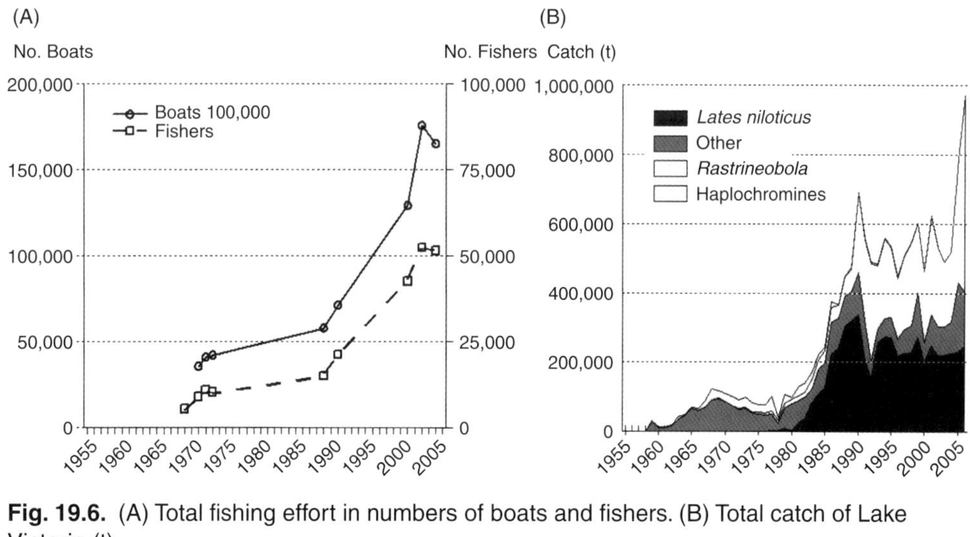

Fig. 19.6. (A) Total fishing effort in numbers of boats and fishers. (B) Total catch of Lake Victoria (t).

rate of increase in numbers of boats, gears and fishers has accelerated. Similar patterns are found in all riparian countries. The accelerated growth in effort started 5 years later in Tanzania than in Uganda and Kenya, but now seems to stabilize in all countries. From around 1998, gear use shifted, with large increases in gill nets - in particular of the mesh sizes (5–6 in) and long-lines used in the Nile perch fishery and in particular in Tanzania and Uganda (Mkumbo *et al.*, 2007), whereas the numbers in other gears stabilized or decreased. The use of long-lines is associated with a more offshore effort allocation indicating a shift of the Nile perch fishery to deeper waters (O. Mkumbo, personal communication, Montreal, 2007).

State changes 2: catch rates and length-based indicators

Overall catch rates from experimental trawlers as well as from the fishery - the latter mainly comprising the inshore waters up to 20–30 m depth - seem to indicate a rather limited influence of the fishery on stock dynamics (Fig. 19.5). The average annual abundance of Nile perch in the experimental trawl surveys has fluctuated since 1993 in Uganda with a mean of 147 ± 51 kg/h (n = 1406 hauls, between 17 and 252 hauls/year) and since 1996 in Tanzania with a mean of 234 ± 76 kg/h (n = 785 hauls, between 15 and 232 hauls/year), but in both lake regions no significant changes in the stock sizes have occurred over the past decade despite the increase in effort (Fig. 19.6A). Even when excluding the non-exploited Nile perch smaller than 40 cm - constituting 95% of the number of fish in the experimental trawl catches - there are no significant changes in the relative abundance over the past decade (Fig. 19.7, $CPUE \geq 40$ cm) although large fluctuations in biomass have taken place over shorter time periods as has also been observed in acoustic surveys (Getabu *et al.*, 2003).

Catch rates from the fishery are less equivocal as the reliability of the catch statistics for Uganda and Tanzania has been questionable since the mid-1990s (Cowx *et al.*, 2003). For Kenya no significant changes were observed in annual average catch per boat between 1985 and 2003, but during the past 5 years a short-term decreasing trend has been observed (Fig. 19.8). During the last hydroacoustic survey in 2007, Nile perch were not observed inside Winam Gulf in Kenya, which again was filled with water hyacinth (Williams *et al.*, 2005; R. Kayanda, personal communication, České Budějovice, 2007). These recent trends have lead to much concern and speculation about overfishing (e.g. Balirwa *et al.*, 2003; Njiru *et al.*, 2007) while the alternative hypothesis that the changes are due to hyper-eutrophication and highly deteriorated water quality in the Kenyan waters is hardly considered, nor is it regarded as a fisheries management problem. Similarly, in Tanzania catches per boat showed no trend between 1985 and 1999 (Fig. 19.8), but since then the catch and effort recording system has collapsed and no reliable figures on Nile perch catches are available. Since around 1995, catches per fishermen in both countries and in all data sets vary between 2 and 4 t/year, which is comparable to the average individual catch in other African freshwater fisheries (Jul-Larsen *et al.*, 2003; Fig. 19.9).

Trend analysis in the length-based indicators compared with the expected trends (steady-state and only effort as driver) equally shows a limited influence

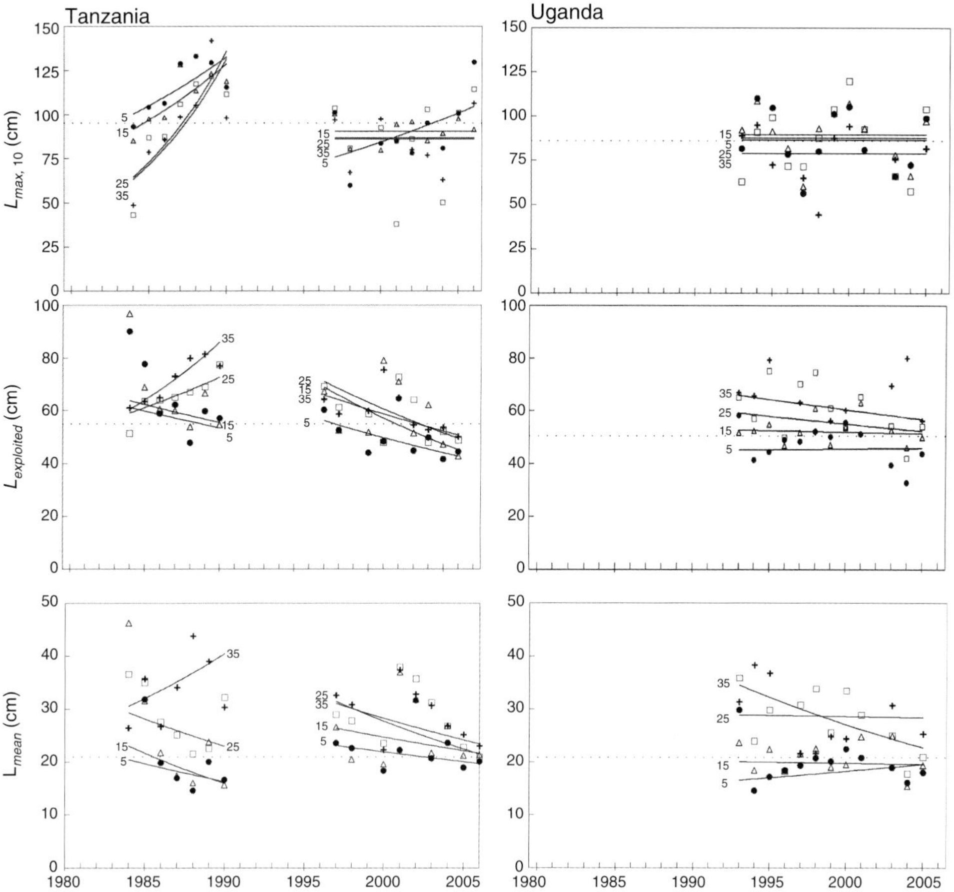

Fig. 19.7. Trends in length-based indicators of Nile perch caught in experimental trawls in Tanzania (1984–1990 and 1997–2006) and in Uganda (1993–2005) at four depth ranges resulting from an ANOVA with a separate-slopes model. When trends are non-significant ($p < 0.05$) the mean over the period and depth category examined are shown. When differences between depths are non-significant the overall trend or the mean over the period is shown. Data points represent annual mean catches by depth. Analyses were carried out on the total number of standardized 30 min hauls/year and depth category (see further Appendix 2 and text). Dotted lines represent the mean of an indicator over the period(s) examined by country. $L_{max,\ 10}$, $L_{exploited}$ ($L_{5\%}$) and L_{mean} are respectively the mean length of ten largest specimens, the length of the exploited population defined as the 5% largest fish in the catch and the overall mean length of the catch. S_{asd} and I_{asd} are the slope and intercept of the abundance-size distribution; $CPUE_{\geq 40\,cm}$ and $CPUE_{<40\,cm}$ are catch rates of specimen larger and smaller than 40 cm (see text for extended definitions).

of the fishery (Table 19.2, Fig. 19.7 and Appendix 2). In addition, two general observations can be made from these analyses. First, overall means of the three length indicators (L_{mean}, $L_{max,\ 10}$, $L_{5\%}$) calculated from opposite parts of the lake are highly similar, which indicate that processes acting on Nile perch population size structure are similar across the lake. Second, there are large inter- and

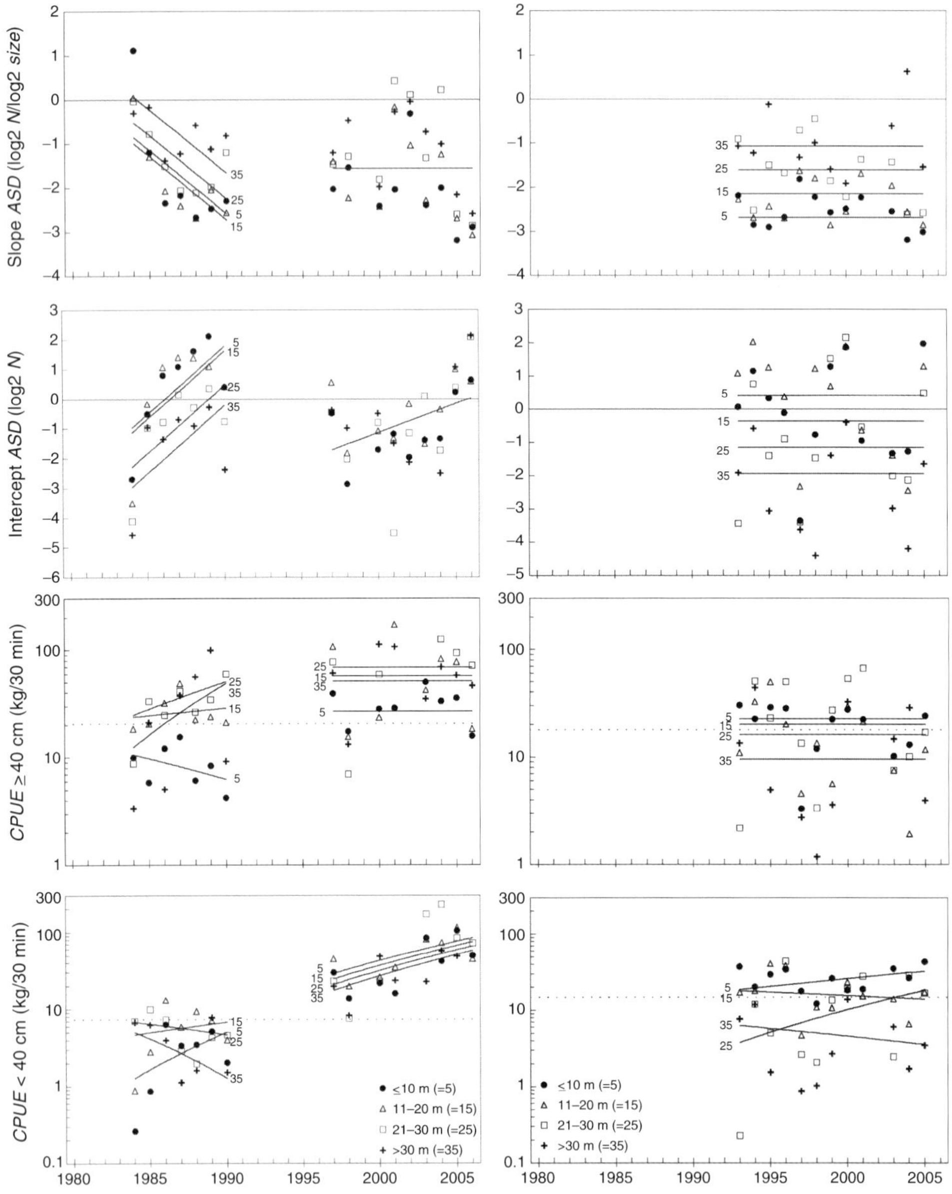

Fig. 19.7. Continued

multi-annual fluctuations in all indicators, but in particular those that show changes in stocks of large Nile perch ($L_{max,\ 10}$ and $CPUE_{\geq 40\,cm}$) that would be expected to show lower inter-annual variation. The observed large inter-annual fluctuations in these indicators could be the result of fluctuating habitat availability due to fluctuations in the extent of the deep-water anoxic layer (see later).

The period of the shift to a Nile perch-dominated ecosystem is covered by the Tanzanian data set between 1984 and 1990. The mean length of Nile perch

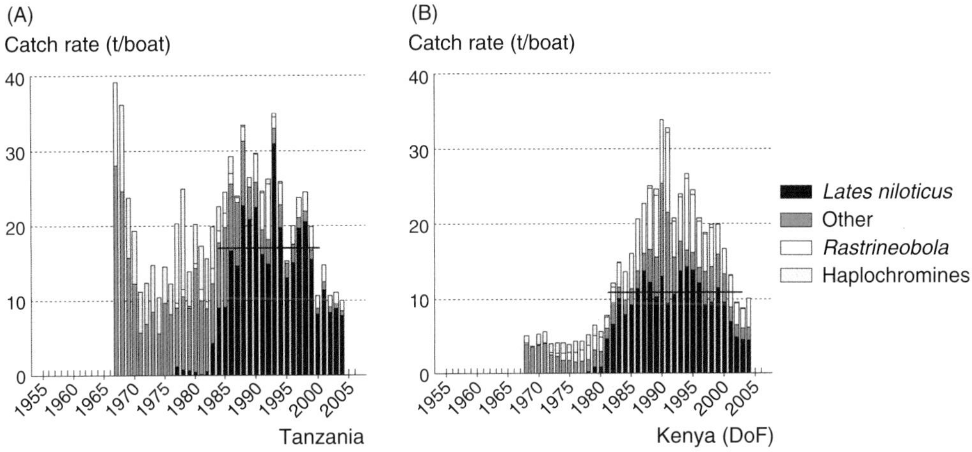

Fig. 19.8. Developments in catch rates of the fishery (t/year/boat) in Tanzania and Kenya. Note that in Tanzania catch rates of all species categories including *Lates niloticus* are severely underestimated after 1999. (From CEDRS Tanzania and Kenya Department of Fisheries.) Black line: mean catch of Nile perch over the years indicated by start and end points of the line. Tanzania: p = ns, mean = 17.9 t/year/boat; Kenya: p = ns, mean = 10.5 t/year/boat.

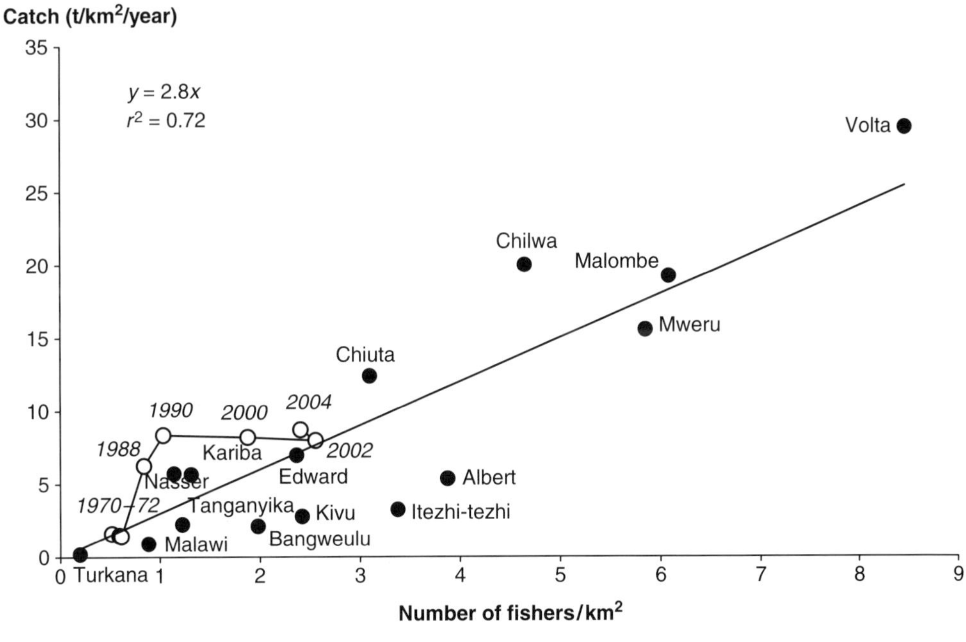

Fig. 19.9. Catch rates plotted versus effort density in 14 African lakes (data from the period 1989–1992). The trend line indicates an average yield of about 3 t/fisher/year irrespective of water body. Superimposed is the development in Lake Victoria between 1970 and 2004, which shows how productivity has increased over time concurrently with the increase in effort. After an initial boom in production in the late 1980s and early 1990s, the catch rates are now approaching the overall mean again. (Adapted from Jul-Larsen *et al*., 2003.)

(L_{mean}) up to 30 m depth, as well as the mean length of the exploited population ($L_{5\%}$) up to 20 m depth decreased. This would be consistent with the expected effects of fishing as exploitation on Nile perch increased rapidly over the same period. The decreasing abundance of Nile perch larger than 40 cm ($CPUE_{\geq 40\,cm}$) at ≤10 m and the relatively stable abundance at 11–20 m would point in the same direction because effort would be highest close to shore. However, mean length at depths >30 m, the length of the exploited population ($L_{5\%}$) at depths >20 m, the catch rate of the exploited population at depths >10 m and all other indicators are not consistent with the expectations. The size of large spawners ($L_{max,\,10}$) increases at all depths and in particular in deeper waters (>20 m) that also show a strong increase in the mean length (L_{mean}) and in length- and catch rates of the exploited population ($L_{5\%}$, $CPUE_{\geq 40\,cm}$). Catch rates of the unexploited part of the population ($CPUE_{<40\,cm}$) increase rapidly in shallow waters <10 m, are relatively stable in waters between 11 and 30 m, and decrease in deep waters, indicating that in most areas of the lake except deeper waters recruitment increases (i.e. increased abundance of juveniles) or is stable. Increasing intercepts (I_{asd}) and decreasing slopes (S_{asd}) of the abundance-size distributions at all depths, together with an increasing mean length of large spawners ($L_{max,\,10}$) and a stable or increasing mean length and abundance of the exploited population ($L_{5\%}$, $CPUE_{\geq 40\,cm}$), indicate an expanding overall abundance of Nile perch population – in particular juveniles and sub-adults – over this period. In conclusion, the decade immediately following the Nile perch boom can be characterized as a period of a rapidly expanding Nile perch population at all depths, with some evidence of fishing limited to inshore waters ≤10 m.

During the last period covered in Tanzanian waters, between 1997 and 2006, mean length and average exploited length decrease at all depths, but for both indicators the decrease is stronger in waters deeper than 20 m, where exploitation is less heavy compared to the shallower waters. Under steady-state conditions these observations would be interpreted as signs of the impacts of fishing. However, contrary to expectation, the size of the large spawners remains the same at all depths at around 88 cm and increases by almost 30% from 77 to 99 cm in depths ≤10 m. Also the catch rate of the exploited part of the population ($CPUE_{\geq 40\,cm}$) remains the same at all depths during this period – being the lowest in shallow waters. Furthermore, the abundance of the unexploited population ($CPUE_{<40\,cm}$) has increased by 2.9–3.3 times compared to 1997, which indicates that Nile perch stocks have no recruitment problem. Last, the stable slope but increasing intercept of the abundance-size distribution strongly suggests that the overall productivity of the Nile perch stocks has increased. The observed decreasing mean lengths and average exploited lengths are therefore not a result of lower abundance of large Nile perch but are due to a larger abundance of small Nile perch.

The Ugandan time series of size and abundance indicators between 1996 and 2006 largely corroborate the conclusions for the Tanzanian waters over the same period. They even indicate less evidence for fishing impacts. All length indicators are stable in shallow waters and the mean length in shallow waters <10 m even show a slight but significant increase of 15% (from 16 to 19 cm). As in Tanzania, catch rates of the exploited part of the population ($CPUE_{\geq 40\,cm}$) are stable at all depths, while catch rates of the unexploited part of the population ($CPUE_{<40\,cm}$) remain stable or increase. Slope and intercept of the abundance-size distribution remain the same. Thus, the increase in fishing effort in this part of the lake seems to be largely balanced by a concomitant increased productivity of the Nile perch stocks. Decreases in length are only observed in the less heavily fished deeper waters: a decrease of 13% in average exploited length (from 60 to 52 cm) in the 21–30 m depth stratum and a decrease of 15% (65 to 53cm) in waters >30 m, and a steep decrease in the mean length of 30% (36–22 cm) in waters >30 m.[5]

Mkumbo and Ezekiel (1999) observed that the size (*TL*) at first maturity (Lm_{50}) in Tanzania had decreased from 76–80 cm (males) and 86–90 cm (females) - quite high for Nile perch in comparison with other stocks (Hopson, 1972; Ogutu-Ohwayo, 1990) - in the late 1980s to 53 cm and 55 cm, respectively. Though they mention changed limnological conditions as a possible cause, they attribute this change mainly to increased fishing pressure on large Nile perch. Similarly, Njiru *et al.* (2007) observe a steady decrease in Nile perch size at maturity (from 60 cm males and 62 cm females to 31 and 22 cm, respectively) in the Kenyan waters between 1998 and 2005 and attribute this to increasing exploitation. Also decreased size of first maturity in Nile tilapia and *Rastineobola* in Kenya is attributed to intensive fishing (Manyala and Ojouk, 2007; Ojouk *et al.*, 2007). The observed decreases in size at maturity, however, could just as well be a function of decreased water quality and lowered ambient oxygen (Pauly, 1984) as has been observed both under natural conditions and experimentally for Nile tilapia (Kolding, 1993; Kolding *et al.*, 2008).

Human pressures leading to eutrophication: demographic and land-use change

The Lake Victoria basin has undergone a rapid increase in population and agricultural production in the past century, especially after the Second World War. From 1960 onwards, the population in a region of 100 km from the lake shore (UNEP, 2006), rose from approximately 10 million to close to 40 million (Fig. 19.3). This strong regional growth has resulted in urbanization, increased deforestation and increased agricultural production associated with increased utilization of fertilizers and biomass burning (Verschuren *et al.*, 2002; Mwanuzi *et al.*, 2005). Tamatamah *et al.* (2005) estimate that 55% of the external deposition of phosphorus on the Lake is from wet and dry atmospheric deposition as the result of the widespread practice of biomass burning and ensuing dust before planting in the rainy season. This has been shown to occur in Lake Malawi (Bootsma *et al.*, 1996, 1999) and Lake Tanganyika (Langenberg *et al.*, 2003) as well. The accelerated land-use processes have increased the supply of nutrients to the lake over the last 50 years and caused it to become increasingly eutrophic (Hecky, 1993; Verschuren *et al.*, 2002).

[5] The northern part of the lake has numerous islands providing protection from winds, which results in more persistent stratification and less aeration of the deeper water column (R. Hecky, personal observation, Montreal, 2007).

System drivers: eutrophication, phytoplankton production and anoxia

Between 1969 and 1993, nutrient loadings to the lake from the surrounding catchment area and the primary productivity in the lake have increased by a factor of 2 along with a six- to tenfold increase in algal biomass in both nearshore and offshore environments (Hecky, 1993; Mugidde, 1993). As a result of the increased algal biomass, light penetration is reduced (Mugidde, 1993; Kling *et al.*, 2001; Fig. 19.10). At present, optimal nutrient concentrations to support fisheries therefore may have been exceeded as there is evidence of light limitation on algal productivity (Silsbe, 2004; Silsbe *et al.*, 2006). The diatom populations collapsed in the mid-1980s and were replaced by cyanobacteria as the dominant group of algae (Kling *et al.*, 2001; Verschuren *et al.*, 2002). The increasing abundance of cyanobacteria reduced light penetration and the depth reduction of the benthic algal distribution that supported the diverse littoral haplochromine community has lead to altered food webs and caused seasonal deep-water hypoxia (Hecky *et al.*, 1994; Silsbe, 2004; Fig. 19.11). The increased eutrophication has resulted in a lake environment with favourable conditions for water hyacinth (*E. crassipes*) along the shores, high concentrations of cyanobacteria all over the lake, and much reduced O_2 concentrations in the deeper parts.

Seasonal and decadal fluctuations and trends in rainfall, water levels and wind stress contribute to changing input of nutrients from the catchments. Water-level fluctuations have a seasonal range of 1.21 m, but have a large variation between years. A sudden level rise in the early 1960s after unusually heavy rains (Fig. 19.12; Flohn, 1987; Sene and Plinston, 1994) caused extensive flooding of the shoreline, drowning shoreline swamps and releasing nutrients from flooded soils and decomposing vegetation (Welcomme, 1970; Azza, 2006). Although water levels since 1964 have been receding with a rate of 0.032 m/year, water levels in the following decades remained – with decadal fluctuations – at the

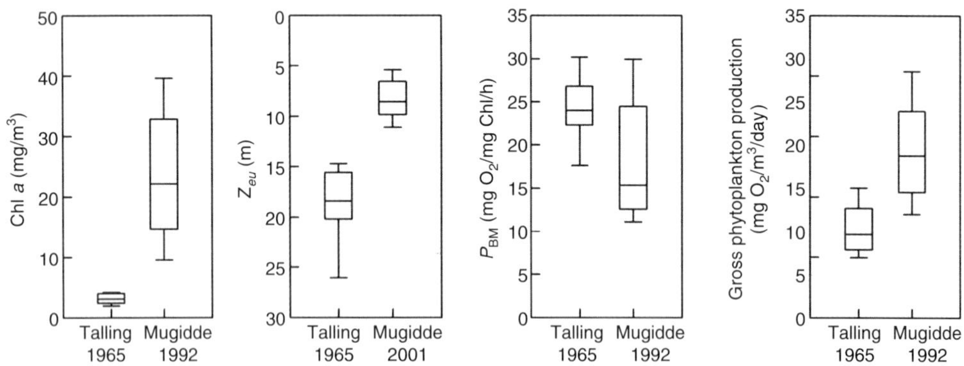

Fig. 19.10. Comparison of historic and modern-day measurements of gross phytoplankton production and its constituents. Gross phytoplankton production (PPG; right panel) is a product of the phytoplankton biomass (chl; left panel), the euphotic depth (Z_{eu}; middle left panel) and the photosynthetic irradiance parameters (P_{BM} and α_B ($mgO_2/mgChl/mol/m^2$: not shown)). (From Silsbe *et al.*, 2005, based on data in Talling, 1965; Mugidde, 1992, 2001.)

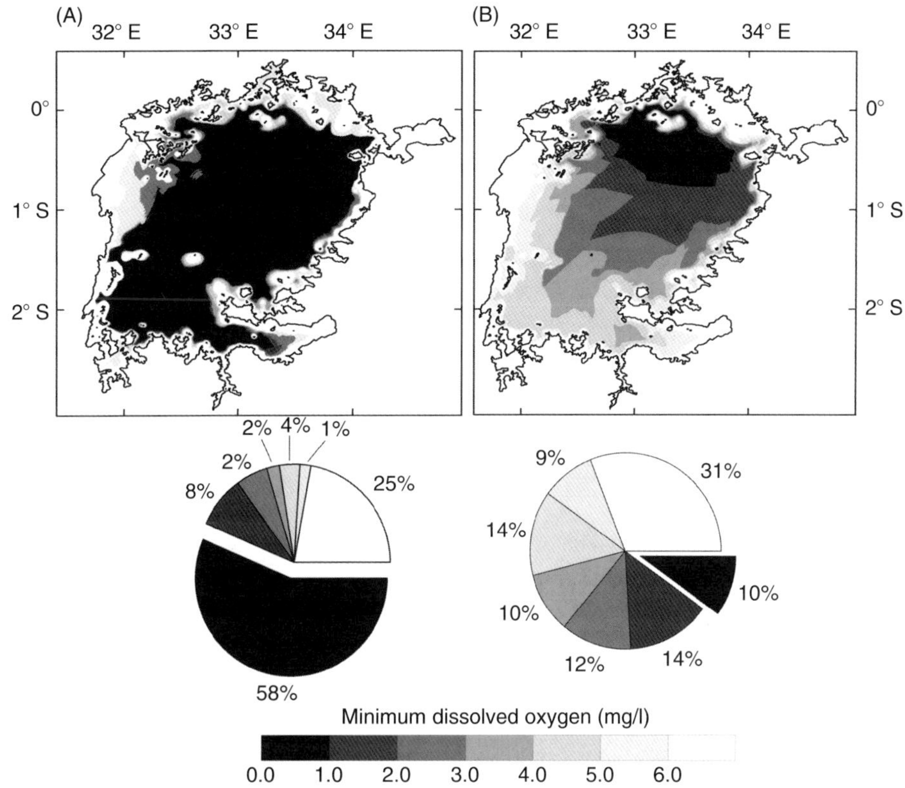

Fig. 19.11. Planimetric distribution of the minimum dissolved oxygen (DO) concentrations in the water column in (A) February 2000 and (B) February 2001 with pie charts for both spatial surveys showing dissolved oxygen concentrations as a percentage of total lake volume. During each survey DO was measured through depth at 52 stations distributed over the entire lake, and the data was spatially interpolated using a Krieging algorithm and a bathymetric map. Both surveys coincided with a period of high annual thermal stratification and deoxygenation. Identical surveys conducted in August 2000 and 2001 when the lake was largely unstratified show no DO below 4 mg/l but with 32% and 36% of total lake volume between 4 and 6 mg/l in each year, respectively.

elevated level maintained by the precipitation in the region. At the same time the average wind stress remained high resulting in increased deep-water mixing. Phosphorus concentrations began to rise post Second World War (Fig. 19.11) and have continued to rise since the early 1960s by approximately 1 μg/l/year to the present (Mwanuzi *et al.*, 2005). The shift to dominance of cyanobacteria also increased the nitrogen loading to the lake (Mugidde *et al.*, 2003). Both processes resulted in increased nutrient availability in Victoria and drove eutrophication. Phosphorus loadings increased dramatically from the early 1950s onwards, increasing the demand for dissolved Si, an essential element for diatom production. The increased diatom productivity resulted in increased sedimentation of Si (Fig. 19.12) and caused dissolved Si concentrations to fall to low levels favouring the succession to cyanobacteria (Hecky, 1993; Kling *et al.*, 2001). Increased

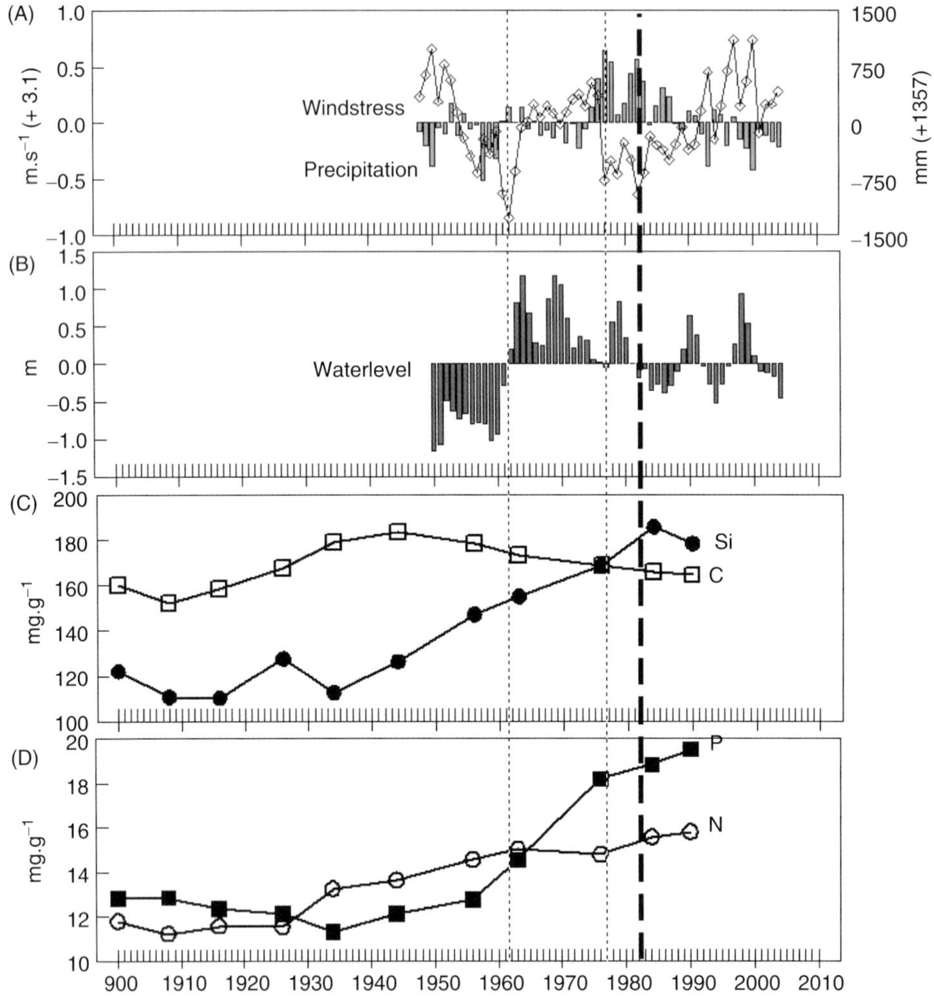

Fig. 19.12. (A) Annual deviations of precipitation and the average annual wind stress over Lake Victoria since 1948. (B) Annual deviation of the long-term maximum relative water level of Lake Victoria measured at Jinja (Uganda) since 1950. (C and D) Nutrient concentrations in sediment cores from 1900 onwards. C = carbon, Si = silica, N = nitrogen, P = phosphorus. Vertical light dotted lines: period of high wind stress and high water levels. Thick dotted line: start of the Nile perch boom in Uganda. (Panels C and D redrawn from Hecky, 1993.)

loading of phosphorus and depletion of Si preceded the Nile perch boom that started in 1982 and may have facilitated the boom by increasing the availability of invertebrate prey (Hecky, 1993).

Seasonal changes in wind stress and wind direction drive the seasonal mixing (turnover) pattern of the lake that can be associated with seasonal fish kills (Ochumba, 1990), while internal seiches can cause deoxygenation of near surface waters (Hecky *et al.*, 1994) at any time during the stratified season. Whether the incidence of these seasonal kills has increased over the past two decades is not known but can be expected (Goudswaard, 2006). However, changes in wind stress and wind direction may have larger-scale habitat effects as well. From the mid-1970s to the mid-1990s, average annual wind stress was low (Fig. 19.12), which may lead to longer and more stable anoxic layers covering larger areas of the lake, thereby reducing the size of suitable habitat for Nile perch. The size of the anoxic layer varies considerably between years (Fig. 19.11); thus, the large bottom trawl-based inter-annual fluctuations in biomass of large Nile perch could therefore be the result of contracting and expanding populations as a result of changed habitat availability.

Physicochemical and optical conditions in the fringing wetland areas have also changed considerably between 1960 and the 1990s. The increased eutrophication initially resulted in favourable conditions for water hyacinth (Azza, 2006) and the habitats formed were thought to enhance the recovery of some haplochromine species as Nile perch appears to avoid the hyacinth mats (Chapman *et al.*, 1996, 2002; Njiru *et al.*, 2002). Currently, however, continued eutrophication seems to degrade euhydrophyte-dominated littoral habitats and impair the functioning of marginal wetlands as fish habitats (Azza, 2006).

Alternative Explanations and the Future of Lake Victoria's Fisheries

From the major changes in the state of the Lake Victoria ecosystem, as well as from the general ecological expectations that are known from similar situations (Jul-Larsen *et al.*, 2003; Kolding and van Zwieten, 2006), a series of logical inferences can be derived from the combined set of ecosystem and fisheries indicators. In particular, there is a need to explain *why there appears to be a limited documentable impact on the exploited stocks* from the increasing fishing pressure on the lake, which is in contradiction with both the expectations (Table 19.2) as well as the predictions derived from various stock assessment methods (Table 19.1).

Comparing the observations with the general expectations, the only possibility for a constant or even increasing biomass under increased exploitation is if production has increased. From the ecosystem drivers, there are clear indications that this has happened. Thus, the chain of events is: the introduction of the Nile perch, together with a general increase in eutrophication, has lead to a loss of biodiversity (Goudswaard, 2006; Witte *et al.*, 2007a) resulting in a dramatic simplification of the ecosystem, *preceded* by accelerated primary productivity (Hecky, 1993; Mugidde, 1993; Silsbe, 2004) and followed by an increased survival of juvenile Nile perch (Table 19.2; Fig. 19.7). Subsequently, this has led to a higher turnover from primary productivity into biomass at the top-predator (Nile perch) level. These processes have driven the Nile perch boom, which has accelerated and subsidized the productivity of the ecosystem through the high turnover of fish biomass and therefore more rapid recycling of nutrients. The demographic changes around the lake, with a population growth that is among the highest in the world, have led to urbanization, industrial development, deforestation and agriculture, which have resulted in further eutrophication of the lake environment. Up till now this eutrophication has caused an increase in productivity of the various trophic levels in the aquatic food web, cascading up to and including Nile perch.

In order to test this, an attempt has been made at estimating the annual production and exploitation rates of Nile perch from Uganda and Tanzania as described above. The results are shown in Fig. 19.13, which corroborate the expected increased productivity curves and the relatively constant fishing mor-

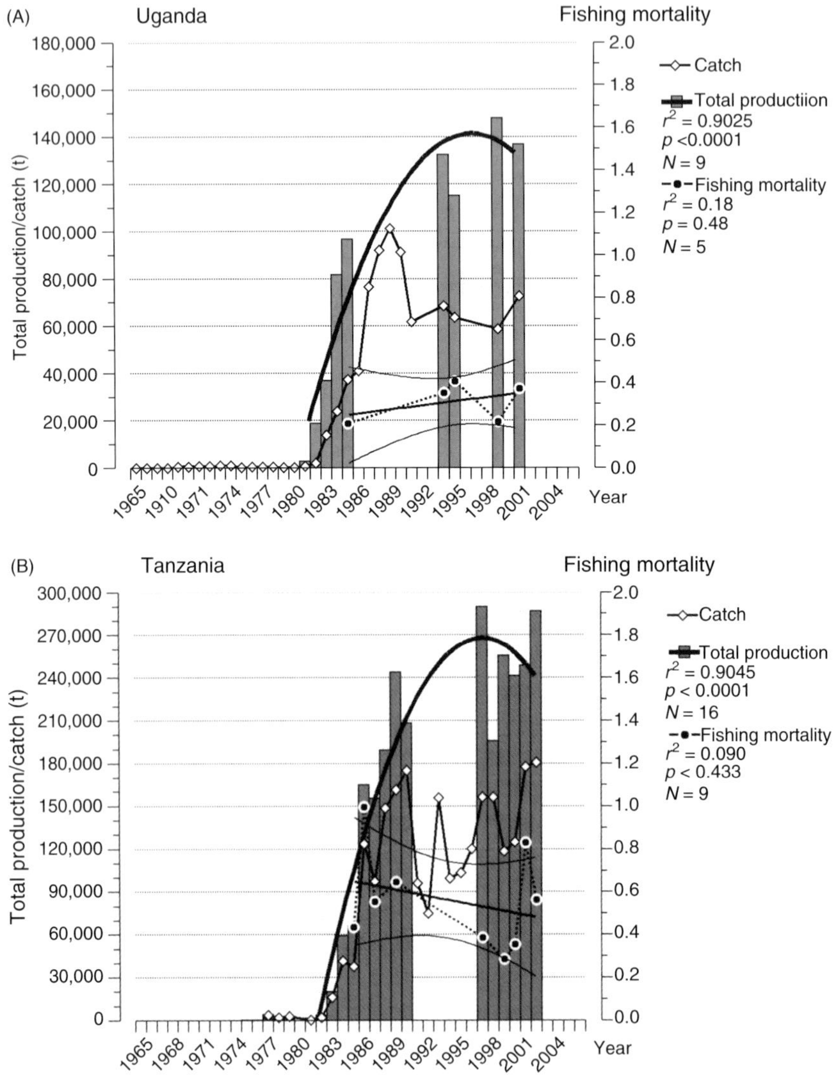

Fig. 19.13. Total reported catch (diamonds + black line), estimated total annual biological production (bars with fitted 2nd order polynomial) and estimated annual fishing mortality (circles + dotted line) for Nile perch in (A) Uganda and (B) Tanzania. Due to the under-reporting of catches broken down by species in Tanzania since 1998, the annual Nile perch catches have been raised proportionally to total reported catch. The total production has in both countries increased logistically with increased eutrophication as predicted, while there are no significant changes in fishing mortality (fitted linear regression). Mean F for Uganda from 1985 = 0.34/year $\rightarrow E = F/Z \approx 0.5$. Mean F for Tanzania from 1997 = 0.48/year $\rightarrow E = F/Z \approx 0.6$.

tality rates. Based on the reported catches, the mean exploitation rates (E) of Nile perch have remained stable around 0.5, which in theory should be optimal for a top predator (Kolding, 1994; Jul-Larsen *et al.*, 2003). The greatly increased production of Nile perch and *Rastrineobola*, as well as developments in the fishing industry and export markets, has led to an increased attraction of labour and investments into the fisheries resulting in a considerably increased fishing effort. Concurrently, the increased biological productivity of the lake has, up until the present, been sufficient to absorb the increase in fishing pressure as evidenced by the overall stability in the fishing mortality and standing biomass of Nile perch (Figs19.5 and 19.13) and even increasing biomass of *Rastineobola* (hydroacoustic surveys, LVFO-Hydroacoustic survey report, Jinja, February 2008). The individual return from the fishery (i.e. the catch per unit of effort), averaged over the whole lake, therefore has remained relatively stable. This does not exclude the possibility of local overfishing in heavily exploited areas of the lake, or local decimation from deteriorating water quality as in Winam Gulf (Kenya), which will be seen in the catch rates from the fishery as well.

On the whole, however, we conclude that there are limited, if any, observable *long-term* impacts of the fishery on the Nile perch biomass despite the large increase in fishing effort. This conclusion is in direct contrast to the view taken by Getabu *et al.* (2003) who suggest, based on hydroacoustic surveys that Nile perch biomass declined by about 50% between 1999 and 2001, which they attributed to fishing pressure although they also mention that environmental factors might be important. They also noted an increase in *Rastrineobola* stocks, which is attributed to a reduction in Nile perch *predation*, which again is due to fishing. However, short-term intra-decadal fluctuations have been observed several times for Nile perch stock (Fig. 19.5) and can be expected for *Rastrineobola* independent of predation because of the high turnover rate of the species. The inter-annual fluctuations in the abundance of large Nile perch ($CPUE_{>40\,cm}$, $L_{max,10}$) are larger than could be expected for this long-living species. These fluctuations may be indicative for inter-annual or longer-term fluctuations in the size of the anoxic layer (Fig. 19.11) regulating the dispersion of large fish within the available habitat, particularly the deep bottom areas. If so, then fluctuating, abundance indices have little to do with the impacts of fishing or recruitment, but are governed by the spatial distribution of Nile perch in relation to the oxycline and its vertical position at the time of trawl surveys.

The increase in fishing effort is expected to continue until the increased fish productivity is fully absorbed by the fishery, which will happen, judged from comparison with other small-scale African fisheries (Jul-Larsen *et al.*, 2003; FAO, 2004), when the average return to the individual small-scale fishermen stabilizes around 3 t/year (Fig. 19.9). Further increases in numbers of fishermen will then only occur if: (i) the average income from the fishery is maintained due to increase in prices (actually happening); or (ii) other targets, usually lower in the food web, are chosen. The latter will be indicated by a decrease in average trophic level of the whole fishery, the so-called fishing down the food web (Pauly *et al.*, 1998), which in our view is a healthy sign from an ecological perspective if the downward trend in overall trophic level is a result of utilizing all trophic levels for fishing (Jul-Larsen *et al.*, 2003; Kolding and van Zwieten, 2006), a process that can be described as fishing through the food web (Essington *et al.*, 2005). There are already indications of increased exploitation of lower trophic species such as cichlids and *Rastrineobola* in the most recent annual catch statistics (Fig. 19.6B).

Will the increase in fish production continue: what happens under hyper-eutrophication?

With increased eutrophication, the overall biological fish productivity can be expected to level off and eventually decrease, as the water quality deteriorates (Ryder *et al.*, 1974; Kolding and van Zwieten, 2006). The shape of the nutrient-productivity relationship is dependent on the individual species' capacity to endure deteriorating water quality and changes in the phytoplankton community (Fig. 19.14). Species susceptible to anoxic conditions, such as Nile perch (Fish, 1956; Schofield and Chapman, 2000), will be affected first, while more hardy species such as *Rastineobola* (seen via ROV in hypoxic waters in Lake Victoria; Hecky, Waterloo, 2006, personal communication), Nile tilapia, catfishes or *Protopterus* will continue to flourish (Njiru *et al.*, 2002) and some haplochromines may adapt (Witte *et al.*, 2007a). Early warning signs of the effects on fish stocks and the fishery of deteriorating water quality will be first observed in the relatively closed bays and heavily populated/urbanized bays as Mwanza and Winam Gulf.[6] Indications for this will be reduced catch rates starting with Nile perch, species change and movements of fishermen out of the affected regions into still productive offshore areas. Without good spatially explicit monitoring systems following fish stocks and fishing patterns, these developments may therefore initially not be observed in overall catches and catch rates of fishermen.

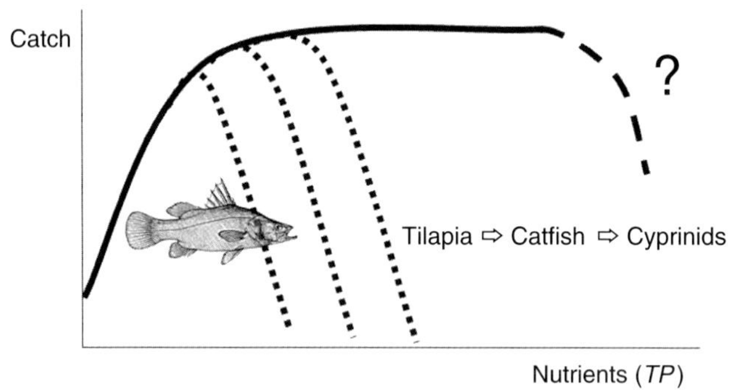

Fig. 19.14. Hypothetical model showing the development of community fisheries yield (catch) under increasing eutrophication represented by increased phosphorus loading (*TP*). Species that cannot adapt to higher loadings (increasing hypoxia) start declining after reaching a species-specific maximum (dotted lines). Compare with Fig. 19.13.

[6] Nile perch is now being reported to have more or less disappeared from Winam Gulf (R. Kayanda and I. Cowx, personal communication, České Budějovice, 2007).

That the productivity of Nile perch may be levelling off is corroborated in Fig. 19.13, which indicates that the top of the dome may have been reached, at least for Nile perch. Silsbe *et al.* (2006) concluded that the ideal nutrient concentrations in Lake Victoria to support a productive fishery, while maintaining an acceptable level of water quality, have actually now been exceeded. The estimated production trajectories appear to support this conclusion. In addition, the overall catch rates plotted versus effort density (Fig. 19.9) is again approaching the average trend of approximately 3 t/fisher/year observed in 14 other African lakes. Further nutrient enrichment will therefore most likely seriously affect the Nile perch fishery as well as fish biodiversity while other - more hardy - species like Nile tilapia, *Rastrineobola*, *Clarias* and *Protopterus* are likely to continue to grow. If indeed the top of the productivity curve for Nile perch as a function of eutrophication has now been reached, it will be of utmost importance to consider this aspect if catches should begin to decline. Theoretically, yields, under steady-state, follow a similar dome-shaped curve with increased effort, as the productivity-nutrient curve. A future decline in the Nile perch fishery may therefore easily, but also erroneously, be blamed on the fishery, whereas the real cause would be increased eutrophication. Ignoring the effects of increased nutrient loadings as a driver in the Lake Victoria ecosystem may therefore have severe consequences, not only for the application of assessment models, but for implementation of appropriate fishery management measures. Presently, all management efforts are recommended, and focused, on regulating the fishery, but if the overriding driver is the nutrient loads, as the evidence presented here indicates, then all these activities could be considered largely futile.

Conclusions

The overall conclusion is that Lake Victoria is undergoing rapid and profound changes in nearly all indicators from both bottom-up and top-down processes. It will be a major challenge not only to continue the monitoring of these changes, but far more so in trying to manage to control them. Effective management requires consensus on causes and effects of observed changes and political willingness to act on these changes. So far, most of the various activities in terms of research and management on the lake have operated in isolation only regarding their 'own' set of indicators from their respective disciplines. Fisheries management for example seems to have been based purely on classical fisheries theory with effort as the only driver, or preconceived truisms on gear regulations inherited from elsewhere as has been described for other southern African fisheries (Malasha, 2003). Unfortunately, the background information for these actions are from single species stock assessments, based on steady-state models, which cannot handle increased productivity or deteriorating water quality, and therefore uniformly indicate overfished stocks of Nile perch that are in danger of collapse (e.g. Pitcher and Bundy, 1995; Mkumbo *et al.*, 2002; Getabu *et al.*, 2003; Cowx, 2005; Matsuishi *et al.*, 2006; Njiru *et al.*, 2007). Little or no attention has been paid to the implications for fisheries management from the parallel work of the limnologists and ecologists on the lake.

The ubiquitous notion of overfishing, as is quoted and feared in nearly every written source on the fisheries of Lake Victoria, starting from descriptions of the fishery on *O. esculentes* in the 1920s onwards (Graham, 1929; Garrod, 1961; Fryer, 1973; and citations above), should be re-evaluated. Lake Victoria is not in a steady-state, which explains why classical approaches to fisheries assessments have all provided misleading results, as the underlying assumptions have been violated. Not only does this illustrate the danger of applying stock assessment models uncritically, but it also highlights the strange tendency of faithfully believing in model predictions (a limit is reached and a collapse is imminent), in spite of contradictory facts (the stocks have not changed). Bottom-up processes rather than the fishery seem to drive the Lake Victoria fishery. The production-nutrient relationship, however, is dome-shaped which means that production eventually will taper off (Downing and Plante, 1993) and subsequently decline as a result of deteriorating water quality (Ryder *et al.*, 1974). According to Silsbe *et al.* (2006), the primary productivity seems now to have reached a maximum and a continued nutrient enrichment of Lake Victoria will not increase gross phytoplankton production as it is already light-limited over most of the lake's surface area. Further, nutrient enrichment and ensuing deterioration of the water quality will most likely seriously affect the fish biodiversity, as already suggested for Lake Victoria and documented in Lake Erie and elsewhere (Seehausen *et al.*, 1997; Ludsin *et al.*, 2001; Egerston and Downing, 2004), and the Nile perch fishery, as anoxic waters become more extensive and rise to shallower depths. The observed changes in biomass and size structure in the deeper less exploited and unfished parts of the lake indicate that other processes than fishing may have a significant impact on the Nile perch stocks. Whether these changes are linked to the increase in eutrophication and the subsequent increase in the area of deoxygenated water remains to be established, but that such links exist is highly likely and should not be excluded when assessing future changes in Nile perch stocks.

The present status of the Lake Victoria-exploited stocks appears to a very large extent, and perhaps even exclusively, to be bottom-up driven in which case the present fisheries management concentrating on limiting fishery activities will have limited or no effect. In fact, if there are no visible effects from increased fishing on the exploited stocks as shown here, then there will be no detectable effects from management regulations either. The implications of the increased eutrophication in terms of biodiversity and fishery productivity have been largely ignored among the fishery researchers – at least in terms of management information and advice. Instead changes, both in the lake environment and in important stocks, are generally traced to the fishing activities only. It is therefore a paradox that while the observed resilience of the stocks has caused controversy about the actual status, the unified consensus on the management solutions appears to be a continuation of traditional gear and effort regulations (van der Knaap *et al.*, 2002; Matsuishi *et al.*, 2006). A renewed implementation of existing fishing regulations and measures through co-management instead of the largely failed enforcement of fishing regulations is now being revived through beach management units (BMUs). Attempts at certifying the Nile perch trade (Scholz, 2007) also are largely based on measures to adhere to slot sizes and associated mesh-size regulations. Despite these incentives to adhere to regulations, it is more or less agreed that overfishing is a fact and that therefore the effectiveness of management is low, with the ubiquitous conclusion that regulations should be reinforced (e.g. Njiru *et al.*, 2007). This prevailing view of overfishing is partly attributed to (misleading) assessment results and partly to (again misleading) decreased catches in the official statistics. From the late 1990s to 2006, the catch assessment surveys largely deteriorated while effort was intensely monitored by three consecutive biannual frame surveys. However, if catch rates and particularly yields in the future will begin to fall, while effort continues to grow, it will be a big open question whether the decrease is due to environmental degradation or due to overfishing or both.

As eutrophication appears to be the single most important driver of the productivity, the main threat to the fishery, in our view, is hyper-eutrophication rather than overexploitation. Fisheries management of Lake Victoria therefore needs to reset its priorities and integrate environmental information. The potentially very strong impact of the increased eutrophication on the fishery, and the emergent signs that a maximum in the gross primary production may have been reached, would warrant a much higher management commitment on the eutrophication issue from both environmental and fisheries perspectives. Even if fisheries managers are mandated to manage the fishery and not the environment, then this is not a reason to ignore the potential effects of the changing environment on the stocks that they are required to manage. Actually, with increased eutrophication the fishery could even intensify and diversify to lower trophic levels in order to remove as much organic matter as possible. In fact, the shift down the food chain has already begun and *Rastrineobola* now constitutes the greatest proportion of the fish catch while the biomass is still increasing. Thus, the fishery on *Rastrineobola* and recently also *Caridina* shows the start of such a 'fishing down' process and should be considered a positive development, although the net economic and social effects of more lower valued fish compared to potential declines in the higher valued Nile perch remains to be determined.

Towards an ecosystem approach to management

An integrated assessment of ecosystems is a demanding challenge. The development of an ecosystem-based management in a huge system like Lake Victoria requires a basic comprehension of the main processes taking place at the scale of the Lake and its catchments, as is attempted here. Such an understanding will provide an information framework for regional and local management concerns. Over the past decade several large projects funded by the World Bank and the EU have been dedicated at gathering the information to arrive at this understanding. Considerable amounts of data and information were collected on many aspects of the ecosystem. Unfortunately, analyses from different disciplinary components were rarely integrated and information was not made available in a historic context (time series). Several documents referred to 'lack of prioritization of research' indicating either a lack in the basic understanding of main interacting processes, or at least that research has not lead to the necessary framework to assist in focus and prioritization (see references in Kolding *et al.*, 2005). As the extensive literature of Lake Victoria shows, numerous short-term isolated studies with descriptions of states and processes are not sufficient to allow comprehension. This reductionistic approach will rather lead to information overload, barring the emergence of meaningful insights and strategies for management instead of leading to a pragmatic basic understanding of ecosystems (Peters, 1991). Knowledge about variations in (isolated) processes cannot give directives on whether systems states are 'good' or 'bad'. The reductionistic perspective does not give directionality in time and therefore bars value assessments (Steele, 1998). Conversely, a holistic approach at a higher scale than observations on individual processes (Kolding, 1994) that combines an analysis of key processes operational in Lake Victoria, with indicators of important phenomena arranged in time series, will aid in real comprehension of the system, as shown here. An assessment along these lines will give guidance to the choice of important indicators and the development of monitoring systems that will form the basis in ecosystem-based management (Choi *et al.*, 2005). Solutions to address the relative impacts of eutrophication and fishing can be found by making use of the spatial differentiation in trophic conditions on the lake and careful assessment of effort allocation of fishermen in reaction to local conditions. Research on, and monitoring of, the same set of indicators by the three countries can be considered replicates under different conditions of fishing pressure and eutrophication that will be a powerful aid in learning what will, and will not, work in management. The present disciplinary segregation and approach, as well as continued implementation of traditional single-species management regulations, albeit disguised as co-management, will not solve the present problems in Lake Victoria. It is apparent on Lake Victoria and other aquatic ecosystems that the environmental and fisheries research and management activities need integration, and that a holistic Ecosystem Approach to Fisheries (EAF) management is adopted. A systematic updating of a comprehensive set of indicators, together with a rehabilitation of the historic data and information, will be a necessary tool in developing the information base for such an ecosystem approach.

Acknowledgements

The ideas in this chapter have been gradually developed during a series of workshops with participants from the three riparian fisheries research institutions (TAFIRI, KMFRI and FIRRI) and from the Departments of Fisheries from Kenya, Tanzania and Uganda under the auspices of LVEMP. We gratefully acknowledge the inspiring discussions and sharing of information with the many researchers and managers involved, although we concede that our views may not be shared by all. It has been a major work to reformat and reconstruct the trawl data series of Tanzania and Uganda back in time and we are deeply grateful for the help and assistance from Professor Bwathondi, Kees Goudswaard, John Okaronon, Enock Mlaponi and Samuel Bassa.

References

Abila, R.O. and Jansen, E.G. (1997) Socio-economics of the Lake Victoria fisheries: from local to global markets. The fish exporting and fishmeal industries of Lake Victoria - structure strategies and social-economic impacts in Kenya. IUCN, Nairobi, Kenya No. 2.

Allen, K.R. (1971) Relationship between production and biomass. *Journal of the Fisheries Research Board Canada* 28, 1573-1581.

Azza, N.G.T. (2006) The dynamics of shoreline wetlands and sediments of northern Lake Victoria. UNESCO-IHE PhD thesis, Leiden, Taylor & Francis Balkema ISBN 0-415-42649-9, 170 pp.

Balirwa, J.S., Chapman, C.A., Chapman, L.J., Cowx, I.G., Geheb, K., Kaufman, L., Lowe-McConnel, R.H., Seehausen, O., Wanink, J.H., Welcomme, R.L. and Witte, F. (2003) Biodiversity and fishery sustainability in the Lake Victoria basin: an unexpected marriage. *Bioscience* 55, 703-715.

Barel, C.D.N., Dorit, R., Greenwood, P.H., Fryer, G., Hughes, N., Jackson, P.B.N., Kawanabe, H. and Lowe-McConnel, R.H. (1985) Destruction of fisheries in Africa's lakes. *Nature* 315, 19-20.

Baskin, Y. (1992) Africa's troubled waters: fish introductions and a changing physical profile muddy Lake Victoria's future. *BioScience* 42, 476-481.

Beddington, J.R. and Cooke, I.G. (1983) The potential yield of fish stocks. FAO Fisheries Technical Paper No. 242, pp. 1-47.

Bootsma, H.A., Bootsma, M.J. and Hecky, R.E. (1996)The chemical composition of precipitation and its significance to the nutrient budget of lake Malawi. In: Johnson, T.C. and Odada, E.O. (eds) *The Limnology, Climatology and Paleoclimatology of the East African Lakes*. Gordon and Breach Publishers, Amsterdam, The Netherlands, pp. 103-140.

Bootsma, H.A., Mwita, J., Mwichande, B., Hecky, R.E., Kihedu, J. and Mwambungu, J. (1999)The atmospheric deposition of nutrients on lake Malawi/Nyasa. In: Bootsma, H.A. and Hecky, R.E. (eds) *Water Quality Report for the SADC/GEF Lake Malawi/Nyasa*. SADC/GEF Project Report, Biodiversity Conservation Project, Burlington, Ontario, Canada, pp. 85-111.

Budeba, Y.L. and Cowx, I.G. (2007) Contribution of *Caridina nilotica* (Roux) in the Dagaa fishery in Lake Victoria, Tanzania. *Aquatic Ecosystem Health & Management* 10, 381-391.

Chapman, L.J., Chapman, C.A., Ogutu-Ohwayo, R., Chandler, M., Kaufman, L. and Keiter, A.E. (1996) Refugia for endangered fishes from an introduced predator in Lake Nabugabo, Uganda. *Conservation Biology* 10, 554-561.

Chapman, L.J., Chapman, C.A., Nordlie, F.G. and Rosenberger, A.E. (2002) Physiological refugia: swamps, hypoxia tolerance and maintenance of fish diversity in the Lake Victoria region. *Comparative Biochemistry and Physiology* A 133, 421-437.

Choi, J.S., Frank, K.T., Petrie, B.D. and Leggett, W.C. (2005) Integrated assessment of a large marine ecosystem: a case study of the devolution of the eastern Scotian shelf, Canada. *Oceanography and Marine Biology: An Annual Review* 43, 47-67.

Cowx, I. (ed.) (2005) Review of the exploitation pressures on the fisheries resources of Lake Victoria. LVEMP National Secretariat, Entebbe, Uganda, 125 pp.

Cowx, I.G., van der Knaap, M., Muhoozi, L.I. and Othina, A. (2003) Improving fishery catch statistics for Lake Victoria. *Aquatic Ecosystem Health & Management* 6, 299-310.

Degnbol, P. and Jarre, A. (2004) Review of indicators in fisheries management - A development perspective. *South African Journal of Marine Science* 26, 303-326.

Downing, J.A. and Plante, C. (1993) Production of fish populations in lakes. *Canadian Journal of Fisheries and Aquatic Science* 50, 110-120.

Egerston, C.J. and Downing, J.A. (2004) Relationship of fish catch and composition to water quality in a suite of agriculturally eutrophic lakes. *Canadian Journal of Fisheries and Aquatic Science* 61, 1784-1796.

Essington, T.E., Beaudreau, A.H. and Wiedenmann, J. (2005) Fishing through marine food webs. *PNAS* 103, 3171-3175.

FAO (2004) African freshwaters: are small-scale fisheries a problem? The State of World Fisheries and Aquaculture, SOFIA, FAO, Rome, Italy, pp. 131-137.

Fish, G.R. (1956) Some aspects of the respiration of six species of fish from Uganda. *Journal of Experimental Biology* 33, 186–195.

Flohn, H. (1987) East-African rains of 1961/62 and the abrupt change of the White Nile discharge. *Paleoecology of Africa* 18, 3–18.

Fryer, G. (1973) The Lake Victoria fisheries: some facts and fallacies. *Biological Conservation* 5, 304–308.

Fryer, G. and Iles, T.D. (1972) *The cichlid fishes of the great lakes of Africa: their biology and evolution*. Oliver and Boyd, Edinburgh, 642 pp.

Garrod, D.J. (1961) The rational exploitation of the *Tilapia esculenta* stock of the North Buvuma island area, Lake Victoria. *East African Agriculture and Forestry Journal* 27, 69–71.

Getabu, A. (2003) Assessment of fish stocks in Lake Victoria using a combination of acoustic and trawling methods. PhD thesis, University of Hull, Hull, UK, 397 pp.

Getabu, A., Tumwebaze, R. and MacLennan, D.N. (2003) Spatial distribution and temporal changes in the fish populations of Lake Victoria. *Aquatic Living Resources* 16, 159–165.

Gislason, H. and Rice, J. (1998) Modelling the response of size and diversity spectra of fish assemblages to changes in exploitation. *ICES Journal of Marine Science* 56, 362–370.

Goldschmidt, T., Witte, F. and Wanink, J. (1993) Cascading effects of the introduced Nile perch on the detritivorous/phytoplanktivorous species in the sublittoral areas of Lake Victoria. *Conservation Biology* 7, 686–700.

Goudswaard, P.C. (2006) Causes and effects of the Lake Victoria ecological revolution. PhD thesis, Leiden University, The Netherlands, 177 pp.

Goudswaard, P.C., Witte, F. and Wanink, J.H. (2006) The shrimp *Caridina nilotica* in Lake Victoria (East Africa), before and after the Nile perch increase. *Hydrobiologia* 563, 31–34.

Goudswaard, P.C., Witte, F. and Katunzi, E.F.B. (2008) The invasion of an introduced predator, Nile perch (*Lates niloticus*, L.) in Lake Victoria (East Africa): chronology and causes. *Environmental Biology of Fishes* 81, 127–139.

Graham, M. (1929) The Victoria Nyanza and its fisheries. A report on the fishing survey of Lake Victoria 1927–1928. Crown Agents, London, 255 pp.

Greenwood, P.H. (1974) Cichlid fishes of Lake Victoria, East Africa. The biology and evolution of a species flock. Trustees of the British Musum (Natural History), London, 134 pp.

Hecky, R.E. (1993) The eutrophication of Lake Victoria. *Verhandlungen der Internationalen Verein Für Limnologie* 25, 39–48.

Hecky, R.E., Bugenyi, F.W.B., Ochumba, P., Talling, J.F., Mugidde, R., Gophen, M. and Kaufman, L. (1994) The deoxygenation of Lake Victoria. *Limnology and Oceanography* 39, 1476–1481.

Hopson, A.J. (1972) A study of Nile perch (*Lates niloticus* L. Pisces: Centropomidae) in Lake Chad. Overseas Research Publication No 19, London, 93 pp.

Jackson, P.B.N. (1971) The African Great Lakes: food source and world treasure. *Biological Conservation* 5, 302–304.

Johnson, T.C., Scholtz, C.A., Talbot, M.R., Kelts, K., Rickets, R.D., Ngobi, G., Beuning, K., Sseemmanda, I. and McGill, J.W. (1996) Late Pleistocene desiccation of Lake Victoria and rapid evolution of cichlid fishes. *Science* 273, 1091–1093.

Jul-Larsen, E., Kolding, J., Nielsen, J.R., Overa, R. and van Zwieten, P.A.M. (2003) Management, co-management or no management? Major dilemmas in southern African freshwater fisheries. Part 1: Synthesis Report. FAO Fisheries Technical Paper 426/1. FAO, Rome, Italy.

Kanamitsu, M., Ebisuzaki, W., Woollen, J., Yang, S.-K., Hnilo, J.J., Fiorino, M. and Potter, G.L. (2002) NCEP-DEO AMIP-II Reanalysis (R-2). *Bulletin of the American and Meteorological Society* 83, 1631–1643.

Katunzi, E.F.B., van Densen, W.L.T., Wanink, J.H. and Witte, F. (2006) Spatial and seasonal patterns in the feeding habits of juvenile *Lates niloticus* (L.), in the Mwanza Gulf of Lake Victoria. *Hydrobiologia* 568, 121–133.

Kaufman, L. (1992) Catastrophic changes in species-rich freshwater ecosystems. The lessons of Lake Victoria. *BioScience* 42, 846–858.

Kaufman, L. and Schwartz, J. (2002) Nile perch population dynamics in Lake Victoria: implications for management and conservation. In: Ruth, M. and Lindhom, J. (eds) *Dynamic*

Modeling for Marine Conservation. Springer, New York, pp. 257–313.

Kitchell, J.F., Schindler, D.E., Ogutu-Ohwayo, R. and Reinthal, P.N. (1997) The Nile perch in Lake Victoria: interactions between predation and fisheries. *Ecological Applications* 7, 653–664.

Kolding, J. (1993) Population dynamics and life history styles of Nile tilapia (*Oreochromis niloticus*) in Fergusons Gulf, Lake Turkana, Kenya. *Environmental Biology of Fishes* 37, 25–46.

Kolding, J. (1994) Plus ça change, plus c'est la même chose. On the ecology and exploitation of fish in fluctuating tropical freshwater systems. Dr Scient thesis, Department of Fisheries and Marine Biology, University of Bergen, Bergen, Norway, ISBN 82-7744-011-1. 197 pp.

Kolding, J. and van Zwieten, P.A.M. (2006) Improving productivity in tropical lakes and reservoirs. Challenge Program on Water and Food – Aquatic Ecosystems and Fisheries Review Series 1, Theme 3 of CPWF, c/o WorldFish Center, Cairo, Egypt, 139 pp.

Kolding, J., van Zwieten, P.A.M., Manyala, J., Okedi, J., Mgaya, Y.D. and Orach-Meza, F. (2005) Lake Victoria Environmental Management Program (LVEMP): Regional Synthesis Report on Fisheries Research and Management. States, trends and processes. LVEMP National Secretariat, Entebbe, Uganda, 126 pp.

Kolding, J., Haug, L. and Stefansson, S. (2008) Effect of oxygen on growth and reproduction in Nile tilapia (*Oreochromis niloticus*). *Canadian Journal of Fisheries and Aquatic Science* 65(7), 1413–1424.

Kudhongania, A.W. and Chitamwebwa, D.B.R (1995) Impact of environmental change, species introductions and ecological interactions on the fish stocks in Lake Victoria. In: Pitcher, T.J. and Hart, P.J.B (eds) *Impact of species changes in African Lakes* (Fish and Fisheries series). Chapman and Hall, 19–32 pp.

Kudhongania, A.W. and Cordone, A.J. (1974) Batho-spatial distribution pattern and biomass estimate of the major demersal fishes in Lake Victoria. *African Journal of Tropical Hydrobiology and Fisheries* 3, 15–32.

Kudhongania, A.W., Twongo, T. and Ogutu-Ohwayo, R. (1992) Impact of the Nile perch on the fisheries of Lakes Victoria and Kyoga. *Hydrobiologia* 232, 1–10.

Langenberg, V.T., Nyamushahu, S., Roijackers, R. and Koelmans, A.A. (2003) External nutrient sources for Lake Tanganyika. *Journal of Great Lakes Research* 29(Suppl. 2), 169–180.

Ligtvoet, W. and Witte, F. (1991) Perturbation through predator introduction: effects on the food web and fish yields in Lake Victoria (East Africa). In: Horwood, E. and Ravera, O. (eds) *Terrestrial and Aquatic Ecosystems. Perturbation and Recovery*. New York, pp. 263–268.

Ludsin, S.A., Kershner, M.W., Blocksom, K.A., Knight, R.L. and Stein, R.A. (2001) Life after death in Lake Erie: nutrient controls drive fish species richness, rehabilitation. *Ecological Applications* 11, 731–746.

Malasha, I. (2003) The emergence of colonial and post-colonial fisheries regulations: the case of Zambia and Zimbabwe. In: Jul-Larsen, E., Kolding, J., Overå, R., Raakjær Nielsen, J. and van Zwieten, P.A.M. (eds) *Management, Co-management or No-management? Major Dilemmas in Southern African Freshwater Fisheries*. FAO Fisheries Technical Paper 426/2, FAO, Rome, Italy, pp. 253–266.

Manyala, J.O. and Ojouk, J.E. (2007) Survival of the Lake Victoria *Rastineobola argentea* in a rapidly changing environment: biotic and abiotic interactions. *Aquatic Ecosystem Health & Management* 10, 407–415.

Matsuishi, T., Muhoozi, L., Mkumbo, O.C., Budeba, Y., Njiru, M., Asila, A., Othina, A. and Cowx, I.G. (2006) Are the exploitation pressures on the Nile perch fisheries resources of Lake Victoria a cause for concern? *Fisheries Management and Ecology* 13, 53–71.

Mkumbo, O.C. (2002) Assessment and management of Nile perch (*Lates niloticus* L.) stocks in the Tanzanian waters of Lake Victoria. PhD thesis, University of Hull, Hull, UK, 276 pp.

Mkumbo, O.C. and Ezekiel, C.N. (1999) Distribution and abundance of fish stocks in Lake Victoria, Tanzania. In: Tweddle, D. and Cowx, I.G. (eds) *Lake Victoria Fisheries Research Project Phase II*. Report on fourth FIDAWOG workshop held at Kisumu, 16–20 August 1999, pp. 27–46.

Mkumbo, O.C., Ezekiel, C.N., Budeba, Y.L. and Cowx, I.G. (2002) Analysis of exploitation patterns of Nile perch in Lake Victoria. In: Cowx, I.G. (ed.) *Management and Ecology of Lake and Reservoir Fisheries*. Fishing News Books, Blackwell Science, Oxford, UK, pp. 84–95.

Mkumbo, O.C., Nsinda, P., Ezekiel, C.N., Cowx, I.G. and Aeron, M. (2007) Towards sustainable exploitation of Nile perch consequential to regulated fisheries in Lake Victoria. *Aquatic Ecosystem Health & Management* 10, 449–457.

Mugidde, R. (1992) Changes in phytoplankton productivity and biomass in Lake Victoria (Uganda). MSc thesis, University of Manitoba, Winnipeg, Canada.

Mugidde, R. (1993) The increase in phytoplankton primary productivity and biomass in Lake Victoria (Uganda). *Verhandlungen der Internationalen Verein Für Limnologie* 25, 846–849.

Mugidde, R. (2001) Nutrient status and phytoplankton nitrogen fixation in Lake Victoria, East Africa. PhD thesis, University of Waterloo, Waterloo, Canada.

Mugidde, R., Hecky, R.E., Hendzel, L. and Taylor, W.D. (2003) Pelagic nitrogen fixation in Lake Victoria, Uganda. *Journal of Great Lakes Research* 29, 76–88.

Mwanuzi, F.L., Abuodha, J.O.Z., Muyodi, F.J. and Hecky, R.E. (2005) Lake Victoria Regional Water Quality Synthesis Report. Lake Victoria Environmental Management Program (LVEMP), National Secretariat, Entebbe, Uganda.

Nicholson, S.E. (1998) Historical fluctuations of Lake Victoria and other lakes in the Northern Rift Valley of East Africa. In: Lehman, J.T. (ed.) *Environmental Change and Response in East African Lakes*. Kluwer, Dordrecht, The Netherlands, pp. 7–36.

Njiru, A. Othina, A.N., Getabu, A., Tweddle, D. and Cowx, I.G. (2002) Is the invasion of water hyacinth, *Eichhornia crassipes* Solms (Mart.), a blessing to Lake Victoria fisheries? In: Cowx, I.G. (ed.) *Management and Ecology of Lake and Reservoir Fisheries*. Fishing News Books, London, pp. 255–263.

Njiru, M., Waithaka, E., Muchiri, M., van der Knaap, M. and Cowx, I.G. (2005) Exotic introductions to the fishery of Lake Victoria: what are the management options? *Lakes & Reservoirs: Research and Management* 10, 147–155.

Njiru, M., Nzungi, P., Getabu, A., Wakwabi, E., Othina, A., Jembe, T. and Wekesa, S. (2007) Are fisheries management, measures in Lake Victoria successful? The case of Nile perch and Nile tilapia fishery. *African Journal of Ecology* 45, 315–323.

Ochumba, P.B.O. (1990) Massive fish kills within the Nyanza Gulf of Lake Victoria, Kenya. *Hydrobiologia* 208, 93–99.

Ogutu-Ohwayo, R. (1990) The decline of the native fishes of Lake Victoria and Kyoga (East Africa) and the impact of introduced species, especially the Nile perch, *Lates niloticus* and the Nile tilapia, *Oreochromis niloticus*. *Environmental Biology of Fishes* 27, 81–96.

Ojouk, J.E., Njiru, M., Ntiba, M.J. and Mavuti, K.M. (2007) The effect of overfishing on the life-history strategies of Nile tilapia, *Oreochromis niloticus* (L.) in the Nyanza Gulf of Lake Victoria, Kenya. *Aquatic Ecosystem Health & Management* 10, 443–448.

Okaronon, J. (2004) Current status of the fish stocks of Lake Victoria (Uganda). PhD thesis, University of Nairobi, Kenya.

Page, L.N. (2006) The Nile perch fishery in Lake Victoria: management proposals for fishery sustainability and biodiversity conservation. BA thesis, Reed College, Portland, Oregon, 68 pp.

Pauly, D. (1982) Studying single species dynamics in multi-species context. In: Pauly, D. and Murphy, G.I. (eds) *Theory and Management of Tropical Fisheries*. ICLARM, Manila, The Philippines, pp. 33–70.

Pauly, D. (1984) A mechanism for the juvenile-to adult transition in fishes. *Journal du Conseil International pour l'Exploration de la mer* 41, 280–284.

Pauly, D., Christensen, V., Dalsgaard, J., Froese, R. and Torres, F.J. Jr. (1998) Fishing down marine food webs. *Science* 279, 860–863.

Peters, R.H. (1991) *A Critique for Ecology*. Cambridge University Press, Cambridge, 366 pp.

Pitcher, T.J. and Bundy, A. (1995) Assessment of the Nile perch fishery in Lake Victoria. In: Pitcher, T.J. and Hart, P.B.J. (eds) *The Impact of Species Change in African Lakes*. Fish and Fisheries Series (18), Chapman & Hall, London, pp. 163–179.

Pringle, R.M. (2005). The origin of the Nile perch in Lake Victoria. *BioScience* 55, 780–787.

Rabour, C.O., Gichuki, J. and Moreau, J. (2003) Growth, mortality and recruitment of Nile perch *Lates niloticus* (L. Centropomidae) in the Nyanza Gulf of Lake Victoria: an evaluation update. *Naga* 26, 8–12.

Reynolds, J.E., Gréboval, D.F. and Mannini, P. (1995) Thirty years on: the development of the Nile perch fishery in Lake Victoria. In: Pitcher, T.J. and Hart, P.J.B. (eds) *The Impact of Species Changes in African Lakes*. Fish and Fisheries Series (18), Chapman & Hall, London, pp. 181–211.

Rice, J. and Rochet, M.-J. (2005) A framework for selecting a suite of indicators for fisheries management. *ICES Journal of Marine Science* 62, 516–527.

Ryder, R.A., Kerr, S.R., Loftus, K.M. and Regier, H.A. (1974) The morphoedaphic index, a fish yield estimator – review and evaluation. *Journal of the Fisheries Research Board Canada* 31, 663–688.

Schindler, D.E., Kitchell, J.F. and Ogutu-Ohwayo, R. (1998) Ecological consequences of alternative gill net fisheries for Nile Perch in Lake Victoria. *Conservation Biology* 12, 56–64.

Schofield, P.J. and Chapman, L.J. (2000) Hypoxia tolerance of introduced Nile perch: implications for survival of indigenous fishes in the Lake Victoria basin. *African Zoology* 35, 35–42.

Scholz, U. (2007) The dilemma of the Nile perch ecolabelling could be a strategy to secure long-term market access of a fishing sector that secures the livelihoods of around 150,000 fishers in the Nile-perch fishery. *SAMUDRA Report* 48, 10–12.

Seehausen, O., van Alphen, J.J.M. and Witte, F. (1997) Cichlid fish diversity threatened by eutrophication that curbs sexual selection. *Science* 277, 1808–1811.

Sene, K.J. and Plinston, D.T. (1994) A review and update of the hydrology of Lake Victoria in East Africa. *Hydrological Sciences Journal* 39, 47–63.

Silsbe, G.M. (2004) Phytoplankton production in Lake Victoria, East Africa. MSc thesis, University of Waterloo, Ontario, Canada.

Silsbe, G.M., Hecky, R.E., Guildford, S.J. and Mugidde, R. (2006) Variability of chlorophyll a and photosynthetic parameters in a nutrient-saturated tropical great lake. *Limnology and Oceanography* 51, 2052–2063.

Shin, Y.-J. and Cury, P. (2004) Using an individual-based model of fish assemblages to study the response of size spectra to changes in fishing. *Canadian Journal of Fisheries and Aquatic Science* 61, 414–431.

Shin, Y.-J., Rochet, M.-J., Jennings, S., Field, J.G. and Gislason, H. (2005) Using size-based indicators to evaluate the ecosystem effects of fishing. *ICES Journal of Marine Science* 62, 384–396.

Stager, J.C. and Johnson, T.C. (2007) The late Pleistocene desiccation of Lake Victoria and the origin of its endemic biota. *Hydrobiologia*, online DOI 10.1007/s10750-007-9158-2

Steele, J.H. (1998) Regime shifts in marine ecosystems. *Ecological Applications* 8, 33–36.

Talling, J.F. (1965) The photosynthetic activity of phytoplankton in East African lakes. *Internationale Revue der gesamten Hydrobiologie* 50, 1–32.

Tamatamah, R.A., Hecky, R.E. and Duthie, H.C. (2005) The atmospheric deposition of phosphorus in Lake Victoria (East Africa). *Biogeochemistry* 73, 325–433.

Tufte, E.R. (2006) *Beautiful Evidence*. Graphics Press, Connecticut, 213 pp.

Tumwebaze, R. (1997) Application of hydroacoustics in fish stock assessment of Lake Victoria. M.Phil thesis, University of Bergen, Bergen, Norway.

Tumwebaze, R., Getabu, A., Bayona, J., MacLennan, D. and Cowx, I.G. (2002) Fisheries of Lake Victoria: an underwater perspective. In: Cowx, I.G. (ed.) *Management and Ecology of Lake and Reservoir Fisheries*. Fishing News Books, Blackwell Science, Oxford, pp. 70–83.

Twongo, T., Bugenyi, F.W.B. and Wanda, F. (1995) The potential for further proliferation of water hyacinth in lakes victoria, Kyoga and Kwania and some urgent aspects for research. *African Journal of Tropical Hydrobiology and Fisheries* 6, 1–10.

UNEP (2006) Africa's Lakes: Atlas of Our Changing Environment. Division of Early Warning and Assessment (DEWA). United Nations Environment Programme (UNEP), Nairobi, Kenya. Earthprint.com ISBN: 92 807 2694. 396 pp.

van der Knaap, M., Ntiba, N.J. and Cowx, I.G. (2002) Key elements of fisheries management on Lake Victoria. *Aquatic Ecosystem Health & Management* 5, 245–254.

Verschuren, D., Johnson, T.C., Kling, H.J., Edgington, D.N., Leavitt, P.R., Brown, E.T., Talbot, M.R. and Hecky, R.E. (2002) History and timing of human impact on Lake Victoria, East Africa. *Proceedings of the Royal Society London B* 269, 289–294.

Videler, J. and Wardle, C.S. (1991) Fish swimming stride by stride: speed limits and endurance. *Reviews of Fish Biology and Fisheries* 1, 23–40.

Walters, C. and Kitchell, J.F. (2001) Cultivation/depensation effects on juvenile survival and recruitment: implications for the theory of fishing. *Canadian Journal of Fisheries and Aquatic Science* 58, 39–50.

Walters, C., Christensen, V. and Pauly, D. (1997) Structuring dynamic models of exploited ecosystems from trophic mass-balance assessments. *Reviews in Fish Biology and Fisheries* 7, 139–172.

Wanink, J. (1998) The pelagic cyprinid *Rastrineobola argentea* as a crucial link in the distrupted ecosystem of Lake Victoria. PhD thesis, University of Wageningen, The Netherlands, ISBN 90-5651-042-8. 288 pp.

Welcomme, R.L. (1970) Studies on the effects of the abnormal high water levels on the ecology of fish in certain shallow regions of Lake Victoria. *Journal of Zoology*. 160, 405–436.

Welcomme, R.L. (1988) International introductions of inland aquatic species. FAO Fisheries Technical Paper 294:1–318, FAO, Rome, Italy.

Welcomme, R.L. (1999) A review of a model for qualitative evaluation of exploitation levels in multi-species fisheries. *Fisheries Management and Ecology* 6, 1–19.

Williams, A.E., Duthie, H.C. and Hecky, R.E. (2005) Water hyacinth in Lake Victoria: why did it vanish so quickly and will it return. *Aquatic Botany* 81, 300–314.

Witte, F. (1984) Ecological differentiation in Lake Victoria haplochromines: comparison of cichlid species flocks in African lakes. In: Echelle, A.A. and Kornfield, I. (eds) *Evolution of Fish Species Flocks*. Orono Press, Maine, pp. 155–167.

Witte, F. and Goudswaard, P.C. (1985) Aspects of the haplochromine fishery in southern Lake Victoria. In: CIFA. Report of the 3rd session of the sub-committee for the Development and Management of the Fisheries in Lake Victoria, 4–5 October 1984, Jinja. FAO fisheries report 335, pp. 81–88.

Witte, F., Goldschmidt, P.T., Goudswaard, P.C., Ligtvoet, W., van Ooijen, M.J.P. and Wanink, J.H. (1992) Species extinction and concomitant ecological changes in Lake Victoria. *Netherlands Journal of Zoology* 42, 214–232.

Witte, F., Goldschmidt, P.T., Goudswaard, P.C., Katunzi, E.F.B., Mkumbo, O.C., Seehausen, O. and Wanink, J.H. (1999) Lake Victoria's ecological changes and their relationships with the riparian societies. In: Kawanabe, H., Coulter, G.W. and Roosevelt, A.C. (eds) *Ancient Lakes, their Cultural and Biological Diversity*. Kenobi productions Ghent, Belgium, pp. 189–202.

Witte, F., Msuku, B.S., Wanink, J.H., Seehausen, O., Katunzi, E.F.B., Goudswaard, P.C. and Goldschmidt, P.T. (2000) Recovery of cichlid species in Lake Victoria: an examination of factors leading to differential extinction. *Reviews in Fish Biology and Fisheries* 10, 233–241.

Witte, F., Wanink, J.H. and Kishe-Machumu, M.A. (2007a) Species distinction and the biodiversity crisis in Lake Victoria. *Transactions of the American Fisheries Society* 136, 1146–1159.

Witte, F., Wanink, J.H., Kishe-Machamu, M.A., Mkumbo, O.C., Goudswaard, P.C. and Seehausen, O. (2007b) Differential decline and recovery of haplochromine trophic groups in the Mwanza Gulf of Lake Victoria. *Aquatic Ecosystem Health & Management* 10, 416–433.

Yin, X. and Nicholson, S.E. (1998) The water balance of Lake Victoria. *Journal of Hydrological Science* 43, 789–812.

Zwieten, P.A.M., van Goudswaard, P.C. and Kapasa, C.K. (2003) Mweru-Luapula is an open exit fishery where a highly dynamic population of fishermen makes use of a resilient resource base. In: Jul-Larsen, E., Kolding, J., Overå, R., Raakjær Nielsen, J. and Zwieten, P.A.M. van (eds) *Management, Co-management or No Management? Major Dilemmas in Southern African Freshwater Fisheries*. Part 2: Case studies. FAO Fisheries Technical Paper. No. 426/2, pp. 1–33.

Appendix 19.1

Forty-one indicators to assess the states of Lake Victoria's fish stocks and the pressures and drivers acting on them compiled by researchers from Uganda, Tanzania and Kenya. (From Kolding *et al.*, 2005.)
Time series data available: + = >10 years; ~ = some points; o = anecdotal

Level	Type of indicators	Indicator	Time series	Source
System	Productivity	1. Lake depth	na	Silsbe, 2004
		2. Wind stress	+	Mwanuzi *et al.*, 2005
		3. Rainfall	+	A. Lotsch – Worldbank; Silsbe, 2004
		4. Water level/water balance	+	Mwanuzi *et al.*, 2005
		5. Secchi depth	~	Mwanuzi *et al.*, 2005; various published literature since 1930 (see references in Kolding *et al.*, 2005)
		6. Temperature	+	Silsbe, 2004 and various published sources (see Kolding *et al.*, 2005)
		7. Nutrients	~	Hecky *et al.*, 1994; Mwanuzi *et al.*, 2005
		8. Oxygen	~	Silsbe, 2004
		9. Primary production (Chl *a*)	~	Mwanuzi *et al.*, 2005
		10. Species composition of microalgae	o	Mwanuzi *et al.*, 2005
		11. Water hyacinth	~	Mwanuzi *et al.*, 2005
	Secondary productivity	12. Lake flies (abundance)	o	No data available; some published observations
		13. *Caridina* (abundance)	~	Hydroacoustic surveys
		14. Zooplankton (abundance)	~	Mwanuzi *et al.*, 2005; Wanink, 1998
Effort	Catch and effort	15. Catch	+	CEDRS Tanzania, Kenya, Uganda
		16. Effort (number of fishermen, boats, types of gear)	+ ~	CEDRS, Tanzania, Kenya, Uganda
		17. Spatial distribution of fishermen	~	Frame surveys Tanzania, Uganda, Kenya
		18. MSY estimates	-	Pitcher and Bundy, 1995
	Social and economic	19. Contribution of the fisheries to GDP	~	National Statistical Bureaus
		20. Contribution of Nile perch to domestic food supply	~	Fisheries Departments, National Statistical Bureaus
		21. Boat owners, crew	~	Frame surveys
		22. Export volumes and values	+	Fisheries Departments, National Statistical Bureaus
		23. Percentage of level of education by age group	~	Various sources (see Kolding *et al.*, 2005)
		24. Total earnings in the fishery	~	Fisheries Departments, National Statistical Bureaus

Continued

Appendix 19.1 Continued

Forty-one indicators to assess the states of Lake Victoria's fish stocks and the pressures and drivers acting on them compiled by researchers from Uganda, Tanzania and Kenya. (From Kolding *et al.*, 2005.)
Time series data available: + = >10 years; ~ = some points; o = anecdotal

Level	Type of indicators	Indicator	Time series	Source
		25. Fish prices	~	Fisheries Departments, National Statistical Bureaus
		26. Per capita fish consumption	~	Fisheries Departments, National Statistical Bureaus
		27. Landing sites and factories	+	Fisheries Departments, National Statistical Bureaus
		28. Feed production (fishmeal reduction *Rastrineobola*/*Caridina*)	o	No data available
		29. Number of processing plants	+	Fisheries Departments, National Statistical Bureaus
	Control and surveillance, co-management, fish quality assurance	30. Number of BMUs	+	Fisheries Departments
		31. Enforcement statistics	~	Fisheries Departments
		32. Number of inspectors at landing sites	+	Fisheries Departments
		33. Number of certified landing beaches	+	Fisheries Departments
Stocks	Biodiversity, community, food web	34. Species composition (experimental trawls)	+ ~	Trawl surveys
		35. Feeding habits of main commercial species	~	Various sources: e.g. for Nile perch Katunzi *et al.*, 2006
	Life history	36. Size indicators (L_{mean}, L_{max}, $L_{frequency}$)	+ ~	Trawl surveys
		37. Length at 50% maturity	~	Various published sources see http://www.fishbase.org
		38. Growth parameter estimates	~	Various published sources see http://www.fishbase.org
		39. Slope/intercept of the biomass size spectrum	+ ~	Trawl surveys
	Stock abundance	40. Experimental catch rates (trawls, gill nets) by species	+ ~	Trawl surveys; experimental gill net surveys (occasionally conducted in Uganda)
		41. Fishery catch rates (by species)	+ ~	CEDRS Tanzania, Uganda, Kenya

Appendix 19.2

Results of the separate slopes model (2). Df = degrees of freedom; MS = mean square; F = F-statistic; scale parameter = square root of the deviance divided by the degrees of freedom; LR = likelihood ratio; ns = not significant; $CPUE < 40$ cm and $CPUE \geq 40$ cm = experimental catch rates for Nile perch smaller and larger than or equal to 40 cm. Slope, Int = slope and intercept of the abundance size distribution; $L_{max,\,10}$ $L_{5\%}$ L_{mean} = maximum, exploited and mean length of Nile perch in the experimental trawler catch. For explanation of the indicators see text.

		Tanzania 1984–1990				Tanzania 1997–2006				Uganda 1993–2005			
Indicator		$CPUE_{<40\,cm}$	$CPUE_{\geq 40\,cm}$	Slope	Intercept	$CPUE_{<40\,cm}$	$CPUE_{\geq 40\,cm}$	Slope	Intercept	$CPUE_{<40\,cm}$	$CPUE_{\geq 40\,cm}$	Slope	Intercept
Df-model		7	7	7	7	7	7	7	7	7	7	7	7
Df-error		1,547	1547	20	20	390	390	28	28	1,086	1,086	38	38
MS-error		6.91	14.1	0.49	1.77	12.5	15	0.96	1.8	11.63	17	2.58	7.4
F		15.01	29.17	4.74	3.54	7.01	4	1.45	1.2	13.96	3.19	7.11	2.77
r^2		0.06[a]	0.12[a]	0.62[b]	0.55[c]	0.11[a]	0.07[a]	0.27 ns	0.22[c]	0.08[a]	0.02[b]	0.57[a]	0.34[c]
Type 1 error analysis													
MS-model	Year	129.3[a]	0.05 ns	9.25[a]	23.8[b]	452.8[a]	2.88 ns	3.58 ns	11.5[c]	28.5 ns	40.7 ns	0.34 ns	0.73 ns
	Depth	19.5[cb]	871.3[a]	1.56[c]	6.35[c]	128.5[c]	100[a]	2 ns	0.72 ns	314.4[a]	90.3[b]	5.58[a]	5.63[b]
	Interaction	64[a]	86.3[a]	0.72 ns	0.35 ns	11.51 ns	40.3 ns	0.02 ns	0.28 ns	55.1[b]	22.4 ns	0.32 ns	1.99 ns
Type 3 error analysis													
MS-model	Year	0.04 ns	101[b]	9.25[a]	23.8[b]	4.77[a]	1.55 ns	3.58 ns	11.5[c]	33.8 ns	53.4 ns	0.32 ns	0.7 ns
	Depth	148.2[a]	799.8[a]	1.56[c]	6.4[c]	34.4[c]	55.3[c]	1.99 ns	0.78 ns	305.1[a]	102.3[a]	5.6[a]	14.8[c]
	Interaction	64[a]	86.3[a]	0.73 ns	0.35 ns	11.5 ns	40.3 ns	0.02 ns	0.28 ns	55.1[b]	22.4 ns	0.32 ns	1.98 ns

Continued

Appendix 19.2 Continued

Results of the separate slopes model (2). Df = degrees of freedom; MS = mean square; *F* = F-statistic; scale parameter = square root of the deviance divided by the degrees of freedom; LR = likelihood ratio; ns = not significant; $CPUE < 40$ cm and $CPUE \geq 40$ cm = experimental catch rates for Nile perch smaller and larger than or equal to 40 cm. Slope, Int = slope and intercept of the abundance size distribution; $L_{max,\,10}$ $L_{5\%}$ L_{mean} = maximum, exploited and mean length of Nile perch in the experimental trawler catch. For explanation of the indicators see text.

		Tanzania 1984–1990			Tanzania 1997–2006			Uganda		
Indicator		$L_{max,\,10}$	$L_{5\%}$	L_{mean}	$L_{max,\,10}$	$L_{5\%}$	L_{mean}	$L_{max,\,10}$	$L_{5\%}$	L_{mean}
Df-model		8	8	8	8	8	8	8	8	8
Df-deviance		272	10,974	210,181	350	9,883	183,764	452	25,154	463,976
Deviance		886	38,434	11,495,901	1,730.8	27,982	865,635	2,247	60,303	2,148,180
Scale		1.81	1.87	2.67	2.22	1.68	2.17	2.23	1.55	2.15
Type 1 Chi square (LR Statistics)										
MS-model	Year	220.4[a]	80.58[a]	3,178[a]	13.46[a]	1,597.4[a]	4,397[a]	2.46 ns	3,756[a]	3,482[a]
	Depth	68.1[a]	1,032.8[a]	19,760[a]	3.54 ns	769.2[a]	4,249[a]	16.42[a]	28.2[a]	67,734[a]
	Interaction	36.5[a]	223.1[a]	955[a]	9.92[c]	70.1[ac]	495[a]	3.5 ns	114[a]	3,220[a]
Type 3 (Chi square Wald statistics)										
MS-model	Year	232.7[a]	23.1[a]	495[a]	13.32[a]	1,816.1[a]	5,195[a]	2.3 ns	3,332[a]	495[a]
	Depth	82.4[a]	494.1[a]	16,263[a]	2.89 ns	460[a]	3,408[a]	15.73[b]	86.7[a]	67,781[a]
	Interaction	36.4[a]	221.4[a]	944[a]	9.87[c]	70.1[a]	498[a]	3.45 ns	115[a]	3205[a]

[a] $p < 0.001$
[b] $p < 0.01$
[c] $p < 0.05$

Index